A Vital Rationalist

A Vital Rationalist

Selected Writings from
Georges Canguilhem

Edited by François Delaporte

Translated by Arthur Goldhammer

with an introduction by Paul Rabinow and

a critical bibliography by Camille Limoges

ZONE BOOKS · NEW YORK

1994

© 1994 Urzone, Inc.
ZONE BOOKS
611 Broadway, Suite 608
New York, NY 10012

All rights reserved.

No part of this book may be reproduced, stored in a retrieval system, or transmitted in any form or by any means, including electronic, mechanical, photocopying, microfilming, recording, or otherwise (except for that copying permitted by Sections 107 and 108 of the U.S. Copyright Law and except by reviewers for the public press) without written permission from the Publisher.

Sources for the excerpts are listed on pp. 480–81.

Printed in the United States of America.

Distributed by The MIT Press,
Cambridge, Massachusetts, and London, England

Library of Congress Cataloging-in-Publication Data

Canguilhem, Georges, 1904–
    A vital rationalist : selected writings from Georges Canguilhem / edited by François Delaporte; translated by Arthur Goldhammer with an introduction by Paul Rabinow and a critical bibliography by Camille Limoges.
    p. cm.
    Includes bibliographical references.
    ISBN 0-942299-72-8
    1. Science–History. 2. Science–Philosophy.
I. Delaporte, François, 1941– . II. Title.
Q125.C34   1993
500–dc20   93-8613
                                                CIP

Contents

Editor's Note *by François Delaporte*   9
Introduction: A Vital Rationalist
  by Paul Rabinow   11

PART ONE   METHODOLOGY

  I   The History of Science   25
  II The Various Models   41
  III The History of the History of Science   49

PART TWO   EPISTEMOLOGY

  IV Epistemology of Biology   67
  V   Epistemology of Physiology
      A Baroque Physiology   91
      An Experimental Science   103
      *The Major Problems of Nineteenth-Century
        Physiology*   115
  VI Epistemology of Medicine
      The Limits of Healing   129
      The New Situation of Medicine   133
      A Medical Revolution   145

PART THREE   HISTORY

    VII   Cell Theory   161
    VIII   The Concept of Reflex   179
    IX   Biological Objects   203

PART FOUR   INTERPRETATIONS

    X   René Descartes   219
    XI   Auguste Comte   237
    XII   Claude Bernard   261

PART FIVE   PROBLEMS

    XIII   Knowledge and the Living
           Science and Life   287
           The Concept of Life   303
    XIV   The Normal and the Pathological   321
           Introduction to the Problem   321
           The Identity of the Two States   327
           Implications and Counterpositions   337
    XV   Normality and Normativity   351

           Critical Bibliography *by Camille Limoges*   385
           Notes   455

Translator's Note
The texts collected here are translated from the French for the first time, but for two exceptions: I have included passages from my translation of Georges Canguilhem's *Ideology and Rationality* (Cambridge, MA: MIT Press, 1988) and from Carolyn Fawcett's translation of *The Normal and the Pathological* (New York: Zone Books, 1989).

Editor's Note

François Delaporte

The texts collected in this volume introduce English-language readers to an especially difficult and complex dimension of George Canguilhem's work, namely his philosophy of biology and medicine. Its primary purpose, then, is to chart the main themes of Canguilhem's thought, which is distinguished by minute attention to developments in biology and medicine over the past fifty years. To achieve this end, importance was given to questions of methodology in the history of science. This in itself was necessary because the object of historical discourse is not scientific discourse as such but the historicity of scientific discourse insofar as it represents the implementation of an epistemological project (*projet de savoir*). If the history of science is the history of a discourse subject to the norm of critical rectification, then it is clearly a branch of epistemology. Canguilhem recognizes that the disciplines whose history he writes give the appearance of a genesis, that is, a process opposed to the diversity of the various forms of pseudo-science. This, in fact, is the source of his interest in epistemological breaks. Studying the history of an activity itself defined by its reference to truth as an epistemological value forces one to focus attention on both the failures and successes of that activity. Taking a macroscopic view of the history of science,

Canguilhem undertook to study the emergence of three disciplines: biology, physiology and medicine. Depending on the subject of study, Canguilhem will sometimes provide a history of theory, sometimes a history of concepts and sometimes a history of biological objects. But the objective is always the same: to describe how ideology and science are at once intertwined and separate. Further, his studies of René Descartes, Auguste Comte and Claude Bernard clearly reveal why, as Louis Althusser once put it, Canguilhem is considered one of the best "teachers of how to read works of philosophy and science." The reader, we assume, will not be surprised that the present work ends with a series of general questions concerning the relation of knowledge to life and of the normal to the pathological. Canguilhem began with error and on that basis posed the philosophical problem of truth and life. For Michel Foucault, this approach constituted "one of the crucial events in the history of modern philosophy."

Introduction: A Vital Rationalist

Paul Rabinow

Georges Canguilhem was born in Castelnaudary in southwestern France on June 4, 1904. Although his father was a tailor, Canguilhem likes to refer to himself, not without a certain twinkle in his eye, as being of peasant stock, rooted in the harmonious, cyclical life of the soil and the seasons, his sensibilities formed by the yearly round of the fruit trees. The story of his sentimental education is a classic one. High marks on national examinations sent him on a journey to Paris to study; once there, he was a great success. After completing his studies at the prestigious Lycée Henri IV, he entered the most elite educational institution in France, the Ecole Normale Supérieure, in 1924. Among his *promotion*, his cohort, were Jean-Paul Sartre, Raymond Aron and Paul Nizan; Maurice Merleau-Ponty entered the Ecole a year later. Already at this time, Canguilhem was interested in themes that he would return to and develop throughout his intellectual life: in particular, a paper on Auguste Comte's theory of order and progress, which Canguilhem submitted for a diploma, displays the beginnings of this persistent interest in the relation of reason and society – an interest he shared with his other distinguished classmates but which Canguilhem developed in a highly original manner. The philosopher Alain's judgment of Canguilhem in 1924 as "lively, resolute and content" still captures the man's spirit almost three-quarters of a century later.[1]

Once he became *agrégé* in philosophy in 1927, the young Canguilhem began his teaching tour of provincial lycées, as was required of all Ecole Normale graduates in repayment to the state for their education. His initial peregrinations ended in 1936 in Toulouse, where he taught at the lycée, while beginning his medical training. In 1940, he resigned from his teaching post, because, as he wrote the Rector of the Académie de Toulouse, he hadn't become an *agrégé* in philosophy in order to preach the doctrine of the Vichy regime.[2] He took advantage of his newly found free time to complete his medical studies. Prophetically, in both a philosophic and political sense, Canguilhem replaced Jean Cavaillès, the philosopher of mathematics — he had been called to the Sorbonne — at the University of Strasbourg, which relocated to Clermont-Ferrand in 1941, when Strasbourg was annexed by the Reich. He participated in the formation of an important resistance group to which he made available his skills. All in all, a life in the century, as the French say: like so many of his compatriots, Canguilhem's life was shaped by the conjuncture of France's enduring institutions and the contingent events of his time.

In 1943, Canguilhem defended his medical thesis, "Essais sur quelques problèmes concernant le normal et le pathologique." The continued timeliness and exceptional durability of this work is attested to by the fact that he updated it twenty years later with significant new reflections, and that it was translated into English decades later as *The Normal and the Pathological*.[3] After the war, he resumed his post at the University of Strasbourg (in Strasbourg), where he remained until 1948. After first refusing the important administrative post of inspecteur général de philosophie at the Liberation, he finally accepted it in 1948, and served until 1955, when he accepted the Chair of History and Philosophy of Sciences at the Sorbonne and succeeded Gaston Bachelard as director of the Institut d'histoire des sciences et des techniques. His reputation as a ferocious examiner lives on in Paris today, as does a deep well of affection for the intellectual and institutional support he provided over the decades.[4]

History and Philosophy of Science
Louis Althusser paid Canguilhem a compliment when he compared him (as well as Cavaillès, Bachelard, Jules Vuillemin and Michel Foucault) to an anthropologist who goes into the field armed with "a scrupulous respect for the reality of real science."[5] The comparison is revealing if not quite an accurate description of Canguilhem's method. More strictly ethnographic studies of laboratory life, like those of Bruno Latour, would come later and would aim not merely at correcting a positivist and idealist understanding of science as a single unified activity achieving a cumulative understanding of nature, but also at dismantling the very idea of science – a position as far from Canguilhem's as one could imagine. Nonetheless, Althusser's statement captures the move, first initiated by Bachelard, away from the static universalism that the French university system had enshrined in its rationalist and idealist approaches to science. For Bachelard, philosophy's new role was to analyze the historical development of truth-producing practices. The philosophy of science became the study of regional epistemologies, the historical reflection on the elaboration of theories and concepts by practicing scientists, physicists, chemists, pathologists, anatomists and so on. The aim was not to attack science but to show it in action in its specificity and plurality.

Canguilhem is clear and adamant that even though philosophy had lost its sovereignty and its autonomy, it still had important work to accomplish. Unlike the task of the scientist, the epistemologist's problem is to establish "the order of conceptual progress that is visible only after the fact and of which the present notion of scientific truth is the provisional point of culmination."[6] Truths are found in the practices of science; philosophy analyzes the plurality of these truths, their historicity, and consequently their provisionality, while affirming – not legislating, as the older French philosophy of science sought to do – their normativity. Epistemology is a rigorous description of the process by which truth is elaborated, not a list of final results. Althusser's encomium takes for granted that science exists and holds a privileged status, but Canguilhem, like Foucault and Pierre Bourdieu, never doubted this: "To take as one's object of inquiry nothing other than sources, inventions, influences, priorities, simultaneities, and successions is at bottom to fail to distinguish between science and other aspects of culture."[7] This assumption – Latour has called it the key symbol of French philosophy and history of science – is the cornerstone of the whole architecture of the house of reason inhabited by Canguilhem.[8] Science, for Canguilhem, is "a discourse verified in a delimited sector of experience."[9] Science is an exploration of the norm of rationality at work. But just as firm as the belief in science is the belief in its historicity and its plurality. There are only diverse sciences at work at particular historical moments: physics is not biology; eighteenth-century natural history is not twentieth-century genetics.

Thus, for Canguilhem, "the history of science is the history of an object – discourse – that *is* a history and *has* a history, whereas science is the science of an object that is *not* a history, that has *no* history."[10] Science, through its use of method, divides nature into objects. These objects are secondary, in a sense, but not derivative; one could say that they are both constructed and discovered. The history of science performs a similar set of operations on scientific objects. The object of historical discourse is "the historicity of scientific discourse, in so much as that history effectuates a project guided by its own internal norms but traversed by accidents interrupted by crises, that is to say by moments of judgment and truth."[11] These truths are always contestable and in process, as it were, but no less "real" on account of their contingency. The history of science is not natural history: it does not identify the science with the scientist, the scientists with their biographies, or sciences with their results, nor the results with their current pedagogical use. The epistemological and historical claims assumed by this notion of the history of science are magisterial and run counter to much of contemporary *doxa* in the social studies of science. The texts gathered in this volume provide the evidence for Canguilhem's position. François Delaporte has arranged them in a conceptual and pedagogical fashion with such clarity that it would be fruitless and inappropriate to burden them with extended commentary. Indeed, they provide a kind of coherent "book," which, except for his second doctoral dissertation,[12] Canguilhem himself never wrote; he preferred, after 1943, the essay form crammed with precise, almost aphoristic, sentences, many with the density of kryptonite.

The Normal and the Pathological

Although Canguilhem published in the late 1930s a philosophical treatise on ethics and epistemology, *Traité de logique et de morale*, intended as an unconventional textbook for advanced lycée students, the work for which he is best known starts with his medical thesis, where he investigates the very definition of the normal and the pathological. This work signaled a major reversal in thinking about health. Previously, medical training in France had privileged the normal; disease or malfunction was understood as the deviation from a fixed norm, which was taken to be a constant. Medical practice was directed toward establishing scientifically these norms and, practice following theory, toward returning the patient to health, reestablishing the norm from which the patient had strayed.

As François Dagognet, the philosopher of biology, has crisply observed, Canguilhem launched a frontal attack on "that edifice of normalization" so essential to the procedures of a positivist science and medicine.[13] He did so by re-posing the question of the organism as a living being that is in no preestablished harmony with its environment. It is suffering, not normative measurements and standard deviations, that establishes the state of disease. Normativity begins with the living being, and with that being comes diversity. Each patient whom a doctor treats presents a different case; each case displays its own particularity. One of Canguilhem's famous aphorisms drives this point home: "An anomaly is not an abnormality. Diversity does not signify sickness." With living beings, normality is an activity, not a steady state. The result, if one follows Canguilhem's reasoning, is that "a number, even a constant number, translates a style, habits, a civilization, even the underlying vitality of life."[14] The recent discovery that human body temperature has a much wider range of normality than was previously assumed demonstrates this point. Normality – and this is one of Canguilhem's constant themes – means the ability to adapt to changing circumstances, to variable and varying environments. Illness is a reduction to constants, the very norms by which we measure ourselves as normal. Normality equals activity and flexibility. Hence there is no purely objective pathology; rather, the basic unit is a living being that exists in shifting relations with a changing environment. Arguing for a dramatic reversal, Canguilhem maintained that illness ultimately is defined by the very terms that had defined *health*, namely stable norms, unchanging values.¹⁵ Life is not stasis, a fixed set of natural laws, set in advance and the same for all, to which one must adhere in order to survive. Rather, life is action, mobility and pathos, the constant but only partially successful effort to resist death, to use Bichat's famous definition: "Life is the collection of functions that resist death."

Canguilhem's work has been a consistent and disciplined historical demonstration, a laying-out of the consequences, of these principles. Life has its specificity: "Life, whatever form it may take, involves self-preservation by means of self-regulation."¹⁶ This specificity can – in fact, *must* – be elaborated perpetually, but it can never be evaded. Canguilhem's punctuate, historical essays are not a philosophy of life, like those of Hans Jonas or Maurice Merleau-Ponty, which seek to fix an understanding of life with a single set of concepts. Rather, Canguilhem's tightly written didactic forays display how the life sciences, including the therapeutic ones, have simultaneously elaborated concepts of life and the ways these concepts must be seen as an integrated part of the phenomenon under study: life and its norms.

Although he has been careful not to turn these explorations into a panegyric of vitalism, Canguilhem demonstrates the constant presence of evaluative notions like "preservation," "regulation," "adaptation" and "normality," in both everyday and scientific approaches to life. "It is life itself, and not medical judgment, that makes the biological normal a concept of value and not a concept of statistical reality."¹⁷ Humanity's specificity lies not in the fact that it is separate from the rest of nature but, rather, in the fact that it has created systematic knowledge and tools to help it cope. This testing, parrying with pathology, this active relation to the environment, this normative mobility and projective ability, humanity's conceptual career, is central to its health. "Being healthy means being not only normal in a given situation but also normative in this and other eventual situations. What characterizes health is the possibility of transcending the norm, which defines the momentary normal, the possibility of tolerating infractions of the habitual norm and instituting new norms in new situations."[18] Life is an activity that follows a norm. But health is not being normal; health is being normative.

In general, reflections on the relationships between concepts and life require clarification of the fact that at least two distinct orders are being investigated. First, there is life as form, life as the "universal organization of matter" (*le vivant*), and second, there is life as the experience of a singular living being who is conscious of his or her life (*le vécu*). By "life" – in French – one could mean either *le vivant*, the present participle of the verb "to live" (*vivre*), or the past participle *le vécu*. Canguilhem is unequivocal on this point: the first level of life, form, controls the second, experience. Although it is only the first level, the power and form-giving dimensions of life, which constitutes the explicit subject matter of his work, the presence of the second is frequently felt nonetheless.[19] For all its declarative clarity, the claim of priority only thinly masks the keen awareness of suffering and searching – in a word, pathos – which is the experiential double, the constant companion, of Canguilhem's insistent conceptualism. The pathos of existence is always close at hand for this physician cum philosopher cum pedagogue.

In fact, a not-so-latent existentialism, albeit of a distinctive and idiosyncratic sort, shadows Canguilhem's conception of medicine. One easily hears echoes of Sartre and Merleau-Ponty's early themes, transposed to a different register and played with a distinctive flair. Canguilhem's variants of "to freedom condemned" and "the structure of comportment" are composed in a different key. His individual is condemned to adapt to an environment and to act using concepts and tools that have no preestablished affinities with his surrounding world. "Life becomes a wily, supple intelligence of the world, while reason, for its part, emerges as something more vital: it finally develops a logic that is more than a mere logic of identity."[20] Reason and life are intertwined, not opposed, but neither legislates the other.

A New Understanding of Life: Error

It has become a commonplace to say that Canguilhem's recognition by an English-speaking public, beyond a few specialists in the history of the life sciences, follows in the wake of the success of one of his favorite students and friends, Michel Foucault. While not exactly false, such an appreciation remains insufficient unless we also ask what it was in Canguilhem's work which so interested Foucault. And, even further, are these problems the most pertinent for an American audience? Canguilhem's work, it is worth underlining, is relevant for diverse reasons. The question to be asked then is, Why read him today? The answer lies partially in another frequent commonplace. Canguilhem's predecessor, Bachelard, invented a method for a new history of the "hard sciences" of chemistry, physics and mathematics; his student, Foucault, worked on the "dubious sciences" of Man; Canguilhem himself has spent his life tracing the liniments of a history of the concepts of the sciences of life. Let us suggest that today it is the biosciences — with a renewed elaboration of such concepts of norms and life, death and information — that hold center stage in the scientific and social arena; hence the renewed relevance of Georges Canguilhem.

In his 1966 essay "Le Concept et la vie," Canguilhem analyzed the contemporary revolution under way in genetics and molecular biology. The essay, a historical tour de force, traces the concept of life as form (and experience) as well as knowledge of that form, from Aristotle to the present. Canguilhem demonstrates the continuity of problematization and the discontinuity of answers in the history of the concept of life. This historical reconstruction provides the groundwork for an analysis of our contemporary conceptualization of life. Canguilhem frames James D. Watson and Francis Crick's discovery of the structure of the double helix as an information system, one in which the code and the (cellular) milieu are in constant interaction. There is no simple, unidirectional causal relation between genetic information and its effects. The new understanding of life lies not in the structuring of matter and the regulation of functions, but in a shift of scale and location – from mechanics to information and communication theory.[21] In an important sense, the new understanding of life as information rejoins Aristotle insofar as it posits life as a logos "inscribed, converted and transmitted" within living matter.[22] However, we have come a long way since Aristotle. The telos of life most commonly proposed today is more an ethological one, seeing behavior as determined and humans more as animals, than a contemplative one that assigns a special place to reflection and uncertainty. From sociobiologists to many advocates of the Human Genome Project, the code is the central dogma.

Canguilhem rejects this telos. If *homo sapiens* is as tightly programmed as the ethologists (or many molecular biologists) think, then how, Canguilhem asks, can we explain error, the history of errors and the history of our victories over error? Genetic errors are now understood as informational errors. Among such errors, however, a large number arise from a maladaption to a milieu. Once again he reintroduces the theme of normality as situated action, not as a pregiven condition. Mankind makes mistakes when it places itself in the wrong place, in the wrong relationship with the environment, in the wrong place to receive the information needed to survive, to act, to flourish. We must move, err, adapt to survive. This condition of "erring or drifting" is not merely accidental or external to life but its fundamental form. Knowledge, following this understanding of life, is "an anxious quest" (*une recherche inquiète*) for the right information. That information is only partially to be found in the genes. Why and how the genetic code is activated and functions, and what the results are, are questions that can be adequately posed or answered only in the context of life, *le vivant*, and experience, *le vécu*.

Conclusion

Michel Foucault, in an essay dedicated to Canguilhem, "La Vie, l'expérience et la science," characterized a division in French thought between subject-oriented approaches, which emphasize meaning and experience, and those philosophies which take as their object knowledge, rationality and concepts.[23] The rhetorical effect was marvelous. While everyone had heard of Sartre and Merleau-Ponty, few people beyond a small circle of specialists had actually read the work of Cavaillès on the philosophy of set theory in mathematics or Canguilhem on the history of the reflex arc.[24] The irony was made more tantalizing by allusions to the unflinching and high-stakes activities in the resistance of one side of the pair (Cavaillès was killed by the Nazis after forming the resistance network that Canguilhem joined), while the others lived in Paris, writing pamphlets. Foucault was revealing to us a hidden relationship of truth and politics, indicating another type of intellectual, one for whom totality and authenticity bore different forms and norms. However, there is a certain insider's humor involved; twenty years earlier, Canguilhem had employed the same distinctions, applying them to Cavaillès during the 1930s while mocking those who deduced that a philosophy without a subject must lead to passivity and inaction. Cavaillès, who had made the philosophic journey to Germany during the 1930s and warned early on of the dangers brewing there, did not, Canguilhem tells us, hesitate when the war finally came.[25] Rather than writing a moral treatise to ground his actions, he joined the resistance while finishing his work on logic as best he could. Truth and politics were distinct domains for these thinkers of the concept; one was ethically obliged to act in both domains while never losing sight of the specificity of each. Cavaillès's example of rigorous thought and principled action, while still compelling today (especially given the misunderstanding and moralizing about French thought rampant across the Rhine, the Channel and the Atlantic), would seem to demand a renewed conceptualization. The rise and ephemeral glory of structuralism and Althusserianism have shown that removing the humanist subject in the social sciences by itself guarantees neither an epistemological jump from ideology to science nor more effective political action (any more than reinserting a quasi-transcendental subject will provide such guarantees). While Canguilhem's work enables one to think and rethink such problems, it obviously does not offer any readymade answers for the future. Deploying readymade solutions from the past, when history has moved on, concepts changed, milieus altered, would, Canguilhem has taught us, constitute a major error — an error matched in its gravity only by those seeking to annul history, blur concepts and homogenize environments. Living beings are capable of correcting their errors, and Canguilhem's work offers us tools to begin, once again, the process of doing so.

PART ONE

Methodology

CHAPTER ONE

The History of Science

The Object of Historical Discourse
[1] When one speaks of the "science of crystals," the relation between science and crystals is not a genitive, as when one speaks of the "mother of a kitten." The science of crystals is a discourse on the nature of crystal, the nature of crystal being nothing other than its identity: a mineral as opposed to an animal or vegetable, and independent of any use to which one may put it. When crystallography, crystal optics and inorganic chemistry are constituted as sciences, the "nature of crystal" just is the content of the science of crystals, by which I mean an objective discourse consisting of certain propositions that arise out of a particular kind of work. That work, the work of science, includes the formulation and testing of hypotheses, which, once tested, are forgotten in favor of their results.

When Hélène Metzger wrote *La Genèse de la science des cristaux*, she composed a discourse about discourses on the nature of crystal.[1] But these discourses were not originally the same as what we now take to be the correct discourse about crystals, the discourse that defines what "crystals" are as an object of science. Thus, the history of science is the history of an object – discourse – that *is* a history and *has* a history, whereas science is the science of an object that is *not* a history, that has *no* history.

The object "crystal" is a given. Even if the science of crystals must take the history of the earth and the history of minerals into account, that history's time is itself a given. Because "crystal" is in some sense independent of the scientific discourse that seeks to obtain knowledge about it, we call it a "natural" object.[2] Of course, this natural object, external to discourse, is not a scientific object. Nature is not given to us as a set of discrete scientific objects and phenomena. Science constitutes its objects by inventing a method of formulating, through propositions capable of being combined integrally, a theory controlled by a concern with proving itself wrong. Crystallography was constituted as soon as the crystalline species could be defined in terms of constancy of face angles, systems of symmetry, and regular truncation of vertices. "The essential point," René Just Haüy writes, "is that the theory and crystallization ultimately come together and find common ground."[3]

The object of the history of science has nothing in common with the object of science. The scientific object, constituted by methodical discourse, is secondary to, although not derived from, the initial natural object, which might well be called (in a deliberate play on words) the pre-text. The history of science applies itself to these secondary, nonnatural, cultural objects, but it is not derived from them any more than they are derived from natural objects. The object of historical discourse is, in effect, the historicity of scientific discourse. By "historicity of scientific discourse" I mean the progress of the discursive project as measured against its own internal norm. This progress may, moreover, meet with accidents, be delayed or diverted by obstacles, or be interrupted by crises, that is, moments of judgment and truth.

The history of science was born as a literary genre in the eighteenth century. I find that insufficient attention has been paid to a significant fact about the emergence of this genre: it required no fewer than two scientific and two philosophical revolutions as its preconditions. One scientific revolution occurred in mathematics, in which Descartes's analytic geometry was followed by the infinitesimal calculus of Leibniz and Newton; the second revolution, in mechanics and cosmology, is symbolized by Descartes's *Principles of Philosophy* and Newton's *Principia*. In philosophy, and, more precisely, in the theory of knowledge, that is, the foundations of science, Cartesian innatism was one revolution and Lockeian sensualism the other. Without Descartes, without a rending of tradition, there would be no history of science. [*Etudes*, pp. 16–17]

[2] Was Bernard Le Bouvier Fontenelle mistaken when he looked to Descartes for justification of a certain philosophy of the history of science? From the denial that authority holds any validity in science, Fontenelle reasoned, it follows that the conditions of truth are subject to historical change. But does it then make sense to propose a historicist reading of a fundamentally antihistoricist philosophy? If we hold that truth comes only from the evidence and the light of nature, then truth, it would seem, has no historical dimension, and science exists *sub specie aeternitatis* (hence the Cartesian philosophy is antihistoricist). But perhaps Fontenelle deserves credit for noticing an important but neglected aspect of the Cartesian revolution: Cartesian doubt refused to comment on prior claims to knowledge. It not only rejected the legacy of ancient and medieval physics but erected new norms of truth in place of the old. Hence, it rendered all previous science obsolete and consigned it to the surpassed past [*le passé dépassé*]. Fontenelle thus realized that when Cartesian philosophy killed tradition – that is, the unreflective continuity of past and present – it provided at the same time a rational foundation for a possible history, for an emergent consciousness that

the evolution of humankind has meaning. If the past was no longer judge of the present, it was, in the full sense of the word, witness to a movement that transcended it, that dethroned the past in favor of the present. As Fontenelle was well aware, before the Moderns could speak about the Ancients, even to praise them, they had to take their distance. [*Etudes*, p. 55]

[3] According to Descartes, however, knowledge has no history. It took Newton, and the refutation of Cartesian cosmology, for history – that is, the ingratitude inherent in the claim to begin anew in repudiation of all origins – to appear as a dimension of science. The history of science is the explicit, theoretical recognition of the fact that the sciences are critical, progressive discourses for determining what aspects of experience must be taken as real. The object of the history of science is therefore a nongiven, an object whose incompleteness is essential. In no way can the history of science be the natural history of a cultural object. All too often, however, it is practiced as though it were a form of natural history, conflating science with scientists and scientists with their civil and academic biographies, or else conflating science with its results and results with the form in which they happen to be expressed for pedagogical purposes at a particular point in time. [*Etudes*, pp. 17-18]

The Constitution of Historical Discourse
[4] The historian of science has no choice but to define his object. It is his decision alone that determines the interest and importance of his subject matter. This is essentially always the case, even when the historian's decision reflects nothing more than an uncritical respect for tradition.

Take, for example, the application of probability to nineteenth-century biology and social science.[4] The subject does not fall within the boundaries of any of the nineteenth century's mature sciences; it corresponds to no natural object, hence its study cannot fall back on mere description or reproduction. The historian himself must create his subject matter, starting from the current state of the biological and social sciences at a given point in time, a state that is neither the logical consequence nor the historical culmination of any prior state of a developed science – not of the mathematics of Pierre-Simon Laplace or the biology of Charles Darwin, the psychophysics of Gustav Fechner, the ethnology of Frederick Taylor or the sociology of Emile Durkheim. Note, moreover, that Adolphe Quêtelet, Sir Francis Galton, James McKeon Catell and Alfred Binet could develop biometrics and psychometrics only after various nonscientific practices had provided raw material suitable for mathematical treatment. Quêtelet, for example, studied data about human size; the collection of such data presupposes a certain type of institution, namely, a national army whose ranks are to be filled by conscription, hence an interest in the standards for selecting recruits. Binet's study of intellectual aptitudes presupposes another type of institution, compulsory primary education, and a concomitant interest in measuring backwardness. Thus, in order to study the particular aspect of the history of science defined above, one must look not only at a number of different sciences bearing no intrinsic relation to one another but also at "nonscience," that is, at ideology and political and social praxis. Our subject, then, has no natural theoretical locus in one or another of the sciences, any more than it has a natural locus in politics or pedagogy. Its theoretical locus must be sought in the history of science itself and nowhere else, for it is this history and only this history that constitutes the specific domain in which the theoretical issues posed by the development of scientific practice find their resolution.[5] Quêtelet, Gregor Mendel, Binet and Théodore Simon established new and unforeseen relations between mathematics and practices that were originally nonscientific, such as selection, hybridization and orientation. Their discoveries were answers to questions they asked themselves in a language they had to forge for themselves. Critical study of those questions and those answers is the proper object of the history of science. Should anyone wish to suggest that the concept of history proposed here is "externalist," the foregoing discussion should suffice to dispose of the objection.

The history of science can of course accommodate various kinds of objects within the specific theoretical domain that it constitutes: there are always documents to be classified, instruments and techniques to be described, methods and questions to be interpreted, and concepts to be analyzed and criticized. Only the last of these tasks confers the dignity of history of science upon the others. It is easy to be ironic about the importance attached to concepts, but more difficult to understand why, without concepts, there is no science. The history of science is interested in, say, the history of instruments or of academies only insofar as they are related, in both their uses and their intentions, to theories. Descartes needed David Ferrier to grind optical glass, but it was he who provided the theory of the curves to be obtained by grinding.

A history of results can never be anything more than a chronicle. The history of science concerns an axiological activity, the search for truth. This axiological activity appears only at the level of questions, methods and concepts, but nowhere else. Hence, time in the history of science is not the time of everyday life. A chronicle of inventions or discoveries can be periodized in the same way as ordinary history. The dates of birth and death listed in scientific biographies are dates from the ordinary calendar, but the advent of truth follows a different timetable in each discipline; the chronology of verification has its own viscosity, incompatible with ordinary history. Dmitry Mendeleyev's periodic table of the elements accelerated the pace of progress in chemistry, and eventually led to an upheaval in atomic physics, while other sciences maintained a more measured pace. Thus, the history of science, a history of the relation of intelligence to truth, generates its own sense of time. Just how it does this depends on how the progress of science permits this history to reconstitute the theoretical discourse of the past. A new discovery may make it possible to understand a discourse that was not understood when it was first enunciated, such as Mendel's theory of heredity, or it may demolish theories once considered authoritative. Only contact with recent science can give the historian a sense of historical rupture and continuity. Such contact is established, as Gaston Bachelard taught, through epistemology, so long as it remains vigilant.

The history of science is therefore always in flux. It must correct itself constantly. The relation between Archimedes' method of exhaustion and modern calculus is not the same for today's mathematician as it was for Jean Etienne Montucla, the first great historian of mathematics. This is because no definition of mathematics was possible before there was mathematics, that is, before mathematics had been constituted through a series of discoveries and decisions. "Mathematics is a developmental process [*un devenir*]," said Jean Cavaillès. The historian of mathematics must take his provisional definition of what mathematics is from contemporary mathematicians. Many works once relevant to mathematics in an earlier period may therefore cease to be relevant in historical perspective; from a newly rigorous standpoint, previously important works may become trivial applications. [*Etudes*, pp. 18–20]

Recursion and Ruptures
[5] In establishing such a close connection between epistemology and the history of science I am, of course, drawing on the

inspirational teachings of Gaston Bachelard.[6] The fundamental concepts of Bachelard's epistemology are by now well known, so well known, perhaps, that they have been disseminated and discussed, especially outside France, in a vulgarized, not to say sanitized, form, devoid of the polemical force of the original. Among them are the notions of new scientific spirit, epistemological obstacle, epistemological break [*rupture*], and obsolete or "official" science....

To my mind, the best summary of Bachelard's research and teaching can be found in the concluding pages of his last epistemological work, *Le Matérialisme rationnel*.[7] Here the notion of epistemological discontinuity in scientific progress is supported by arguments based on the history and teaching of science in the twentieth century. Bachelard concludes with this statement: "Contemporary science is based on the search for true [*véritable*] facts and the synthesis of truthful [*véridique*] laws." By "truthful" Bachelard does not mean that scientific laws simply tell a truth permanently inscribed in objects or intellect. Truth is simply what science speaks. How, then, do we recognize that a statement is scientific? By the fact that scientific truth never springs fully blown from the head of its creator. A science is a discourse governed by critical correction. If this discourse has a history whose course the historian believes he can reconstruct, it is because it *is* a history whose meaning the epistemologist must reactivate. "Every historian of science is necessarily a historiographer of truth. The events of science are linked together in a steadily growing truth.... At various moments in the history of thought, the past of thought and experience can be seen in a new light."[8] Guided by this new light, the historian should not make the error of thinking that persistent use of a particular term indicates an invariant underlying concept, or that persistent allusion to similar experimental observations connotes affinities of method or approach. By observing these rules he will avoid the error of, for instance, seeing Pierre Louis Moreau de Maupertuis as a premature transformist or geneticist.[9] [*Ideology and Rationality*, pp. 10-12]

[6] When Bachelard speaks of a norm or value, it is because in thinking of his favorite science, mathematical physics, he identifies theory with mathematics. His rationalism is built on a framework of mathematism. In mathematics one speaks not of the "normal" but of the "normed." In contrast to orthodox logical positivists, Bachelard holds that mathematics has epistemological content, whether actual or potential, and that progress in mathematics adds to that content. On this point he agrees with Jean Cavaillès, whose critique of logical positivism has lost nothing of its vigor or rigor. Cavaillès refutes Rudolph Carnap by showing that "mathematical reasoning is internally coherent in a way that cannot be rushed. It is by nature progressive."[10] As to the nature of this progress, he concludes,

> One of the fundamental problems with the doctrine of science is precisely that progress is in no way comparable to increasing a given volume by adding a small additional amount to what is already there, the old subsisting with the new. Rather, it is perpetual revision, in which some things are eliminated and others elaborated. What comes after is greater than what went before, not because the present contains or supersedes the past but because the one necessarily emerges from the other and in its content carries the mark of its superiority, which is in each case unique.[11]

Nevertheless, the use of epistemological recursion as a historical method is not universally valid. It best fits the disciplines for the study of which it was originally developed: mathematical physics and nuclear chemistry. Of course, there is no reason why one cannot study a particularly advanced specialty and then abstract rules for the production of knowledge which may, with caution, be extrapolated to other disciplines. In this sense, the method cannot be generalized so much as it can be broadened. Yet it cannot be extended to other areas of the history of science without a good deal of reflection about the specific nature of the area to be studied. Consider, for example, eighteenth-century natural history. Before applying Bachelardian norms and procedures to the study of this subject, one must ask when a conceptual cleavage[12] occurred whose effects were as revolutionary as were those of the introduction of relativity and quantum mechanics into physics. Such a cleavage is barely perceptible in the early Darwinian years,[13] and, to the extent that it is visible at all, it is only as a result of subsequent cataclysms: the rise of genetics and molecular biology.

Hence, the recurrence method must be used judiciously, and we must learn more about the nature of epistemological breaks. Often, the historian in search of a major watershed is tempted to follow Kant in assuming that science begins with a flash of insight, a work of genius. Frequently the effects of that flash are said to be all-embracing, affecting the whole of a scientist's work. But the reality is different. Even within one man's work we often find a series of fundamental or partial insights rather than a single dramatic break. A theory is woven of many strands, some of which may be quite new while others are borrowed from older fabrics. The Copernican and Galilean revolutions did not sweep away tradition in one fell swoop. Alexandre Koyré has located what he considers to be the decisive "mutation" in Galileo's work, the decisive change in thinking that made him unable to accept medieval mechanics and astronomy.[14] For Koyré, the elevation of mathematics – arithmetic and geometry – to the status of key to intelligibility in physics indicated a rejection of Aristotle in favor of Plato. Koyré's argument is sufficiently well known that I need not discuss it in detail. But in painting a quite accurate picture of Galileo as an Archimedean as much as a Platonist, is not Koyré abusing the freedom of the recurrence method?[15] And is he not somewhat overstating the case in saying that the change in Galileo's thinking marked a total repudiation of Aristotelianism? Is not Ludovico Geymonat right to point out that Koyré's interpretation neglects all that Galileo preserved from Aristotelian tradition even as he was proposing that mathematics be used to bolster logic?[16] Thus, Koyré is himself challenged on the very point on which he challenged Pierre Duhem when he wrote, "The apparent continuity in the development of physics from the Middle Ages to the present [a continuity that Jean-Paul Caverni and Pierre Maurice Duhem have so assiduously stressed] is illusory.... No matter how well the groundwork has been laid, a revolution is still a revolution."[17] [*Ideology and Rationality*, pp. 13-15]

Science and Scientific Ideologies

What is scientific ideology?
[7] Scientific ideology, unlike a political class ideology, is not false consciousness. Nor is it false science. The essence of false science is that it never encounters falsehood, never renounces anything, and never has to change its language. For a false science there is no prescientific state. The assertions of a false science can never be falsified. Hence, false science has no history. By contrast, a scientific ideology does have a history. A scientific ideology comes to an end when the place that it occupied in the encyclopedia of knowledge is taken over by a discipline that operationally demonstrates the validity of its own claim to scientific status, its own "norms of scientificity." At that point, a certain form of nonscience is excluded from the domain of science. I say "nonscience" rather than use Bogdan Suchodolski's term "anti-

science" simply in order to take note of the fact that, in a scientific ideology, there is an explicit ambition to be science, in imitation of some already constituted model of what science is. This is a crucial point. The existence of scientific ideologies implies the parallel and prior existence of scientific discourses. Hence, it also presupposes that a distinction has already been made between science and religion.

Consider the case of atomism. Democritus, Epicurus and Lucretius claimed scientific status for their physics and psychology. To the antiscience of religion they opposed the antireligion of science. Scientific ideology neglects the methodological requirements and operational possibilities of science in that realm of experience it chooses to explore; but it is not thereby ignorance, and it does not scorn or repudiate the function of science. Hence, scientific ideology is by no means the same thing as superstition, for ideology has its place, possibly usurped, in the realm of knowledge, not in the realm of religious belief. Nor is it superstition in the strict etymological sense. A superstition is a belief from an old religion that persists despite its prohibition by a new religion. Scientific ideology does indeed stand over [*superstare*] a site that will eventually be occupied by science. But science is not merely overlain; it is pushed aside [*deportare*] by ideology. Therefore, when science eventually supplants ideology, it is not in the site expected. [*Ideology and Rationality*, pp. 32-34]

How scientific ideologies disappear and appear
[8] For another, I hope convincing, example of the way in which scientific ideologies are supplanted by science, consider the Mendelian theory of heredity. Most historians of biology believe that Maupertuis was the forerunner of modern genetics because in his *Vénus physique* he considered the mechanisms by which normal and abnormal traits are transmitted. He also used the calculus of

probabilities to decide whether the frequency of a particular abnormality within a particular family was or was not fortuitous, and explained hybridization by assuming the existence of seminal atoms, hereditary elements that combined during copulation. But it is enough to compare the writings of Maupertuis and Mendel to see the magnitude of the gap between a science and the ideology that it replaces. The facts that Mendel studies are not those gleaned by a casual observer; they are obtained through systematic research – research dictated by the nature of Mendel's problem, for which there is no precedent in the pre-Mendelian literature. Mendel invented the idea of a *character*, by which he meant not the elementary agent of hereditary transmission but the element of heredity itself. A Mendelian character could enter into combination with *n* other characters, and one could measure the frequency of its appearance in successive generations. Mendel was not interested in structure, fertilization or development. For him, hybridization was not a way of establishing the constancy or inconstancy of a global type; it was a way of decomposing a type, an instrument of analysis, a tool for separating characters that made it necessary to work with large samples. Hence, Mendel was interested in hybrids despite his repudiation of an age-old tradition of hybrid research. He was not interested in sexuality or in the controversy over innate versus acquired traits or over preformation versus epigenesis. He was interested only in verifying *his* hypothesis via the calculation of combinations.[18] Mendel neglected everything that interested those who in reality were not his predecessors at all. The seventeenth-century ideology of hereditary transmission is replete with observations of animal and plant hybrids and monsters. Such curiosity served several purposes. It supported one side or the other in the debates between preformationists and epigenesists, ovists and animalculists. As a result, it was useful in resolving legal questions concerning the subordination of the sexes, paternity, the purity of bloodlines and the legitimacy of the aristocracy. These concerns were not unrelated to the controversy between innatism and sensualism. The technology of hybridization was perfected by agronomists in search of advantageous varieties, as well as by botanists interested in the relations between species. Only by isolating Maupertuis's *Vénus physique* from its context can that work be compared with the *Versuche über Pflanzenhybriden*. Mendel's science is not the end point of a trail that can be traced back to the ideology it replaced, for the simple reason that that ideology followed not one but several trails, and none was a course set by science itself. All were, rather, legacies of various traditions, some old, others more recent. Ovism and animalculism were not of the same age as the empirical and mythological arguments advanced in favor of aristocracy. The ideology of heredity[19] was excessively and naively ambitious. It sought to resolve a number of important theoretical and practical legal problems without having examined their foundations. Here the ideology simply withered away by attrition. But the elimination of its scientific underpinnings brought it into focus as an ideology. The characterization of a certain set of observations and deductions as an ideology came after the disqualification of its claim to be a science. This was accomplished by the development of a new discourse, which circumscribed its field of validity and proved itself through the consistency of its results.

[9] Instructive as it is to study the way in which scientific ideologies disappear, it is even more instructive to study how they appear. Consider briefly the genesis of a nineteenth-century scientific ideology, evolutionism. The work of Herbert Spencer makes an interesting case study. Spencer believed that he could state a universally valid law of progress in terms of evolution from the simple to the complex through successive differentiations.

Everything, in other words, evolves from more to less homogeneity and from lesser to greater individuation: the solar system, the animal organism, living species, man, society, and the products of human thought and activity, including language. Spencer explicitly states that he derived this law of evolution by generalizing the principles of embryology contained in Karl-Ernst von Baer's *Über Entwickelungsgeschichte der Thiere* (1828). The publication of the *Origin of Species* in 1859 confirmed Spencer's conviction that his generalized theory of evolution shared the scientific validity of Darwin's biology. But he also claimed for his law of evolution the support of a science more firmly established than the new biology: he claimed to have deduced the phenomenon of evolution from the law of conservation of energy, which he maintained could be used to prove that homogeneous states are unstable. If one follows the development of Spencer's work, it seems clear that he used von Baer's and, later, Darwin's biology to lend scientific support to his views on social engineering in nineteenth-century English industrial society, in particular, his advocacy of free enterprise, political individualism and competition. From the law of differentiation, he deduced that the individual must be supported against the state. But perhaps this "deduction" was contained in the principles of the Spencerian system from the very beginning.

The laws of mechanics, embryology and evolution cannot validly be extended beyond the domain proper to each of these sciences. To what end are specific theoretical conclusions severed from their premises and applied out of context to human experience in general, particularly social experience? To a practical end. Evolutionist ideology was used to justify industrial society as against traditional society, on the one hand, and the demands of workers, on the other. It was in part antitheological, in part antisocialist. Thus, evolutionist ideology was an ideology in the

Marxist sense: a representation of nature or society whose truth lay not in what it said but in what it hid. Of course, evolutionism was far broader than Spencer's ideology. But Spencer's views had a lasting influence on linguists and anthropologists. His ideology gave meaning to the word *primitive* and salved the conscience of colonialists. A remnant of its legacy can still be found in the behavior of advanced societies toward so-called underdeveloped countries, even though anthropology has long since recognized the plurality of cultures, presumably making it illegitimate for any one culture to set itself up as the yardstick by which all others are measured. In freeing themselves from their evolutionist origins, contemporary linguistics, ethnology and sociology have shown that an ideology disappears when historical conditions cease to be compatible with its existence. The theory of evolution has changed since Darwin, but Darwinism is an integral part of the history of the science of evolution. By contrast, evolutionist ideology is merely an inoperative residue in the history of the human sciences. [*Ideology and Rationality*, pp. 34–37]

Chapter Two

The Various Models

The Positivist Tradition

[10] Events completely extrinsic to science and logic, portrayed conventionally if at all in standard histories of scientific research, yield an account that claims, if only in ritual fashion, to trace the logical development of a scientific idea. This would be surprising only if there were no distinction between science and the history of science. In that case, a biologist could write a history of his work in exactly the same way as he would write a scientific paper, relying on exactly the same criteria he would use in evaluating the truth of a hypothesis or the potential of a particular line of research. But to proceed in this way is to treat hypotheses and research programs not as projects but as objects. When a scientific proposition is judged to be true, it takes on a retroactive validity. It ceases to be part of the endless stream of forgotten dreams, discarded projects, failed procedures and erroneous conclusions – things, in short, for which someone must shoulder the responsibility. The elimination of the false by the true – that is, the verified – appears, once it is accomplished, to be the quasi-mechanical effect of ineluctable, impersonal necessity. Importing such norms of judgment into the historical domain is, therefore, an inevitable source of misunderstanding. The retroactive effect

of the truth influences even one's assessment of the respective contributions of various investigators to a scientific discovery (an assessment that only a specialist is competent to make), because the tendency is to see the history of the subject in the light of today's truth, which is easily confused with eternal truth. But if truth is eternal, if it never changes, then there is no history: the historical content of science is reduced to zero. It should come as no surprise that it was positivism, a philosophy of history based on a generalization of the notion that theory ineluctably succeeds theory as the true supplants the false, that led to science's contempt for history. Over time, a research laboratory's library tends to divide into two parts: a museum and a working reference library. The museum section contains books whose pages one turns as one might examine a flint ax, whereas the reference section contains books that one explores in minute detail, as with a microtome. [*Formation du réflexe*, pp. 155–56]

[11] Eduard Jan Dijksterhuis, the author of *Die Mechanisierung des Weltbildes*, thinks that the history of science is not only science's memory but also epistemology's laboratory. This phrase has been quoted frequently. The idea, which has been accepted by numerous specialists, has a less well known antecedent. Pierre Flourens, referring in his eulogy of Georges Cuvier to the *Histoire des sciences naturelles* published by Magdelaine de Saint-Agy, states that the history of science "subjects the human mind to experiment...makes an experimental theory of the human spirit." Such a conception is tantamount to modeling the relation between the history of science and the science of which it is the history on the relation between the sciences and the objects of which they are sciences. But experimentation is only one of the ways in which science relates to objects, and it is not self-evident that this is the relevant analogy for understanding history's relation to its object. Furthermore, in the hands of its recent champion, the meth-

odological statement has an epistemological corollary, namely, that there exists an eternal scientific method. In some periods this method remains dormant, while in others it is vigorous and active. Gerd Buchdahl has characterized this corollary as naive,[20] and one would be inclined to agree if he were willing to apply the same description to the empiricism or positivism underlying his own view. It is no accident that I attack positivism at this point in the argument: for after Flourens but before Dijksterhuis, Pierre Lafitte, a confirmed disciple of Auguste Comte, compared the history of science to a "mental microscope."[21] The use of such an instrument, Lafitte suggests, reveals hidden truths: the understanding of science is deepened through discussion of the difficulties scientists faced in making their discoveries and propagating their results. The image of the microscope defines the context as the laboratory, and there is, I think, a positivist bias in the idea that history is simply an injection of duration into the exposition of scientific results. A microscope merely magnifies otherwise invisible objects; the objects exist whether or not one uses the instrument to look for them. The implicit assumption is that the historian's object is lying there waiting for him. All he has to do is look for it, just as a scientist might look for something with a microscope. [*Etudes*, pp. 12-13]

Historical Epistemology
[12] To understand the function and meaning of the history of science, one can contrast the image of the laboratory with that of a school or tribunal, that is, an institution where judgment is passed on either the past of knowledge or knowledge of the past. But if judgment is to be passed, a judge is essential. Epistemology provides a principle on which judgment can be based: it teaches the historian the language spoken at some point in the evolution of a particular scientific discipline, say, chemistry. The

historian then takes that knowledge and searches backward in time until the later vocabulary ceases to be intelligible, or until it can no longer be translated into the less rigorous lexicon of an earlier period. Antoine-Laurent Lavoisier, for example, introduced a new nomenclature into chemistry. Hence, the language spoken by chemists after Lavoisier points up semantic gaps in the language of earlier practitioners. It has not been sufficiently noticed or admired that Lavoisier, in the "Discours préliminaire" to his *Traité élémentaire de chimie*, assumed full responsibility for two decisions that left him open to criticism: "revising the language spoken by our teachers" and failing to provide "any historical account of the opinions of my predecessors." It was as though he understood the lesson of Descartes, that to institute a new branch of knowledge is in effect to sever one's ties to whatever had presumptively usurped its place.

There are in fact two versions of the history of science: the history of obsolete knowledge and the history of sanctioned knowledge, by which I mean knowledge that plays an active [*agissant*] role in its own time. Without epistemology it is impossible to distinguish between the two. Gaston Bachelard was the first to make this distinction.[22] His decision to recount the history of scientific experiments and concepts in the light of the latest scientific principles has long since demonstrated its worth.

Alexandre Koyré's idea of the history of science was basically similar to Bachelard's. True, Koyré's epistemology was closer to Emile Meyerson's than to Bachelard's, and more keenly attuned to the continuity of the rational function than to the dialectics of rationalist activity. Yet it was because he recognized the role of epistemology in doing history of science that he cast his *Etudes galiléennes* and *The Astronomical Revolution* in the form that he did.

Is the dating of an "epistemological break" a contingent or subjective judgment? To see that the answer is no, one need only note that Koyré and Bachelard were interested in different periods in the history of the exact sciences. Furthermore, these periods were not equally equipped to deal mathematically with the problems of physics. Koyré began with Copernicus and ended with Newton, where Bachelard began. Koyré's epistemological observations tend to confirm Bachelard's view that a "continuist" history of science is the history of a young science. Koyré believed, for instance, that science is theory and that theory is fundamentally mathematization. (Galileo, for example, is more Archimedean than Platonist.) He also held that error is inevitable in the pursuit of scientific truth. To study the history of a theory is to study the history of the theorist's doubts. "Copernicus... was not a Copernican."

The history of science thus claims the right to make judgments of scientific value. By "judgment," however, I do not mean purge or execution. History is not an inverted image of scientific progress. It is not a portrait in perspective, with transcended doctrines in the foreground and today's truth way off at the "vanishing point." It is, rather, an effort to discover and explain to what extent discredited notions, attitudes or methods were, in their day, used to discredit other notions, attitudes or methods – and therefore an effort to discover in what respects the discredited past remains the past of an activity that still deserves to be called scientific. It is as important to understand what the past taught as it is to find out why we no longer believe in its lessons. [*Etudes*, pp. 13-14]

Empiricist Logicism
[13] It is easy to distinguish between what Bachelard calls "normality"[23] and what Thomas Kuhn calls "normal science."[24] The two epistemologies do share certain points in common: in particular, the observation that scientific textbooks overemphasize

the continuity of scientific research. Both stress the discontinuous nature of progress. Nevertheless, while the fundamental concepts share a family resemblance, they do not really belong to the same branch. This has been noted by Father François Russo, who, despite reservations about the claims of superiority to which epistemological historians are sometimes prone, argues that Kuhn is mistaken about the nature of scientific rationality as such.[25] Though ostensibly concerned to preserve Karl Popper's emphasis on the necessity of theory and its priority over experiment, Kuhn is unable to shake off the legacy of logical positivism and join the rationalist camp, where his key concepts of "paradigm" and "normal science" would seem to place him. These concepts presuppose intentionality and regulation, and as such they imply the possibility of a break with established rules and procedures. Kuhn would have them play this role without granting them the means to do so, for he regards them as simple cultural facts. For him, a paradigm is the result of a choice by its users. Normal science is defined by the practice in a given period of a group of specialists in a university research setting. Instead of concepts of philosophical critique, we are dealing with mere social psychology. This accounts for the embarrassment evident in the appendix to the second edition of the *Structure of Scientific Revolutions* when it comes to answering the question of how the truth of a theory is to be understood. [*Ideology and Rationality*, pp. 12-13]

Internalism and Externalism

[14] How does one do the history of science, and how should one do it? This question raises another: *what* is the history of science a history *of*? Many authors apparently take the answer to this second question for granted, to judge by the fact that they never explicitly ask it. Take, for example, the debates between what English-speaking writers call internalists and externalists.[26] Exter-

nalism is a way of writing the history of science by describing a set of events, which are called "scientific" for reasons having more to do with tradition than with critical analysis, in terms of their relation to economic and social interests, technological needs and practices, and religious or political ideologies. In short, this is an attenuated or, rather, impoverished version of Marxism, one rather common today in the world's more prosperous societies.[27] Internalism (which externalists characterize as "idealism") is the view that there is no history of science unless one places oneself within the scientific endeavor itself in order to analyze the procedures by which it seeks to satisfy the specific norms that allow it to be defined as science rather than as technology or ideology. In this perspective, the historian of science is supposed to adopt a theoretical attitude toward his specimen theories; he therefore has as much right to formulate models and hypotheses as scientists themselves.

Clearly, both the internalist and externalist positions conflate the object of the history of science with the object of a science. The externalist sees the history of science as a matter of explaining cultural phenomena in terms of the cultural milieu; he therefore confuses the history of science with the naturalist sociology of institutions and fails to interpret the truth claims intrinsic to scientific discourse. The internalist sees the facts of the history of science, such as instances of simultaneous discovery (of modern calculus, for example, or the law of conservation of energy), as facts whose history cannot be written without a theory. Thus, a fact in the history of science is treated as a fact of science, a procedure perfectly compatible with an epistemology according to which theory rightfully takes priority over empirical data. [*Etudes*, pp. 14–15]

CHAPTER THREE

The History of the History of Science

A History of Precursors
[15] Every theory is rightly expected to provide proofs of practical efficacy. What, then, is the practical effect for the historian of science of a theory whose effect is to make his discipline the place where the theoretical questions raised by scientific practice are studied in an essentially autonomous manner? One important practical effect is the elimination of what J.T. Clark has called "the precursor virus."[28] Strictly speaking, if precursors existed, the history of science would lose all meaning, since science itself would merely appear to have a historical dimension.

Consider the work of Alexandre Koyré. Koyré contrasted, on epistemological grounds, the "closed world" of antiquity with the "infinite universe" of modern times. If it had been possible for some ancient precursor to have conceived of "the infinite universe" before its time, then Koyré's whole approach to the history of science and ideas would make no sense.[29]

A precursor, we are told, is a thinker or researcher who proceeded some distance along a path later explored all the way to its end by someone else. To look for, find and celebrate precursors is a sign of complacency and an unmistakable symptom of incompetence for epistemological criticism. Two itineraries cannot be compared unless the paths followed are truly the same.

In a coherent system of thought, every concept is related to every other concept. Just because Aristarchus of Samos advanced the hypothesis of a heliocentric universe, it does not follow that he was a precursor of Copernicus, even if Copernicus invoked his authority. To change the center of reference of celestial motions is to relativize high and low, to change the dimensions of the universe – in short, to constitute a system. But Copernicus criticized all astronomical theories prior to his own on the grounds that they were not rational systems.[30] A precursor, it is said, belongs to more than one age: he is, of course, a man of his own time, but he is simultaneously a contemporary of later investigators credited with completing his unfinished project. A precursor, therefore, is a thinker whom the historian believes can be extracted from his cultural milieu and inserted into others. This procedure assumes that concepts, discourses, speculations and experiments can be shifted from one intellectual environment to another. Such adaptability, of course, is obtained at the cost of neglecting the "historicity" of the object under study. How many historians, for example, have looked for precursors of Darwinian transformism among eighteenth-century naturalists, philosophers and even journalists?[31] The list is long.

Louis Dutens's *Recherches sur l'origine des découvertes attribuées aux modernes* (1776) may be taken as an (admittedly extreme) case in point. When Dutens writes that Hippocrates knew about the circulation of the blood, and that the Ancients possessed the system of Copernicus, we smile: he has forgotten all that William Harvey owed to Renaissance anatomy and mechanical models, and he fails to credit Copernicus's originality in exploring the mathematical possibility of the earth's movement. We ought to smile just as much at the more recent writers who hail René Antoine

Ferchault de Réaumur and Maupertuis as precursors of Mendel without noticing that the problem that Mendel set himself was of his own devising, or that he solved it by inventing an unprecedented concept, the independent hereditary character.[32]

So long as texts and other works yoked together by the heuristic compression of time have not been subjected to critical analysis for the purpose of explicitly demonstrating that two researchers sought to answer identical questions for identical reasons, using identical guiding concepts, defined by identical systems, then, insofar as an authentic history of science is concerned, it is completely artificial, arbitrary and unsatisfactory to say that one man finished what the other started or anticipated what the other achieved. By substituting the logical time of truth relations for the historical time of these relations' invention, one treats the history of science as though it were a copy of science and its object a copy of the object of science. The result is the creation of an artifact, a counterfeit historical object – the precursor. In Koyré's words:

> The notion of a "forerunner" is a very dangerous one for the historian. It is no doubt true that ideas have a *quasi* independent development, that is to say, they are born in one mind, and reach maturity to bear fruit in another; consequently, the history of problems and their solutions can be traced. It is equally true that the historical importance of a doctrine is measured by its fruitfulness, and that later generations are not concerned with those that precede them except in so far as they see in them their "ancestors" or "forerunners." It is quite obvious (or should be) that no-one has ever regarded himself as the "forerunner" of someone else, nor been able to do so. Consequently, to regard anyone in this light is the best way of preventing oneself from understanding him.[33]

A precursor is a man of science who, one knows only much later, ran ahead of all his contemporaries but before the person whom one takes to be the winner of the race. To ignore the fact that he is the creature of a certain history of science, and not an agent of scientific progress, is to accept as real the condition of his possibility, namely, the imaginary simultaneity of "before" and "after" in a sort of logical space.

In making this critique of a false historical object, I have sought to justify by counterexample the concept I have proposed according to which the history of science defines its object in its own intrinsic terms. The history of science is not a science, and its object is not a scientific object. To do history of science (in the most operative sense of the verb "to do") is one of the functions (and not the easiest) of philosophical epistemology. [*Etudes*, pp. 20-23]

A History in the Service of Politics

[16] It was in 1858 that a new polemic, initiated this time by George Prochaska's growing renown, resulted in Descartes's name being brought into the history of the reflex for the first time. The occasion was an article by A.L. Jeitteles, a professor of medicine at Olmütz, entitled *Who Is the Founder of the Theory of Reflex Movement?*[34] Jeitteles summarized Marshall Hall's first paper, said a few words about Hall's priority over Johannes Müller, acknowledged the great value of both men's work, yet claimed that the impetus for research into reflex action came from elsewhere, from an earlier time, and from another source. "It was none other than our eminent, and today insufficiently honored, compatriot, George Prochaska, who richly deserves to be preserved in the eternally grateful memory of our Czech fatherland, so rich in superior men of every kind." Jeitteles asserted that Prochaska was the true founder of the theory of reflex movement, quoted excerpts

from *De functionibus systematis nervosi*, and concluded that the entire theory of the reflex action inherent in the spinal cord was there "preformed and preestablished" (*präformirt und prästabilirt*). Although not interested in investigating whether Hall and Müller, who may not have known Prochaska's work directly, might have been influenced by word of it filtered through "the scientific milieu of his contemporaries and epigones" (*in die gleichzeitige und epigonische wissenschaftliche Welt transpirirte*), Jeitteles asks how this work could have been ignored for so long. His answer, which seems judicious to me, is that Albrecht von Haller's authority is a sufficient explanation. The theory of irritability, of a strength inherent in the muscle, diverted attention from the intrinsic functions of the spinal cord. This only makes Prochaska's merit all the more apparent: rather than rehearse the ideas of the period, his work contradicted them. The final lines of the article are an appeal to some generous historian to revive the great Prochaska as a model for future generations. Jeitteles thought that the man to do this was the current occupant of Prochaska's chair at the venerable and celebrated University of Prague, the "illustrious forerunner of all German universities." That man was the distinguished physiologist Jan Purkinje (1787–1869).

The impetuosity of this plea, which naturally and pathetically combines a claim for the originality of a scholar with an affirmation of the cultural values of an oppressed nationality, is equaled only by the brutality and insolence of the reply it received from an official representative, not to say high priest, of German physiology. Emile Du Bois-Reymond (1818–1896), Müller's student and successor in the chair of physiology at the University of Berlin – who became a member of the Berlin Academy of Sciences in 1851 and who was already celebrated not only for his work in neuromuscular electrophysiology but also for his numerous professions of philosophical faith in the universal validity of mechanistic determinism and the inanity of metaphysical questions[35] — summarily dismissed Prochaska and gave Descartes credit for having had the genius to anticipate both the word and the idea of "reflex." In a commemorative address delivered at the time of Müller's death in 1858, Du Bois-Reymond stated that he had found (*wie ich gefunden habe*) that Descartes, roughly a century and a half before Prochaska, had correctly described reflex movement (*erstens beschrieb...Descartes...die Reflexbewegungen völlig richtig*); he had used the same analogy (with reflection) to describe the phenomenon; and he also deserved credit for the law of peripheral manifestation of sense impressions.[36] The passages that precede and follow these lines on Descartes give a clear indication of Du Bois-Reymond's intention. It was, first of all, to protect Müller's "copyright," as it were: Müller may not have known about Descartes, but Prochaska was another matter. If Prochaska was not the father of the notion of reflex, then he himself fell under the shadow of the judgment proposed in his name against his successors. Furthermore, Descartes was, according to Du Bois-Reymond, a self-conscious mechanist physiologist, a theorist of the animal-machine, and therefore deserving of the same admiration extended to Julien Offray de La Mettrie, the theorist of the man-machine.[37] By contrast, Prochaska was a vague and inconsistent thinker in whose mind the notion of reflex was associated with that of *consensus nervorum*, an anatomical myth of animist inspiration.[38] Indeed, if Prochaska had formulated the principle of the reflection of sense impressions in 1784, he failed to mention it in his *Physiologie oder Lehre von der Natur des Menschen* in 1820.[39] Finally, Prochaska did not know what he was doing the first time he had the opportunity to describe correctly the reflection of sense impressions. As for Müller's contemporaries, the only author who might justly be credited with priority over Müller was Hall, and that was a priority of two months.[40] It may be that in diminishing Prochaska, Du Bois-Reymond was really trying to discredit a group of biologists manifestly guilty in his eyes of the sin of metaphysics, namely, the *Naturphilosophie* school.

Du Bois-Reymond's 1858 text was published in 1887 in the second volume of his *Reden* along with explanatory notes. The notes concerning the passages of Descartes on which Du Bois-Reymond based his comments are particularly valuable for our purposes;[41] some of the relevant passages are from Article 13 of *The Passions of the Soul*, where the palpebral reflex is described. I must point out that Du Bois-Reymond makes no distinction between a description and a definition, and that it is rather disingenuous of him to reproach Prochaska, as he does in one note, for having used the same example as Descartes. It would be laughable to maintain that Charles Scott Sherrington should not have studied the "scratch reflex" because it meant borrowing from Thomas Willis. In any case, Prochaska was an ophthalmologist and, strictly speaking, had no need of Descartes to know that there is such a thing as involuntary occlusion of the eyelids.[42] The second text of Descartes's cited by Du Bois-Reymond is Article 36 of *The Passions of the Soul*. Although it does contain the expression "*esprits réfléchis*" (reflected spirits), this expression, unique in Descartes's work, is used to explain the mechanism of a form of behavior that is not a reflex in the strict sense of the word. If, in fact, Du Bois-Reymond is right to contend that Prochaska did not know what he was doing when he devoted page after page of his *Commentation* of 1784 to the "reflection" of sensory into motor impressions, what are we to say, applying the same criterion of judgment, about an author who uses a pair of words only once? [*Formation du réflexe*, pp. 138-40]

[17] We therefore impute to Du Bois-Reymond, at his request, full responsibility for his historical discovery. If I have dwelt on the details of this controversy, it is because it enables

us at last to establish the precise origin of the widely accepted view that paternity of both the word "reflex" and some rudimentary version of the idea can be traced back to Descartes, a view that Franklin Fearing, as we have seen, repeats several times, but whose origins he never examines.[43] Along with the origin of the assertion, we have discovered its meaning. As for the circumstances, Du Bois-Reymond's address was meant as a rebuke to a Czech professor insufficiently persuaded of the superiority of German civilization. But as far as its scientific implications are concerned, this address can be attributed to a concern – a concern, that is, on the part of a physiologist for whom "scientism" did duty for philosophy – to discover, in Descartes's alleged anticipation of a discovery that was beginning to justify a mechanistic interpretation of a whole range of psychophysiological phenomena, a guarantee and, in a sense, an authentication of the use that people now proposed to make of it. It was not so much for reasons of pure physiology as for reasons of philosophy that Descartes was anointed a great physiologist and illustrious precursor. [*Formation du réflexe*, pp. 141-42]

[18] In the history of the concept of the reflex, very different circumstances and motivations account for the appearance of Descartes, Willis, Jean Astruc and Prochaska, with Johann August Unzer generally being left shrouded in shadow. Prochaska's name came up in the course of a polemic between Marshall Hall and certain of his contemporaries, a polemic that gradually turned into what is commonly called a settling of scores. The story belongs, along with countless other tales of rivalry between scientific coteries, to the anecdotal history of science. Descartes's name came up in the course of a diatribe against one dead man for the apparent purpose of honoring another. In fact, it was a matter of liquidating an opposition, or even – when one looks at it closely – two oppositions. One culture, speaking through the

voice of one of its official representatives, defended its political superiority of the moment against another culture. One philosophy of life, constrained within the framework of a biological research method, treated another philosophy as a mythology allegedly incapable of fostering effective scientific research. It was mechanism against vitalism. [*Formation du réflexe*, p. 155]

A Canonical History

[19] An emperor's wish to glorify and justify new academic institutions led to a new departure in the history of science. In 1807 Napoleon I ordered a report on the progress that had been made in science since 1789. Georges Cuvier, as permanent secretary of the Institut pour les Sciences Physiques et Naturelles since 1803, was assigned responsibility for the *Report* that was eventually published in 1810, while Jean Baptiste Joseph Delambre was made responsible for a similar report on the mathematical sciences. The authorities could pride themselves on having found a new Bernard Le Bouvier Fontenelle, a man capable of supplementing the yearly analyses of the work of the academy with eulogies of deceased academicians. But anyone who would examine the history of a life devoted to research must consider other, similar research contemporary with, or prior to, that of his subject. And when one has received a Germanic education – an education that was, in Henry Ducrotay de Blainville's words, "encyclopedic and philological"[44] – one could conceive of giving a "course in the history of natural science." And when one had chosen, as Cuvier had toward the end of his studies at the Caroline Academy in Stuttgart, to study "cameralistics," or the science of administration and economics,[45] it was only natural to devote space to technology in one's report to the emperor and to adumbrate a theory of the social status of modern science in the 1816 *Réflexions sur la marche actuelle des sciences et sur leurs rapports avec la société*, as well

as in the *Discours sur l'état de l'histoire naturelle et sur ses accroissements depuis le retour de la paix maritime* (1824). The reader of volume three of the *Histoire des sciences naturelles* is not surprised, then, to find that the first lecture is devoted to a reminder, inspired by the Marquis de Condorcet in the *Esquisse* of 1794, of the debt that modern science owes to the technological innovations of the fourteenth and fifteenth centuries: alcohol, clear glass, paper, artillery, printing, the compass. In the same lecture, Cuvier, a Protestant and the official within the ministry of the interior responsible for overseeing non-Catholic religious worship, could not help noticing the encouragement and support that men of learning had found in the Reformation: freedom of thought and the gradual emancipation of philosophy from subservience to theological doctrine.

Blainville and François Louis Michel Maupied's *Histoire des sciences de l'organisation et de leurs progrès, comme base de la philosophie* is constructed on the basis of diametrically opposed judgments. To be sure, the chapter devoted to Conrad Gesner recalls the positive contributions of technology to Renaissance science (vol. 2, pp. 134-35), but immediately thereafter the Reformation is denounced for "reviving the unfortunate reactions that we have previously seen arising out of various struggles of the human spirit, applying method without authority to the explication of dogma" (p. 136). Because of the friendship between the principal author and those two cultural agitators, Blainville and Maupied, the work contains numerous passages concerning the relation of the sciences and their teaching to the new social needs of an emerging industrial society, but these excurses almost always end in sermons. [...]

Blainville and Maupied's *Histoire* is also different from, even diametrically opposed to, that of Cuvier when it comes both to determining the method, or ways and means, of the science of living things, and to appreciating the effects of seventeenth-century philosophies on the development of that science. Cuvier thinks that philosophy encourages the sciences if it disposes minds toward observation but discourages the sciences if it disposes minds toward speculation.[46] Wherever Aristotle's method, based on experience, was adopted, the sciences progressed, whereas Descartes chose the opposite path, and the regrettable consequences of that choice lasted until the middle of the eighteenth century, when the sciences were countered by "another philosophy that was a copy of the true Peripateticism and that has been called the *philosophy of the eighteenth century* or of the *skeptics*." A rather sweeping judgment, it might seem, although it was current at the time in one form or another. Blainville and Maupied's judgment is equally broad, as well as considerably more prolix: Descartes, Bacon and all the others (*sic*), they say, are merely the logical consequence, the elaboration, of Aristotle;[47] Bacon's philosophy is nothing but Aristotle's;[48] Descartes worked in an Aristotelian direction;[49] Descartes built on the work of the great Stagirite;[50] and so on. What is the significance of our two historians' fascination with Aristotle? The answer to this question, I think, determines what view the history of science ought to take of Blainville and Maupied's project. The first step toward answering it, moreover, must come from a final comparison with Cuvier's *Histoire*.

The third lecture in Cuvier's third volume is devoted to Leibniz, and Cuvier dwells at length on the great chain of being and on Charles Bonnet's development of this Leibnizian theme. Cuvier states that "physiology does not follow mathematics in admitting unlimited combinations," and that, in order to accept the notion that there exists a continuous chain of beings, as Bonnet and others do, or that beings can be arranged along a single line, one must have a very incomplete view of nature's organization.[51] "I hope,"

Cuvier says, "to have proven that this system is false,"[52] alluding to what he knows he has demonstrated through comparative anatomy and paleontology, namely, that there is no unity of organic gradation, no unity of structural plan, no unity of composition and no unity of type.

Now, if Blainville, for his part, acknowledges five distinct types of creation, he nevertheless argues that they are arranged in a series, each one being the distinct expression of a general plan whose progressive or regressive order, if one looks at the level of the species for gradations and degradations that ought to apply only to genera, does not proceed without apparent hiatus. If the numerous papers, reports and dissertations published by Blainville can be seen as the *a posteriori* of his zoological system, then the *a priori* is described in his *Histoire des sciences de l'organisation* as an *a priori* not of rational intuition but of divine revelation. This affirmation can be read in the Introduction, signed by Blainville himself: "I conceived and carried out my *Histoire de l'organisation* as a possible foundation for philosophy, while at the same time demonstrating that philosophy is one and the same thing as the Christian religion, which is so to speak only an *a priori*, revealed to man by God himself when the state of society required it."[53] And further: "Science in general is knowledge *a posteriori* of the existence of God through his works."[54]

How, then, does knowledge proceed? Through *reading*. The preliminary analysis of zoological notions at the beginning of volume three confirms this unambiguously: "One does not create in science, one reads what is created. The pretension to create is absurd, even in the greatest geniuses."[55] In virtue of this heuristic imperative, the sciences of organization should be able to discover – that is, to read in the structures and functions of living beings – only what the Book of Genesis affirms about the order of those beings' creation, in the waters, in the air and on earth, ending finally with man, proclaimed to be the "master" of all that went before. Now, it so happens that there is a Western philosopher of Greek antiquity who was able to read that order, which was unknown to Eastern mythology: that philosopher was Aristotle, "who understood that there is in nature a collection of groups, and that each group forms a veritable series whose degrees pass imperceptibly from one to the other, from the most imperfect to the one in which life achieved its highest perfection."[56] Aristotle's goal, clearly, was to achieve knowledge of man regarding all those aspects that make him superior to the animals, a being possessing a touch of the divine.[57]

This key to reading the forms of life gives us the key to reading Blainville and Maupied's *Histoire*. That key is the notion of "measure," an absolute term of reference and comparison. "Measure" is a word that recurs frequently in the *Histoire*. The measure of organized beings in their serial disposition is man.[58] And it was because Aristotle made man the measure of animality that Aristotle himself is the measure of truth for the series of investigations that took animals as their object. Through the centuries Aristotle is the measure of the sciences of organization. [...]

Now that we possess the key to the *Histoire des sciences de l'organisation*, we can understand why certain authors were included in the book while others were excluded. Unlike eclectics such as Cuvier (who was frequently characterized as such, both scientifically and politically[59]), Blainville based his choices on an explicit criterion: "In this history a number of eminent men stand as landmarks of scientific progress. I chose them because their own work and the work of their legitimate predecessors pushed science in the right direction and with an impetus appropriate to the age" (vol. 1, pp. viii–ix).

Consequently, the history of the science of organization is governed by the fundamental, which is to say, divine, law of the organization of organisms – the ascending series. Blainville, by always taking the idea of the animal series (which for him was merely the reading of an ontological fact) as the measure of the importance of men and their works, composed his *Histoire* in the image of God creating the series. ["De Blainville," *Revue d'histoire*, pp. 75–82]

[20] All history of science that is not strictly descriptive may be said to be implicitly normative insofar as its author, owing to his culture at that moment, can do nothing to prevent himself from reacting, as would a chemical reagent, with the meanings he thinks he sees emerging on their own from the past. But Blainville and Maupied's *Histoire* is more than normative in this strong sense: it is a canonical history in the strict sense of the word. How else can one characterize a work in which a man of science, such as Blainville, could write in his signed Introduction that he took account "only of those steps that fell on the straight line between the starting point and the end or goal," and that he neglected "the works of individuals who, voluntarily or involuntarily, veered, as it were, to the left"[60] – a work, moreover, in which Jean-Baptiste Lamarck and Lorenz Oken are called "errant naturalists,"[61] a work that claims to profess the views of the "Christian Aristotle"?[62] In virtue of this, the authors write, "As for those lost children who appear in nearly every era of science, who have struck a bold but misplaced blow, or who fired before being ordered to do so, their efforts have almost always been without effect when not positively harmful. We must not speak of them."[63] If the expression "canonical history" seems too severe for characterizing a work written jointly by a scholar who was a legitimist in politics and a priest who would one day serve as a consultant to the *Index*, one can nevertheless say, having noticed that the authors took several quotations from François-René Vicomte de Chateaubriand's *Etudes historiques*,[64] that their *Histoire* is, in its own way and for the nat-

ural sciences, a complement to that author's *Génie du christianisme*. ["De Blainville," *Revue d'histoire*, pp. 90-91]

Part Two

Epistemology

Chapter Four

Epistemology of Biology

Origins of the Concept

[21] Aristotle was the first to attempt a general definition of life: "Of natural bodies [that is, those not fabricated by man], some possess vitality, others do not. We mean by 'possessing vitality' that a thing can nourish itself and grow and decay."[1] Later he says that life is what distinguishes the animate body from the inanimate. But the term "life," like "soul," can be understood in several senses. It is enough that one of them should accord with some object of our experience "for us to affirm that [that object] is alive."[2] The vegetal state is the minimal expression of the soul's functions. Less than this and there is no life; any richer form of life presupposes at least this much.[3] Life, identified with animation, thus differs from matter; the life-soul is the form, or act, of which the living natural body is the content: such was Aristotle's conception of life, and it remained as vigorous throughout the centuries as Aristotelian philosophy itself did. All the medical philosophies that held, down to the beginning of the nineteenth century, that life was either a unique principle or somehow associated with the soul, essentially different from matter and an exception to its laws, were directly or indirectly indebted to that part of Aristotle's system which can equally well be called biology or psychology.

But through the end of the eighteenth century, Aristotle's philosophy was also responsible for a method of studying the nature and properties of living things, especially animals. Life forms were classified according to similarities and differences in their parts (or organs), actions, functions and modes of life. Aristotle gave naturalists reason to look at life forms in a particular way. The method sidestepped the question of life as such. Its aim was to exhibit, without gaps or redundancies, the observable products of what Aristotle had no difficulty imagining as a plastic power. Hence eighteenth-century naturalists such as Comte Buffon and Carolus Linnaeus could describe and classify life forms without ever defining what they meant by "alive." In the seventeenth and eighteenth centuries, the study of life as such was pursued by physicians rather than naturalists, and it was natural for them to associate life with its normal mode, "health." From the mid seventeenth century onward, then, the study of life became the subject of physiology (narrowly construed). The purpose of this study was to determine the distinctive features of the living, not to divine the essence of this remarkable power of nature. [...]

It was a German physician, Georg Ernst Stahl (1660–1734), who more than anyone else insisted that a theory of life was a necessary prerequisite of medical thought and practice. No physician used the term "life" more often. If a doctor has no idea what the purpose of the vital functions is, how can he explain why he does what he does? Now, what confers life – life being the directed, purposeful movement without which the corporeal machine would decompose – is the soul. Living bodies are composite substances with the faculty to impede or resist the ever-present threat of dissolution and corruption. This principle of conservation, of the autocracy of living nature, cannot be passive, hence it must not be material. The faculty of self-preservation is the basis of Stahl's *Theoria medica vera* (1708). Certain careful readers who would later deny his identification of life with the soul still never forgot his forceful definition of life as the power temporarily to suspend a destiny of corruptibility.

In terms less freighted with metaphysics, Xavier Bichat began his *Recherches physiologiques sur la vie et la mort* (1800) with this celebrated maxim: "Life is the collection of functions that resist death." In defining life in terms of a conflict between, on the one hand, a body composed of tissues of specific structure and properties (elasticity, contractility, sensitivity) and, on the other, an environment, or milieu, as Auguste Comte would later call it, governed by laws indifferent to the intrinsic needs of living things, Bichat cast himself as a Stahl purged of theology. [...]

In the very year of Bichat's death, 1802, the term "biology" was used for the first time in Germany by Gottfried Reinhold Treviranus and simultaneously in France by Jean-Baptiste Lamarck (in *Hydrogéologie*); they thereby staked a claim to independence on behalf of the life sciences. Lamarck had long planned to give the title *Biology* to one of his works, having proposed a theory of life very early in his teaching at the Muséum d'Histoire Naturelle in Paris. By studying the simplest organisms, he argued, one could determine what was "essential to the existence of life in a body."[...]

Lamarck conceived of life as a continuous, steady accumulation and assimilation of fluids by solids, initially in the form of a cellular tissue, "the matrix of all organization." Life originates in matter and motion, but its unique power is evident only in the orderly pattern of its effects, the series of life forms, which gradually increase in complexity and acquire new faculties.[4] Life begins with an "act of vitalization," an effect of heat, "that material soul of living bodies."[5] Individuals must die, yet life, particularly in its most advanced animal forms, comes, over time, to bear ever-less resemblance to the inert passivity of inanimate objects. To call Lamarck's theory of life "materialist" is to forget that for him "all the crude or inorganic *composite* matter that one observes in nature" is the residue of organic decomposition, for only living things are capable of chemical synthesis.

Georges Cuvier's conception was very different. Unlike Lamarck, Bichat and Stahl, Cuvier saw life and death not as opposites but as elements of what he called "modes of life." This concept was intended to capture the way in which highly specialized internal organizations could entertain compatible relations with the "general conditions of existence." "Life," Cuvier argued,

> is a continual turbulence, a flow whose direction, though complex, remains constant. This flux is composed of molecules, which change individually yet remain always the same type. Indeed, the actual matter that constitutes a living body will soon have dispersed, yet that matter serves as the repository of a force that will compel future matter to move in the same direction. Thus, the form of a living body is more essential than its matter, since the latter changes constantly while the former is preserved.[6]

Life thus bears a clear relation to death.

> It is a mistake to look upon [life] as a mere bond holding together the various elements of a living body, when it is actually a spring that keeps those elements in constant motion and shifts them about. The relations and connections among the elements are not the same from one moment to the next; in other words, the state or composition of the living body changes from moment to moment. The more active its life is, the more its exchanges and metamorphoses are never-ending. And the instant of absolute rest, which is called total death, is but the precursor of further moments of putrefac-

tion. From this point on it makes sense, therefore, to use the term "vital forces."[7]

Thus, death is present in life, as both universal armature and ineluctable fate of individual components organized into compatible yet fragile systems.

The work of naturalists like Lamarck and Cuvier led, albeit in different ways, to a conceptual and methodological revolution in the representation of the world of living things. Theories of life subsequently found a logical place in the teachings of physiologists who, nevertheless, believed that their experimental methods had exorcised the specter of metaphysics. Thus, for example, Johannes Müller discussed "life" and the "vital organization" of the organism in the introduction to his *Handbuch der Physiologie des Menschens* (1833-34). And Claude Bernard, who recorded his intellectual progress during the most fertile period of his career (1850-60) in his *Cahier de notes*, always regarded the nature of life as the fundamental question of general biology. The careful conclusions he reached are set forth in *Leçons sur les phénomènes de la vie communs aux animaux et aux végétaux* (1878, especially the first three lectures) more systematically than they are in *Introduction à la médecine expérimentale* (1865). Of course, the Bernardian theory of life involved related explanations of two deliberately opposed maxims: life is creation (1865) and life is death (1875).

Having gained eminently scientific status in the nineteenth century, the question "What is life?" became one that even physicists did not disdain to ask: Erwin Schrödinger published a book bearing that title in 1947. At least one biochemist found the question meaningless, however — Ernest Kahane, *La Vie n'existe pas*, 1962. In this historical résumé of how the concept of life has been used in various domains of science, I owe a great deal to the work of Michel Foucault.[8] ["Vie," *Encyclopaedia*, pp. 764a-66a]

Obstacles to Scientific Knowledge of Life

[22] Contemporary French epistemology is indebted to the work of Gaston Bachelard for its interest in what may be described, in general terms, as obstacles to knowledge. In sketching out a psychoanalysis of objective knowledge, Bachelard, if he himself did not propose, at least hinted at the idea that objects of knowledge are not intrinsically complex but rather are enmeshed in psychological complexes. The question of epistemological obstacles does not arise for either classical empiricism or classical rationalism. For empiricists, the senses are simple receptors; the fact that qualities are associated with sensations is ignored. For rationalists, knowledge permanently devalues the senses; the intellect, its purity restored, must never again be sullied. But contemporary anthropology, informed by psychoanalysis and ethnography, takes a very different view: primitive psychic mechanisms impose certain obsessional constraints on the curious yet docile mind, thereby creating certain generalized *a priori* obstacles to understanding. In the life sciences, then, what we hope to discover is the obsessive presence of certain unscientific values at the very inception of scientific inquiry. Even if objective knowledge, being a human enterprise, is in the end the work of living human beings, the postulate that such knowledge exists – which is the first condition of its possibility – lies in the systematic negation, in any object to which it may be applied, of the reality of the qualities which humans, knowing what living means to them, identify with life. To live is to attach value to life's purposes and experiences; it is to prefer certain methods, circumstances and directions to others. Life is the opposite of indifference to one's surroundings. [...]

Science, however, denies the values that life imputes to different objects. It defines objects in relation to one another – in other words, it measures without ascribing value. Its first major historical success was mechanics, based on the principle of inertia, a concept that comes into being when one considers the movement of matter itself abstracted from the ability of living things to impart movement. Inertia is inactivity and indifference. It should come as no surprise, then, that efforts to extend the methods of materialist science to life have repeatedly been met with resistance, right up to the present day. If such resistance often reflects emotional hostility, it may also stem from a reasoned judgment: namely, that it may be paradoxical to attempt to explain a power such as life in terms of concepts and laws based on the negation of that power. [...]

Persistent questions about the origins of life and theories of spontaneous generation may well point to another latent overdetermination. Nowadays it seems to be taken for granted that our fascination with reproduction is all the greater because society shuns and indeed censors our curiosity about the subject. Children's beliefs about sexuality reflect both the importance and mysteriousness of birth. While many historians of biology ascribe belief in spontaneous generation to the lack of evidence or unpersuasiveness of arguments to the contrary, the theory may well point to a nostalgic desire for spontaneous generation – a myth, in short. Freud's dissident disciple Otto Rank argued in *The Trauma of Birth* (1929) that the child's sudden separation from the placental environment is the source of, or model for, all subsequent anxiety.[9] His *Myth of the Birth of the Hero*, which deals with men who somehow avoid the fetal stage, was supposed to lend support to this view by demonstrating the prevalence of birth-denying myths.[10] Without going so far as to claim that all proponents of what has been called "equivocal generation" or "heterogony," whether materialists or creationists, have done nothing more than give shape to a fantasy originating in the traumatized unconscious, one can still argue that the theory of spontaneous generation stems from an overestimation of the value of life. The idea of procreation and birth is in one sense an idea of sequence and priority, and aversion to that idea must be seen as a consequence of the prestige attached to what is original or primordial. If every living thing must be born, and if it can be born only to another living thing, then life is a form of servitude. But if the living can rise to perfection through an ascendantless ascension, life is a form of domination. ["Vie," *Encyclopaedia*, pp. 766a–66b]

Life as Animation

[23] We completely forget that when we speak of animals, animality or inanimate bodies, the terms we use are vestiges of the ancient metaphysical identification of life with the soul and of the soul with breath (*anima* = *anemos*). Thus when man, the only living creature capable of discourse on life, discussed his own life in terms of respiration (without which there is not only no life but no speech), he thought he was discussing life in general. If Greek philosophers prior to Aristotle, especially Plato, speculated about the essence and destiny of the soul, it was Aristotle's *De anima* that first proposed the traditional distinction between the vegetative or nutritive soul, the faculty of growth and reproduction; the animal or sensitive soul, the faculty to feel, desire and move; and the reasonable or thinking soul, the faculty of humanity. In this context, it matters little whether Aristotle thought of these three souls as distinct entities or as merely hierarchical levels, the lesser of which could exist without the greater, whereas the greater could neither exist nor function without the lesser. The important thing is to remember that for the Greeks the word *psychē* meant cool breath. The Jews, moreover, had ideas of life and the soul quite similar to those of the Greeks: "And the Lord God formed man of the dust of the ground and breathed into his nostrils the breath of life; and the man became a living soul" (Gen.

2:7). This is not the place to retrace the history of the schools of Alexandria – the Jewish school with Philo, the Platonic school with Plotinus – whose teachings, coupled with the preaching of Paul (1 Cor. 15), inspired the fundamental themes of early Christian doctrine concerning life, death, salvation and resurrection. Indeed, the cultural eclecticism of Mediterranean civilizations is even responsible for the polysemic connotations (another way of saying "ambiguity") of the term "spirit," from *spirare* – an ambiguity that permitted it to serve equally well in theology, to denote the third person of the Trinity, and in medicine, where, in the phrases "vital spirit" and "animal spirit," it became an anticipatory trope for the so-called nervous influx.

After 1600, the conception of life as an animation of matter lost ground to materialist or merely mechanistic conceptions of the intrinsic life functions, and it was no longer accepted as an objective answer to the question "What is life?" Yet it survived well into the nineteenth century in the form of a medical-philosophical ideology. For evidence of this, one has only to glance at a little-known text, the preface to the thirteenth edition of the *Dictionnaire de médecine* (1873), published by Jean-Baptiste Baillière under the editorship of two positivist physicians, Emile Littré, the author of a celebrated dictionary of the French language, and Charles Robin, a professor of histology at Paris's Faculté de Médecine. [...]

The *Dictionnaire de médecine* in question was a recasting of the 1855 revised edition of Pierre Hubert Nysten's *Dictionnaire* (1814), itself the revised and expanded successor of Joseph Capuron's *Dictionnaire de médecine* (1806). The editors were keen to point out the difference between the materialist ideas they were accused of championing and the positivist doctrine they professed to teach. To that end, they commented on the various definitions of the terms "soul," "spirit," "man" and "death" that Capuron had proposed in 1806 and they (Littré and Robin) had themselves put forward in 1855.

In 1806, "soul" was defined as the "internal principle of all operations of living bodies; more particularly, the principle of life in the vegetal and in the animal. The soul is simply vegetative in plants and sensitive in beasts; but it is simple and active, reasonable and immortal in man."

In 1855, one found a different definition:

> Term which, in biology, expresses, considered anatomically, the collection of functions of the brain and spinal cord and, considered physiologically, the collection of functions of the encephalic sensibility, that is, the perception of both external objects and internal objects; the sum total of the needs and penchants that serve in the preservation of the individual and species and in relations with other beings; the aptitudes that constitute the imagination, language and expression; the faculties that form the understanding; the will, and finally the power to set the muscular system in motion and to act through it on the external world.

In 1863, this definition was subjected to vehement criticism by Anatole Marie Emile Chauffard, who attacked not only Littré and Robin but also Ludwig Büchner (*Kraft und Stoff*, 1855), the high priest of German materialism at the time. In *De la Philosophie dite positive dans ses rapports avec la médecine*, Chauffard celebrated "the indissoluble marriage of medicine and philosophy" and yearned to found "the notion of the real and living being" on "human reason aware of itself as cause and force." Two years later, Claude Bernard wrote, "For the experimental physiologist, there can be no such thing as spiritualism or materialism.... The physiologist and the physician should not think that their role is to discover the cause of life or the essence of diseases."[11] ["Vie," *Encyclopaedia*, pp. 767a–67b]

Life as Mechanism

[24] At the end of the *Treatise on Man* (completed in 1633 but not published until 1662–64), Descartes wrote:

> I should like you to consider that these functions follow from the mere arrangement of the machine's organs every bit as naturally as the movements of a clock or other automaton follow from the arrangement of its counter-weights and wheels. In order to explain these functions, then, it is not necessary to conceive of this machine as having any vegetative or sensitive soul or other principle of movement and life, apart from its blood and its spirits, which are agitated by the heat of the fire burning continuously in its heart – a fire which has the same nature as all the fires that occur in inanimate bodies.[12]

It is fairly well known that Descartes's identification of the animal (including physical or physiological man) with a mechanized or mechanical automaton is the obverse of both his identification of the soul with thought ("For there is within us but one soul, and this soul has within it no diversity of parts"[13]) and his substantial distinction between the indivisible soul and extended matter. If the *Treatise on Man* surpassed even the summary of its contents given in the fifth part of the 1637 *Discourse on Method* as a manifesto supporting an animal physiology purified of all references to a principle of animation of any kind, it was because William Harvey's discovery of the circulation of the blood and publication of the *Exercitatio anatomica de motu cordis et sanguinis in animalibus* (1628) had, in the meantime, presented a hydrodynamic explanation of a life function – an explanation that many physicians, particularly in Italy and Germany, had tried to imitate, offering a variety of artificial models to explain such other functions as muscular contraction or the equilibrium of fish in water. In fact,

Galileo's students and disciples at the Accademia del Cimento, Giovanni Alfonso Borelli (*De motu animalium*, 1680–81), Francesco Redi and Marcello Malpighi, had actually tried to apply Galileo's teaching in mechanics and hydraulics to physiology; Descartes, though, was satisfied to set forth a heuristic program that was more intentional than operational.

One way of explaining how organs like the eye or organ systems like the heart and vessels work is to build what we would now call "mechanical models." This is precisely what the iatromechanics (or iatromathematicians) of the seventeenth and eighteenth centuries tried to do in order to explain muscular contraction, digestion and glandular secretion. Yet the laws of Galilean or Cartesian mechanics cannot by themselves explain the origin of coordinated organ systems, and such coordinated systems are precisely what one means by "life." In other words, mechanism is a theory that tells us how machines (living or not) work once they are built, but it tells us nothing about how to build them.

In practice, mechanism contributed little to subjects such as embryology. The use of the microscope, which became common in the second half of the seventeenth century, made it possible to observe the "seeds" of living things, living things in the earliest stages of development. But Jan Swammerdam's observations of insect metamorphoses and Anthonie van Leeuwenhoek's discovery of the spermatazoid were initially understood to confirm a speculative conception of plant or animal generation, according to which the seed or egg or spermatic animalcule contains, preformed in a miniature that optical magnification reveals, a being whose evolution will proceed until it attains its adult dimensions. The microscopic observation that did most to validate this theory was undoubtedly Malpighi's examination of the yellow of a chicken's egg falsely assumed not to have been incubated.[14]

There is reason to think that Malpighi's belief in mechanism unconsciously structured his perception of phenomena.

Intentionally or not, behind every machine loomed a mechanic or, to use the language of the day, a builder. Living machines implied a mechanic of their own, and that implication pointed toward a *Summus opifex*, God. It was therefore logical to assume that all living machines had been constructed in a single initial operation, and thence that all the germs of all the preformed living things — past, present or future — were, from the moment of creation, contained one inside the other. Under these conditions, the succession of living things only appears to be a history, because a birth is in reality only an unpacking. When less biased or more ingenious observations led to the revival, in a revised form, of the old view that the embryo grows through epigenesis (the successive appearance of anatomical formations not geometrically derivable from antecedent formations[15]), modern embryology was instituted as a science capable of encouraging physiology to free itself from its fascination with mechanism.

Meanwhile, growing numbers of observations by microscopists, naturalists, physicians and others curious about nature helped to discredit mechanism in a different but parallel way. The hidden inner structure of plant and animal parts gradually came to seem prodigiously complicated compared with the macroscopic structures visible through dissection. The discovery of animalcules, henceforth called Protista, opened up previously unsuspected depths in the empire of the living. Whereas seventeenth-century mechanics was a theory of movements and impulses, that is, a science based on data accessible to sight and touch, microscopic anatomy was concerned with objects beyond the manifest and tangible. Availing oneself of that structural microcosm, that "other world" within, one could conceive of ever more minute microcosms embedded one within the other. The microscope enabled the imagination to conceive of structural complexity on a scale never before imagined, much as modern calculus extended the power of Descartes's analytical geometry. As a result, Pascal and Leibniz, unbeknownst to each other, both found mechanism wanting. But Leibniz's critique, unlike Pascal's, provided the foundation for a new conception of living things — biology would henceforth picture life in terms of organism and organization:

> Thus each organic body of a living creature is a kind of divine machine or natural automaton surpassing infinitely every kind of artificial automaton. For a machine made by human art is not a machine in every part of it.... But Nature's machines, living bodies, are machines even in their minutest parts and to infinity. This is what constitutes the difference between nature and art, between the divine art and our human art.[16]

["Vie," *Encyclopaedia*, pp. 767b–68a]

Life as Organization
[25] Once again, it was Aristotle who coined the term "organized body." A body is organized if it provides the soul with instruments or organs indispensable to the exercise of its powers. Until the seventeenth century, then, the paradigm of the organized body was the animal (because it possessed a soul). Of course, Aristotle said that plants too have organs, although of an extremely simple kind, and people did wonder about the organization of the plant kingdom. Microscopic examination of plant preparations led to generalizations of the concept of organization, and it even inspired fantastic analogies between plant and animal structures and functions. Robert Hooke (*Micrographia*, 1667), Marcello Malpighi (*Anatome plantarum*, 1675) and Nehemiah Grew (*The Anatomy of Plants*, 1682) discovered the structure of bark, wood and cortex,

distinguished tubules, vessels and fibers, and compared roots, twigs, leaves and fruits in terms of the membranes or tissues they contained.

The Greek word *organon* referred to both a musician's instrument and an artisan's tool. The human body was compared to a musical organ in more than one seventeenth-century text, including works by Descartes, Pascal, Jacques-Bénigne Bossuet (*Traité de la connaissance de Dieu et de soi-même*) and Leibniz. "Organization," "organic" and "organize" still carried both biological and musical connotations as recently as the nineteenth century (see Emile Littré's *Dictionnaire de la langue française*). For Descartes, the organic "organ" was an instrument that needed no organist, but Leibniz believed that without an organist there could be no structural or functional unity of the "organ" instrument. Without an organizer, that is, without a soul, nothing is organized or organic: "[W]e would never reach anything about which we could say, here is truly a being, unless we found animated machines whose soul or substantial form produced a substantial unity independent of the external union arising from contact."[17] Less celebrated but more of a teacher, the physician Daniel Duncan wrote: "The soul is a skilled organist, which forms its organs before playing them.... It is a remarkable thing that in inanimate organs, the organist is different from the air that he causes to flow, whereas in animate organs the organist and the air that causes them to play are one and the same thing, by which I mean that the soul is extremely similar to the air or to breath."[18]

The concept of organism developed in the eighteenth century, as naturalists, physicians and philosophers sought semantic substitutes or equivalents for the word "soul" in order to explain how systems composed of distinct components nevertheless work in a unified manner to perform a function. The parts of such a system mutually influence one another in direct or mediated fashion. The word "part" seemed ill-suited to denote the "organs" of which the organism could be seen as the "totality" but not the "sum."

Reading Leibniz inspired Charles Bonnet, whose hostility to mechanism had been confirmed by Abraham Trembley's observations on the reproduction of polyps by propagation, and by his own observations on the parthenogenesis of plant lice.

> I am not yet making the difficulty plain enough: it lies not only in how to form *mechanically* an organ that is itself composed of so many different pieces but primarily in explaining, by the laws of mechanics alone, the host of various *relations* that so closely bind all the organic parts, and in virtue of which they all conspire toward the same general goal – by which I mean, they form that unity which one calls an animal, that organized whole which lives, grows, feels, moves, preserves and reproduces itself.[19]

In Germany the text that did most to place "organism" at the top of the late eighteenth century's list of biological concepts was Kant's *Critique of Judgment* (1790). Kant analyzed the concept of an organized being without using the words "life" or "living thing." An organized being is in one sense a machine, but a machine that requires a formative energy, something more than mere motor energy and capable of organizing otherwise inert matter. The organic body is not only organized, it is self-organizing: "In such a natural product as this every part is thought as *owing* its presence to the *agency* of all the remaining parts, and also as existing *for the sake of the others* and of the whole, that is, as an instrument or organ. But this is not enough.... On the contrary the part must be an organ *producing* the other parts – each, consequently, reciprocally producing the others. No instrument of art can answer to this description, but only the instrument of that nature [can]...."[20] In the same period, the physician Carl Friedrich von Kielmeyer, whom Georges Cuvier had met as a fellow student at the Caroline Academy in Stuttgart, delivered a celebrated lecture on the main ideas of an influential approach to zoology and botany, the *Rapport des forces organiques dans la série des différentes organisations* (1793). The organism is defined as a system of organs in a relation of circular reciprocity. These organs are determined by their actions in such a way that the organism is a system of forces rather than a system of organs. Kielmeyer seems to be copying Kant when he says, "Each of the organs, in the modifications that it undergoes at each moment, is to such a degree a function of those that its neighbors undergo that it seems to be both a cause and an effect." It is easy to see why images of the circle and sphere enjoyed such prestige among Romantic naturalists: the circle represents the reciprocity of means and ends at the organ level, the sphere represents the totality, individual or universal, of organic forms and forces.

In France at the beginning of the nineteenth century it was Auguste Comte's biological philosophy, distinct from but not unrelated to Cuvier's biology, that set forth in systematic fashion the elements of a theory of living organization.[21] Arguing that "the idea of life is really inseparable from that of organization," Comte defined the organism as a *consensus* of functions "in regular and permanent association with a collection of other functions." *Consensus* is a Latin translation of the Greek *sympatheia*. Sympathy, wherein the states and actions of the various parts determine one another through sensitive communication, is a notion that Comte borrowed, along with that of synergy, from Paul-Joseph Barthez, who wrote:

> The preservation of life is associated with the sympathies of the organs, as well as with the organism of their functions.... By the

word "synergy," I mean a concourse of simultaneous or successive actions of the forces of diverse organs, that concourse being such that these actions constitute, by their order of harmony or succession, the intrinsic form of a function of health or of a genus of disease.[22]

Comte, of course, imported the concept of consensus into the theory of the social organism, and he later revised and generalized it in his work on social statics. "Consensus" then became synonymous with solidarity in organic systems, and Comte sketched out a series of degrees of organic consensus, whose effects become increasingly stringent as one rises from plants to animals and man.[23] Once consensus is identified with solidarity, one no longer knows which of the two, organism or society, is the model or, at any rate, the metaphor for the other.

It would be a mistake to ascribe the ambiguity of the relation between organism and society solely to the laxity of philosophical language. In the background, one can see the persistence of technological imagery, vividly present from Aristotle's day onward. At the beginning of the nineteenth century, a concept imported from political economy, the division of labor, enriched the concept of organism. The first account of this metaphorical transcription is due to the comparative physiologist Henri Milne-Edwards, who wrote the article on "Organization" for the *Dictionnaire classique des sciences naturelles* (1827). Since the organism was conceived as a sort of workshop or factory, it was only logical to measure the perfection of living beings in terms of the increasing structural differentiation and functional specialization of their parts, and thus in terms of relative complexity. But that complexity required, in turn, an assurance of unity and individualization. The introduction of cell theory in the biology first of plants (around 1825) and later of animals (around 1840) inevitably turned attention toward the problem of integrating elementary individualities and partial life forms into the totalizing individuality of an organism in its general life form.

Such problems of general physiology would increasingly claim the attention of Claude Bernard over the course of his career as a researcher and professor. For proof one need only consult the ninth of his *Leçons sur les phénomènes de la vie communs aux animaux et aux végétaux*. The organism is a society of cells or elementary organisms, at once autonomous and subordinate. The specialization of the components is a function of the complexity of the whole. The effect of this coordinated specialization is the creation, at the level of the elements, of a liquid interstitial milieu that Bernard dubbed the "internal environment," which is the sum of the physical and chemical conditions of all cellular life. "One might describe this condition of organic perfection by saying that it consists in an ever-more noticeable differentiation of the labor of preparing the constitution of the internal environment." As is well known, Bernard was one of the first to discover the constancy of this internal environment, along with a mechanism, which he called "internal secretion," for regulating and controlling that constancy, which has been known ever since as homeostasis. This was the original, and capital, contribution of Bernardian physiology to the modern conception of living organization: the existence of an internal environment, of a constancy obtained by compensating for deviations and perturbations, provides regulated organisms with an assurance of relative independence from variations stemming from the external conditions of their existence. Bernard was fond of using the term "elasticity" to convey his idea of organic life. Perhaps he had forgotten that the paradigmatic machine of his era, the steam engine, was equipped with a regulator when he wrote:

One treats the organism as a machine, and this is correct, but one considers it as a fixed, immutable, *mechanical machine*, confined within the limits of mathematical precision, and this is a serious mistake. The organism is an *organic machine*, that is, a machine equipped with a flexible, elastic mechanism, owing to the special organic processes it employs, yet without violating the general laws of mechanics, physics or chemistry.[24]

["Vie," *Encyclopaedia*, p. 768a–69a]

Life as Information

[26] Cybernetics is the general theory of servomechanisms, that is, of machines constructed so as to maintain certain outputs (products or effects) within fixed or variable limits. Such machines form the heart of self-regulating systems, and it is hardly surprising that self-regulating organic systems, especially those mediated by the nervous system, became models for the entire class. Of course, the analogy between servomechanisms and organisms runs both ways. In a regulated system, not only do the parts interact with one another but a feedback loop connects one or more monitored outputs to one or more regulatory inputs. Thus cybernetic machines, whether natural or man-made, are often described in terms of communications or information theory. A sensor monitors an output for deviations from a fixed or optimum level. When such a deviation is detected, the feedback loop signals the control input so as to convey an instruction from sensor to effector. It is the information content of this signal that is important, not its intrinsic force or magnitude. The feedback information embodies an order in two distinct senses: a coherent structure as well as a command.

An organism can thus be understood as a biological system, an open dynamical system that seeks to preserve its equilibrium and counteract perturbations. Such a system is capable of altering its relation to the environment from which it draws its energy in order to maintain the level of some parameter or to perform some activity.

Claude Shannon's work on communication and information theory and its relation to thermodynamics (1948) appeared to offer a partial answer to an age-old question about life. The second law of thermodynamics, which states that transformations of an isolated system are irreversible, owing to the degradation of energy in the system or, put another way, to the increase of the system's entropy, applies to objects indifferent to the quality of their state, that is, to objects that are either inert or dead. Yet an organism, which feeds, grows, regenerates mutilated parts, reacts to aggression, spontaneously heals certain diseases – is not such an organism engaged in a struggle against the fate of universal disorganization proclaimed by Carnot's principle? Is organization order amidst disorder? Is it the maintenance of a quantity of information proportional to the complexity of the structure? Does not information theory have more to say, in its own algorithmic language, about living things than Henri Bergson did in the third volume of his *Evolution créatrice* (1907)?

In fact, there is a great gulf, an irreducible difference, between current theories of organization through information and Bernard's ideas about individual development or Bergson's ideas about the evolution of species and the *élan vital*. Bernard had no explanation for the evolution of species, and Bergson had no explanation for the stability or reliability of living structures. But the combination of molecular biology with genetics has led to a unified theory of biochemistry, physiological regulation and heritability of specific variations through natural selection, to which information theory has added a rigor comparable to that of the physical sciences.

One question remains, however, within the theory itself, to which no answer is yet in sight: Where does biological information originate? André Lwoff maintains that biological order can arise only out of biological order, a formulation contemporary with the aphorisms *omne vivum ex vivo, omnis cellula ex cellula*. How did the first self-organization come about if communication depends on a prior source of information? One philosopher, Raymond Ruyer, puts the problem this way: "Chance cannot account for antichance. The mechanical communication of information by a machine cannot account for information itself, since the machine can only degrade or, at best, preserve information." Biologists do not regard this question as meaningless: contemporary theories of the origins of life on earth look to a prior chemical evolution to establish the conditions necessary for biological evolution. Within the strict confines of information theory, one young biophysicist, Henri Atlan, has recently proposed an ingenious and complicated response to the question in the form of what he calls a "noise-based principle of order," according to which self-organizing systems evolve by taking advantage of "noise," or random perturbations in the environment. Might the meaning of organization lie in the ability to make use of disorganization? But why always two opposite terms? ["Vie," *Encyclopaedia*, pp. 769a–69b]

Life and Death
[27] Paradoxically, what characterizes life is not so much the existence of the life functions themselves as their gradual deterioration and ultimate cessation. Death is what distinguishes living individuals in the world, and the inevitability of death points up the apparent exception to the laws of thermodynamics which living things constitute. Thus, the search for signs of death is fundamentally a search for an irrefutable sign of life.

August Weismann's theory of the continuity of the germinative plasma as opposed to the mortality of its somatic support (1885), Alexis Carrel's techniques for culturing embryonic tissue (1912), and the development of pure bacterial cultures established the potential immortality of the single-cell organism, which was mortal only by accident, and they lent credence to the idea that the phenomena of aging and natural death after a certain span of years are consequences of the complexity of highly integrated organisms. In such organisms, the potentialities of each component are limited by the fact that other components perform independent functions. Dying is the privilege, or the ransom – in any case, the destiny – of the most highly regulated, most homeostatic natural machines.

From the standpoint of the evolution of species, death marks an end to the reprieve that the pressure of natural selection grants to mutants temporarily more fit than their competitors to occupy a certain ecological context. Death opens up avenues, frees up spaces and clears the way for novel life forms – but this opening is illusory, for one day the bell will toll for today's survivors as well.

From the standpoint of the individual, the genetic heritage is like a loan, and death is the due date when that loan must be repaid. It is as if, after a certain time, it were the duty of individuals to disappear, to revert to the status of inert matter.

Why, then, did Freud's theory of the "death instinct," presented in *Beyond the Pleasure Principle* (1920), meet with so much resistance? In Freud's mind this idea was associated with energeticist concepts of life and of the psychic processes. If a living thing is an unstable system constantly forced to borrow energy from the external environment in order to survive, and if life is in tension with its nonliving environment, what is so strange about hypothesizing the existence of an instinct to reduce that tension to zero, or, put differently, a striving toward death? "If we grant that the existence of a living thing depends on the prior existence of the inanimate objects from which it arises, it follows that the death instinct is in accord with the formula stated earlier according to which every instinct tends to restore a prior state." Perhaps Freud's theory will now be reexamined in light of the conclusions of Atlan's work: "In fact, the only identifiable project in living organisms is death. But owing to the initial complexity of those organisms, perturbations capable of disrupting their equilibrium give rise to still greater complexity in the very *process* of restoring equilibrium."[25]

Finally, one might also wish to understand the reason for, and meaning of, the reactional desire for immortality, the dream of survival — which Bergson calls a "useful theme of mythification" — found in certain cultures. A dead tree, a dead bird, a carcass — individual lives abolished without consciousness of their destiny in death. Is not the value of life, along with the acknowledgment of life as a value, rooted in knowledge of its essential precariousness?

> Death (or the illusion of death) makes men precious and pathetic. Their ghostly condition is moving. Every act may be their last. Not a face they make is not on the point of vanishing like a face in a dream. Everything in mortals has the value of what is irretrievable and unpredictable.[26]

["Vie," *Encyclopaedia*, pp. 769b–69c]

CHAPTER FIVE

Epistemology of Physiology

A Baroque Physiology

Objectives and Methods
[28] In 1554, when the celebrated Jean Fernel (1497–1558) collected his previously published treatises under the title *Universa medicina*, he provided a preface detailing his conception of medicine's constituent elements and its relation to other disciplines. The first of those elements he called *Physiologia*, and under that head he placed his 1542 treatise *De naturali parte mediciniae*. The object of physiology was described as "the nature of the healthy man, of all his forces and all his functions." It scarcely matters here that Fernel's idea of human nature is more metaphysical than positive. The point to be noted is that physiology was born in 1542 as a study distinct from, and prior to, pathology, which itself was prior to the arts of prognosis, hygiene and therapeutics.

Since then the term "physiology" gradually acquired its current meaning: the science of the functions and functional constants of living organisms. The seventeenth century saw the appearance of, among other works, *Physiologia medica* (Basel, 1610) by Theodor Zwinger (1553–1588), *Medicina physiologica* (Amsterdam, 1653) by J.A. Vander-Linden (1609–1664) and *Exercitationes physiologicae* (Leipzig, 1668) by Johannes Bohn (1640–1718). In the eighteenth century, if Frederick Hoffmann (1660–1742) published his *Funda-*

menta Physiologiae as early as 1718, it was undeniably Albrecht von Haller (1708–1777) who made physiology an independent discipline of research and specialized teaching. His eight-volume *Elementa physiologiae* (1757–66) remained a classic for half a century. But it was even earlier, in 1747, that von Haller, after having used his teacher Herman Boerhaave's *Institutiones medicinae* in his courses for nearly twenty years, decided to publish his first textbook, *Primae lineae physiologiae*, in the introduction to which he defined physiology in a way that established the spirit and method of the discipline for a long time to come: "Someone may object that this work is purely anatomical, but is not physiology anatomy in motion?" [*Etudes*, pp. 226–27]

[29] It is easy to understand why anatomy took priority over the study of organ functions. In many cases people felt that the best way to understand the functions of the organs was to inspect their shapes and structures. Structures were macroscopic, and functions, no matter how complex the underlying processes, could be understood by analogies with man-made instruments suggested by superficial structural similarities. From the structure of the eye, for example, it was possible to deduce a few crude notions about the physiology of vision based on knowledge of the construction and use of optical instruments. But the structure of the brain as revealed by dissection implied nothing about its function, because there was no man-made technology or instrument to which it could be compared. When von Haller described the pancreas as "the largest salivary gland," its secretory function could perhaps be compared to that of the parotid, but it was impossible to go further. In his *Eloge* of the surgeon Jean Méry (d. 1722), Fontenelle remembered one of Méry's frequently quoted statements: "We anatomists are like the deliverymen of Paris, who know even the smallest, most out-of-the-way streets but have no idea what goes on inside people's homes."

To find out what went on inside, several options were available: one could monitor comings and goings, introduce spies into the household, or smash the walls partly or totally in order to catch a glimpse of the interior. Méry's statement notwithstanding, physicians had long used such procedures to find out what was going on inside the animal organism. Experimentation through organ ablation was a natural extension of surgical excision. Andreas Vesalius (1514–1564), the founder of modern anatomy, concluded his celebrated *Humani corporis fabrica* (1543) with remarks on the usefulness of vivisection and a discussion of its techniques, in the course of which he reported on experiments with ablation of the spleen and kidney in dogs. In the seventeenth century, the mechanist conception of organic structures encouraged this practice, at once premeditated and blind. If the body is a machine, one should be able to discover the functions of particular parts, of the mechanism's cogs and springs, by destroying parts and observing the disturbance or breakdown of the machine's operation. ["Physiologie," *Encylopaedia*, pp. 1075 b–c]

[30] By 1780, physiology had finally outgrown iatromechanical theories thanks to the work of Antoine-Laurent Lavoisier and Luigi Galvani. Chemistry and physics would supply the new models. Laws, in the Newtonian sense, would replace mechanical theorems. The Newtonian spirit, which had breathed new life into eighteenth-century science, transformed physiology by supplying it not so much with new concepts as with new methods. Tired of rhetorical conflict, physiologists focused on specific properties of the vital functions. But the very influence of Newtonian science still fostered dogmatic attitudes in many minds of philosophical bent.

Vitalism was one reaction to this dogmatism. Far too much ill has been spoken of vitalism. It did not hinder experimentation or the formulation of new concepts in neurophysiology; on the contrary, it encouraged progress. Vitalists, emulating the Newtonian style, focused on the details of biological functions without speculating as to their causes. The so-called Montpellier School of vitalists was led by Théophile de Bordeu (1722-1776) and Paul-Joseph Barthez (1734-1806). There was no more metaphysics in what they called "vital movement" or the "vital principle" than there was in what Haller called "irritability." Barthez's *Nouveaux éléments de la science de l'homme* (1778) was in many ways a system of empirical physiology: "The vital principle in man," Barthez argued, "should be conceived in terms of ideas distinct from our ideas of the attributes of body and soul."

Antoine Augustin Cournot seems to have grasped the originality of vitalist physiology: "Vitalism brings out the analogies that all manifestations of life exhibit in such astonishing variety, and takes them for its guide, but does not pretend that it can penetrate the essence of life."[27] To study this "astonishing variety," eighteenth-century physiologists looked at the whole animal kingdom, from polyp to man, from the frog to the orangutan – that strange missing link that eighteenth-century writers referred to as "jungle man," and whose linguistic ability and intelligence they studied by comparison with the human.

If classicism in biology means rigorous classification combined with mathematical generalization, then the term does not apply to eighteenth-century physiology, which took all living matter for its subject. It tolerated "in-between" categories, as Leibniz called them, and if it generalized at all, it was in imitation of life itself, working endless variations on a small number of themes. It was a picturesque science, curious about minutiae and about nature's intricate ways.

Eighteenth-century physiology stands poised between the doctrinaire dignity of the previous century's medical systems, which bore the weight of earlier dogmas, and the rather frenetic experimentalism of the nineteenth century. It was a fruitful period, as old ideas were exploded by new experience. Bold speculation was the order of the day, and traditional methods gave way to intuition. It would not be long before new techniques, many discovered by chance, revealed which innovative intuitions were sound and which were not.

The period's physiology was as vital as life itself, as men like Lazzaro Spallanzani and Armand Séguin experimented on themselves, while Robert Whytt, René Antoine Ferchault de Réaumur and Stephen Hales performed similar tests on frogs, buzzards and horses. It was, in every sense of the word, a *baroque* physiology. ["Physiologie animale," *Histoire générale*, vol. 2, pp. 618-19]

[31] The eighteenth century was an age not only of enlightenment but also of progress, and of technological progress first and foremost. [...] Inventiveness and applications were the watchwords governing experimentation in physics and chemistry especially. Researchers investigated heat, electricity, changes of physical state, chemical affinities, the decomposition of matter, combustion and oxidation, and their results often spilled over into physiology, raising new problems for further investigation. Electricity joined light and heat in suggesting analogies that could be used to explain "vital forces." The analysis of different kinds of "airs," or gases, gave positive content to the idea of exchanges between organisms and their environment. This "pneumatic" chemistry resolved the once purely speculative rivalry between iatromechanists and iatrochemists in favor of the latter. New instruments such as the thermometer and calorimeter made it possible to measure important biological parameters. It was in 1715 that Daniel Fahrenheit solved the technical problems that had delayed the construction of sensitive, reliable thermometers, and Réaumur followed with further improvements in 1733. In 1780, Lavoisier and Pierre-Simon Laplace built a device for measuring quantities of heat.

Thus, apart from research on the nervous system, most of the major discoveries in eighteenth-century physiology were the work, if not of amateurs, then of men whose primary specialty was not medicine. Among them were such names as Hales, John Boynton Priestley, Lavoisier, Réaumur and Spallanzani. [...] Contemporary texts therefore give a misleading, altogether too academic picture of the state of the discipline. It is odd that when Pierre Jean George Cabanis published his survey of the new physiology in 1804, he mentioned only works and experiments by physicians, even though he was well aware that one of the reasons for the superiority of the new medicine was the contribution of "the collateral sciences, which are constantly providing us with new insights and instruments."[...]

The seventeenth and eighteenth centuries are alike in that both were dominated by a single great discovery. But William Harvey's work nearly inaugurated his century, whereas Lavoisier's nearly closed his. Harvey invented a mechanical model in order to describe one phenomenon; Lavoisier introduced a chemical model to explain another. ["Physiologie animale," *Histoire générale*, vol. 2, pp. 593-94]

Circulation

[32] The work of those referred to as "iatromechanics" (or, equally appropriately, "iatromathematicians") was constantly motivated by an ambition to determine, through measurement and calculation, the laws of physiological phenomena. This was the least contestable of their postulates, moreover. The circulation of the blood and the contraction of the muscles had always been objects of predilection for the physicians of this school.

In *De motu cordis* William Harvey summarized his conclusions as an anatomist and his observations as a vivisectionist. He calculated the weight of the blood displaced by the heart simply in order to show that so large a quantity of blood could not possibly be produced continuously by any organ or be dissipated by the organism. Giovanni Alfonso Borelli was the first to view the circulatory function, by then well established, as an ideal problem to which to apply the laws of hydraulics. He attempted to calculate the force of the systolic contraction. Assuming that the contractive force of a muscle is proportional to its volume and that the volume of the human heart is equal to the combined volume of the masseter and temporal muscles, he determined that the contractive force of the heart is equal to three thousand Roman pounds (1 Roman pound = 11½ ounces). As for the pressure the heart communicates to the blood, an elaborate series of deductions led him to the figure of 135,000 pounds!

In 1718, James Keill (1673-1719) devoted three essays of his *Tentamina medico-physica* to the problems of determining the quantity of blood, its velocity and the force of the heart. He estimated that the blood accounts for 100 pounds of the weight of a 160-pound man; the blood in the aorta travels at a rate of five feet, three inches per hour; and the force of the heart is twelve ounces. (The modern figures are that the weight of the blood is one third of body weight; the velocity of the blood is twenty inches per second; and the work of the contraction wave of the left ventricle is three and a half ounces.)

A skilled experimentalist and a religious zealot, Stephen Hales made an important contribution to circulatory mechanics when he published his *Statical Essays, Containing Haemastatics etc.* (1733). He had already written important works on mathematical botany. His *Vegetable Staticks* (1727) contained illustrations of instruments he had built to measure variations in sap pressure in roots and branches. From there it was but a short step to measuring the pressure of blood in the vessels using a manometer consisting of a long glass tube attached by a cannula to the jugular vein or carotid or crural artery of a horse, dog or sheep. Hales was able to establish that the blood pressure is lower in the veins than it is in the arteries (in the horse, the blood rose to a height of nine feet when the cannula was inserted into the crural artery but to only fifteen inches when inserted into the jugular vein); that it fluctuates with the systole and the diastole; that it is characteristic of a given animal species; and that it is a test of the state of the heart.

Apart from the tentative work of Borelli and Keill, the next work of equally great importance was Jean Poiseuille's *Recherches sur la force du coeur aortique* (1828). Haller knew and spoke of the work of Hales but treated it as a development of ideas of Borelli's, failing to appreciate the novelty of the concept of arterial pressure.

The importance and originality of Hales's research should not, however, detract from the merits of those who, following him and building on his results, made progress toward solving some of the major problems of hemostatics and hemodynamics. Daniel Bernoulli, professor of anatomy at Basel from 1733 to 1751, was the first to explain correctly how to calculate the work done by the heart as the product of the weight of blood expelled times the systolic displacement. He also made comparative studies of the flow of liquids in rigid pipes and in living vessels (*Hydrodynamica*, 1738). His pupil, Daniel Passavant (*De vi cordis*, 1748), used Hales's figures to arrive at a more accurate evaluation of the work of the heart, one close to presently accepted values.

Toward the end of the seventeenth century, researchers began to investigate the causes of the movement of blood in the veins, which are not directly connected to the arteries. Borelli, though admitting the force of the heart, denied that it was sufficient to drive the blood in the veins. Hence, the microscopic examinations by Marcello Malpighi (1661) and Antonie van Leeuwenhoek

(1690) of the capillary circulation in the mesentery of frogs and the tail of tadpoles assumed a very great importance, and so did Cowper's investigations of the mesentery of a cat (1697). Albrecht von Haller (*De motu sanguinis*, 1752) showed that the heart's pulse could be observed simultaneously in both arteries and capillaries, proving that the power of the heart extended to the capillaries. His theory of irritability then enabled him — as the theory of tonicity enabled Georg Ernst Stahl — to argue that the sheath of the capillary can contract independently, imparting an additional circulatory impetus to the blood. Spallanzani also contributed to the solution of this problem in a series of papers, *Sur la circulation observée dans l'universalité du système vasculaire*, *Les Phenomènes de la circulation languissante*, *Les Mouvements du sang indépendants de l'action du coeur* and *La Pulsation des artères* (1773). ["Physiologie animale," *Histoire générale*, vol. 2, pp. 601-603]

Respiration

[33] From Robert Boyles's *Nova experimenta physiomechanica de vi aëris elastica et ejusdem effectibus* (1669), John Mayow concluded, about 1674, that animal respiration involves the fixation of a "spirit" contained in the air. It is the eventual depletion of this spirit from the air in a confined space that renders it unfit to sustain life. In his *Experiments and Observations on Different Kinds of Air* (1774-77), John Boynton Priestley reported that a sprig of mint will release enough dephlogisticated air (oxygen) to support combustion in an inverted bell jar. In 1775, he informed the Royal Society that dephlogisticated air obtained by the same method could sustain the respiration of a mouse.

Lavoisier's first investigations of the "principles" with which metals combine during calcination had much the same aims as Priestley's studies: the analysis, detection and identification of various kinds of gases. The influence of these gases on animal respiration was initially conceived as a kind of chemical test to study the experimental separation of the hypothetical elements of atmospheric air, which had been downgraded from its ancient status as an element. Lavoisier's more systematic studies of the respiration of birds (1775–76) and guinea pigs (1777) enabled him to present to the Académie des Sciences a definitive paper on changes in the blood during respiration (*Mémoire sur les changements que le sang éprouve dans les poumons et sur le mécanisme de la respiration*, 1777).

Using comparative measurements of the volume of gas absorbed and the quantity of heat released by guinea pigs placed in a calorimeter, Lavoisier and Pierre-Simon Laplace were able to generalize all these observations and to state, in 1780, that respiration is nothing other than a slow form of combustion identical to the combustion of carbon. They were wrong, however, in asserting that respiration is the combustion of carbon alone, as Lavoisier was obliged to admit in his 1785 paper *Sur les altérations qu'éprouve l'air respiré*, in which it was shown that respiration produced not only carbon dioxide from the combustion of carbon but also water from the combustion of hydrogen. They were also wrong to describe the lung as the locus and seat of combustion, the heat from which they believed was distributed throughout the organism by the blood.

Finally, after measuring, in collaboration with Séguin, who volunteered to serve as an experimental subject, energy exchanges in human beings, Lavoisier summed up his views in two papers, *Sur la respiration des animaux* (1798) and *Sur la transpiration des animaux* (1790). His declaration of principle is often cited:

> Comparison of these results with earlier ones shows that the animal-machine is controlled by three principal governors: respiration, which consumes hydrogen and carbon and which supplies caloric; transpiration, which fluctuates with the requirements of caloric;

and, last but not least, digestion, which restores to the blood what it has lost by respiration and transpiration.

["Physiologie animale," *Histoire générale*, vol. 2, pp. 595-96]

[34] The end of the debate over the *causes* of animal heat coincided with the beginnings of a debate over the *seat* of the phenomenon. Lavoisier had proposed that carbon and hydrogen in the blood are oxidized in the vessels of the lung by the action of oxygen on a hydrocarbonic fluid secreted therein. Objections to this view were put forward in 1791 by Jean-Henry Hassenfratz, a former assistant of Lavoisier and later a disciple of the mathematician Lagrange, who deserves credit for having first raised them. If all the heat in the organism is first released in the lungs, Hassenfratz asked, why don't the lungs dry out? Or, in any case, why aren't they warmer than the other organs of the body? Isn't it therefore more likely that heat is released in all parts of the body supplied with blood? According to Lagrange, pulmonary blood, in contact with inhaled air, becomes saturated with dissolved oxygen, which then reacts with the carbon and hydrogen in the blood to yield carbon dioxide and water, which are released with the exhaled air. This explanation is roughly correct (except for the fact that oxidation takes place not in the blood but in the cells themselves), but it was not confirmed experimentally until 1837, when Gustav Magnus used a mercury pump to detect the presence of free gases in venous and arterial blood.

Furthermore, the posthumous publication of Jean Sénebier's *Mémoires sur la respiration* (1803) revealed that Spallanzani devoted the last years of his life to systematic experimentation on respiration in vertebrates and invertebrates, from which he, too, concluded, after thousands of experiments, that oxygen is absorbed and carbon dioxide released by all tissues and organs, and that amphibians and reptiles may absorb more oxygen through the skin than through the lungs. In other words, in animals with lungs, the lungs are the organ of expression but not the organ of exercise of a function coextensive with the entire organism. By performing experiments to dissociate the respiration function from the pulmonary organ, Spallanzani, even more than Lavoisier, but using his methods of comparative physiology, laid the groundwork for a general physiology. ["Physiologie animale," *Histoire générale*, vol. 2, pp. 597-98]

An Experimental Science

New Styles in the Age of Laboratories

[35] Relations between the first systematically experimental physiology and the theoretically, that is, mathematically, more advanced physical and chemical sciences followed different patterns in France and Germany. The first year of the nineteenth century witnessed the publication of *Recherches physiologiques sur la vie et la mort* by Xavier Bichat, who strove to preserve the distinctiveness of biology's subject matter and methods in the face of efforts by physicists and chemists to annex physiology to their own disciplines. Bichat, the brilliant founder of general anatomy, or the study of organic tissues, and a tenacious champion of the concept of the "vital properties" of such tissue, had a profound influence on the first French physiologists to embrace methodical experimentation. Although François Magendie and Claude Bernard, unlike Bichat, never doubted the need to use physical and chemical methods to investigate physiological mechanisms, they never ceased to believe in the uniqueness of organic phenomena. This was the distinctive feature, one might even say the national trademark, of French physiology, at a time when physiology in Germany was already being done, like physics and chemistry, in laboratories equipped with steadily improving, in-

dustrially manufactured equipment, while French physiologists soldiered on with nothing more than the rudimentary facilities available to university professors and hospital physicians as perquisites of their positions. This accounts for the undeniable difference in orientation and style of research on either side of the Rhine. When Bernard compared himself to Hermann von Helmholtz and remarked in his notebook that his German colleague found only what he was looking for, he was remarking not only on a difference in spirit but also on a disproportion of available means. For, by this time, new discoveries in physiology were not to be had on the cheap. This was one reason why doctors who came from the United States to study in Europe generally preferred to study with German physiologists, especially Karl Ludwig (1816-1895), rather than with their French counterparts. The first physiology laboratories were established in the United States in the 1870s, and they soon could boast of facilities and equipment superior to the finest European laboratories. As physiology laboratories grew larger and their equipment more complex, it became common for research to be conducted by teams rather than individuals. Researchers were more anonymous, but the discipline as a whole was less dependent on individual strokes of genius. ["Physiologie," *Encyclopaedia*, pp. 1076a-b]

[36] If physics and chemistry exerted growing influence on research in physiology, it was mainly because physiologists found the techniques of those sciences indispensable as research tools, though not necessarily as theoretical models. While Claude Bernard's often-repeated claim that physiology became scientific when it became experimental need not be taken strictly literally, it is certainly true that the radical difference between the physiological experimentation of the nineteenth century and that of the eighteenth century lay in the systematic use of measuring and detection instruments and equipment borrowed or adapted from,

or inspired by, the flourishing sciences of physics and chemistry. To be sure, Ludwig and his school in Germany deserve credit for their persistent interest in physical and chemical methods, as well as their ingenuity in the construction and use of new instruments. Bernard's research seems relatively artisanal by comparison. It was also more narrowly biological, vivisection being its chief technique. But it would be misleading to suggest that there was a fundamental difference of national intelligence or genius between the two countries. Indeed, the history of physiology (not to be confused with the history of physiologists) shows that researchers in both countries learned from each other and exchanged ideas about how to improve experimental methods by borrowing from other disciplines. Ludwig became famous, for example, not only for building the mercury pump for separating blood gases but even more for the construction of the celebrated kymograph (1846). In terms of technological phylogenesis, the ancestor of this instrument was surely the "hemodynamometer" of Jean Poiseuille. Ludwig's genius was to couple Poiseuille's arterial manometer to a graphic recorder. When Etienne-Jules Marey (1830–1904) set out to develop and perfect the graphic method in France, he was therefore indebted indirectly to Poiseuille and directly to Ludwig. [*Etudes*, pp. 231–32]

[37] Even though analytical techniques borrowed from physics and chemistry proved fruitful in physiology, they could not discredit or supplant the method that Claude Bernard called "operative physiology," in which vivisection, resection and ablation are used to disturb the balance of otherwise intact organisms. This traditional method was used by Julien Jean César Legallois and François Magendie early in the nineteenth century and by Pierre Flourens later on. Gustav Théodore Fritsch and Julius Edward Hitzig used galvanic stimulation of the cortex to distinguish between motor and sensory functions in the cerebral

lobes (1870). Friedrich Goltz refused to admit the validity of any other method.

Most of the early work on glands relied on ablation. Charles-Edouard Brown-Séquard used it to study the adrenal function (1856), Moritz Schiff to study the thyroid function (1859 and 1883), and Emile Gley to study the parathyroid function (1891). Before the active principles of the various endocrine secretions could be identified (adrenalin by Takamine in 1901, thyroxin by Edward Calum Kendall in 1914), physiologists tried to demonstrate the chemical actions of glands by means of organ transplants. In 1849, for example, Arnold Adolphe Berthold reversed the effects of castration in a rooster by transplanting testicles into its peritoneal cavity. In 1884, Schiff transplanted a thyroid from one dog to another, the first instance of an operation that had become commonplace by the end of the century.

The techniques of operative physiology were used in conjunction with the new methods of electrophysiology to map the functions of nerve bundles in the spinal cord and to produce an atlas of cerebral functions. Charles Scott Sherrington's discoveries were based on very precise operative techniques involving differential "preparations" (decorticated, decerebrated and decapitated animals). In studying the functions of the sympathetic nervous system, physiologists relied on vivisection long before turning to chemical methods with John Newport Langley. It was vivisection that enabled Claude Bernard in 1854 to demonstrate the role played by the sympathetic system in calorification (regulating the circulatory flow in the capillaries). [...]

Despite the fact that some of its greatest representatives – Bernard, for instance – insisted that physiology was an independent discipline with methods of its own, while others stressed its subordination to physics and chemistry (Karl Ludwig) or mathematics (Hermann von Helmholtz), nineteenth-century physiology was not altogether devoid of unity of inspiration or purpose. It was the science of functional constants in organisms. One sign that it was an authentic science is that from Magendie to Sherrington and Pavlov we find a great many overlapping studies and discoveries and a large number of separate and simultaneous discoveries (sometimes with disputes over priority, sometimes not). The history of physiology enjoyed a relative independence from the history of physiologists. It matters little whether it was Sir Charles Bell or Magendie who "really" discovered the function of the spinal nerve roots, whether Marshall Hall or Johannes Müller first discovered reflex actions, Emile Du Bois-Reymond or Hermann motor currents, or David Ferrier or Hermann Munk the cortical center of vision. As soon as methods and problems become adjusted to each other, as soon as instruments become so highly specialized that their very use implies the acceptance of common working hypotheses, it is true to say that science shapes scientists just as much as scientists shape science. ["Physiologie en Allemagne," *Histoire générale*, vol. 3, pp. 482-84]

Physiology Is Not an Empirical Science

[38] To concentrate solely on the instrumental side of experimentation would be to give a misleading idea of the development of nineteenth-century physiology, though. Some historical sketches and methodological manifestoes give the impression that instruments and the techniques that used them were somehow *ideas*. To be sure, using an instrument obliges the user to subscribe to a hypothesis about the function under study. For example, Emile Du Bois-Reymond's inductive slide physically embodies a certain idea of the functions of nerve and muscle, but it is hardly a substitute for that idea: an instrument is an aid to exploration but of no use in framing questions. Thus, I cannot agree with those historians of physiology, professional as well as amateur, who

would outdo even Claude Bernard's open hostility to theory by ascribing all progress in nineteenth-century physiology to experimentation. The theories that Bernard condemned were systems such as animism and vitalism, that is, doctrines that answer questions by incorporating them. For Bernard, data collection and research were to be distinguished from fruitpicking and stone quarrying: "To be sure," he wrote, "many workers are useful to science though their activities be limited to supplying it with raw or empirical data. Nevertheless, the true scientist is the one who takes the raw material and uses it to build science by fitting each fact into place and indicating its significance within the scientific edifice as a whole."[28] Furthermore, the *Introduction à l'étude de la médecine expérimentale* (1865) is a long plea on behalf of the value of ideas in research, with the understanding, of course, that in science an idea is a guide, not a straitjacket.

While it is true that empirical experimentation enabled Magendie to establish the difference in function between the anterior and posterior roots of the spinal cord in 1822, it must be granted that Sir Charles Bell had not found it unhelpful eleven years earlier to rely on an "idea," namely, his *Idea of a New Anatomy of the Brain* (1811): if two nerves innervate the same part of the body, their effects must be different. The spinal nerves have both motor and sensory functions, hence different anatomical structures. Given that the spinal cord has two roots, each must be a functionally different nerve.

Although the earliest results in the physiology of nutrition came from Justus von Liebig's chemical analyses and Magendie's investigations of the effects of different diets on dogs, the work of William Prout (1785–1850) on saccharides, fats and albumins in the human diet cannot be said to have suffered from the fact that his work was guided by an "idea," namely, that what humans eat, whether in traditional diets or carefully composed menus, reflects an instinctive need to reconstitute that prototype of all diets, milk.

If the work of Hermann von Helmholtz dominated the physiology of the sensory organs in the nineteenth century, it was because he, justly renowned as an inventor of instruments (such as the ophthalmoscope in 1850), was an ingenious experimentalist with a broad mathematical background that he owed to his training as a physicist. When a mathematical mind turns to natural science, it cannot do without ideas. A student of Johannes Müller, whose law of the specific energy of the nerves and sensory organs guided all the period's thinking about psychophysiology, Helmholtz was able to combine his own insistence on measurement and quantification with a philosophical understanding of the unity of nature that he took from his teacher, whose influence is apparent in all of Helmholtz's work on muscular work and heat. If the 1848 paper on the principal source of heat in the working muscle reports data gathered with temperature-measuring instruments specially designed by Helmholtz himself, his 1847 work on the conservation of force, *Über die Erhaltung der Kraft*, was inspired by a certain idea of the unity of phenomena and the intelligence thereof.

In his final lectures at the Muséum, published by Dastre as *Leçons sur les phénomènes de la vie communs aux animaux et aux végétaux* (1878-79), Claude Bernard discussed, along with other key ideas, the unity of the vital functions: "There is only one way of life, one physiology, for all living things." By then, this idea epitomized his life's work; earlier, however, it had surely guided his research. In the 1840s, it had encouraged him to challenge the conclusions reached by Jean-Baptiste Dumas and Jean Baptiste Boussaingault in their *Statique chimique* (1841), much as von Liebig was doing at the same time in Germany. Dumas and Boussaingault had argued that animals merely break down organic compounds, which only plants could synthesize. Bernard, however, described all his work on the glycogenic function of the liver, from the 1848 paper read to the Académie des Sciences to the doctoral thesis of 1853, as a consequence of the assumption that there is no difference between plants and animals with respect to their capacity to synthesize "intermediate principles." Indeed, there is no hierarchy of plant and animal kingdoms; still more radically, Bernard claimed that from the standpoint of physiology there are no kingdoms. He refused to believe that there was something plants could do that animals could not. In answering his critics by rejecting a certain conception of the division of labor among organisms, Bernard may have revealed the (not very mysterious) secret of his success. To be sure, Bernard's belief was a "feeling," not an "argument," as he stated in the *Leçons de physiologie expérimentale appliquée à la médecine* (1855-56). It was not even a working hypothesis concerning the functions of some organ. But even if it was not strictly necessary to hold this belief in order to discover the liver's glycogenic function, the fact that Bernard *did* hold it helped him to embrace an interpretation of his results that most of his contemporaries found disconcerting.

These examples, drawn from various fields of research, show that experimentalists need not pretend to be pure empiricists, working without ideas of any kind, in order to make progress. Bernard observed that the experimentalist who doesn't know what he is looking for won't understand what he finds. The acquisition of scientific knowledge requires a certain kind of lucidity. Scientific discovery is more than individual good fortune or accidental good luck; hence, the history of science should be a history of the formation, deformation and rectification of scientific concepts. Since science is a branch of culture, education is a prerequisite of scientific discovery. What the individual scientist is capable of depends on what information is available; if we forget that, it is easy to confuse experimentation with empiricism. [*Etudes*, pp. 232-35]

Accidents, the Clinic and Socialization

[39] It is impossible to write the history of random events, and if science were purely empirical it would be impossible to write the history of science. One must have a rough sense of periodization to benefit from anecdotal evidence. Research on digestion offers a good example of this. A great deal was learned about digestive physiology in the second half of the nineteenth century, after researchers discovered how to use gastric fistulas to perform the experiments on which today's understanding of digestion is based. After 1890, in particular, Ivan Pavlov made good use of a technique that he himself had helped to perfect. But that technique had been pioneered, simultaneously but quite independently, by Vassili Bassov in 1842 and Nicolas Blondlot in *Traité analytique de la digestion, considérée particulièrement dans l'homme et les animaux vertébrés* (1843).[29] Nearly two centuries earlier, Regner de Graaf had successfully produced a pancreatic fistula in a dog (*Disputatio medica de natura et usu succi pancreatici*, 1664), but no one ever attempted the same operation with other organs. René Antoine Ferchault de Réaumur's experiments in 1752 and Lazzaro Spallanzani's in 1770, both of which had been performed in order to decide between van Helmont's chemical and Borelli's mechanical explanation of digestive phenomena, involved the collection of gastric juices from the esophagus by ingenious but roundabout means; neither man seems to have thought of introducing an artificial fistula into the stomach. The invention of the artificial gastric fistula followed the American physician William Beaumont's publication of his observations of a Canadian hunter, Alexis Saint-Martin, who, after being shot in the stomach, presented with a stomach fistula whose edges adhered to the ab-

dominal walls. Beaumont, having taken the man into his employ, reported his observations of contractions and gastric secretions in a paper entitled "Experiments and Observations on the Gastric Juice and the Physiology of Digestion" (1833). The history of surgery offers few other cases of spontaneous stomach fistulas, and none was observed in any way comparable to Beaumont's. Thus, an accident suggested a method of experiment – one that Bassov and Blondlot would later make systematic use of. It was no accident, however, that this original accident was first patiently exploited and later intentionally reproduced. The chemists of the period were intensely interested in the chemical composition of foodstuffs, and this had led to interest in the chemistry of digestive secretions. The first chemical analyses of gastric juices were undertaken by Prout (1824). However, because physiologists needed to obtain these juices, uncontaminated by food particles, in considerable quantities, they had to figure out how to retrieve the juices at the moment of secretion. They also had to find the right animal to study, one with an appropriate anatomical structure and digestive patterns.

Thus, accidents and unforeseen events sometimes give rise to new techniques of observation and methods of research. One thing leads to another. Similarly, scientific problems sometimes arise in one domain or field of science only to be resolved in another. For example, the history of physiology cannot be entirely divorced from the parallel histories of the clinic and of medical pathology. And it was not always physiology that instructed pathology: relations among the disciplines were complex. Consider for a moment the history of nervous and endocrine physiology in the nineteenth century. Clinical observation revealed functional disorders and disturbances that physiologists at first found difficult to explain, for they could not identify what regulatory mechanism had gone awry. Without the history of clinical work on

Addison's disease or surgery on goiters, it is impossible to make sense of progress in understanding the physiology of the adrenal and thyroid glands. The work of the physiologist Brown-Séquard often began with some medical finding; and in this respect it differed sharply from the work of certain other physiologists, such as Claude Bernard. [*Etudes*, pp. 236-38]

[40] Disease was not the physiologist's only source of scientific challenges. Healthy individuals are neither idle nor inert and cannot be maintained artificially at the beck and call of ingenious or restless experimentalists. The healthy person too is, by definition, capable of carrying out tasks set by nature and culture. In the nineteenth century, the development of industrial societies in Europe and North America led to the socialization, and therefore politicization, of questions of subsistence, diet, hygiene and worker productivity. It is no accident that problems of energy utilization arose around this time, especially in Germany, in regard to both the steam engine and the human organism. The same doctor, Julius Robert von Mayer, who proved that energy could not be destroyed but only converted from one form to another (1842) – from work to heat, or vice versa – also published the results of his research on dietary energetics in 1845. His work confirmed that of von Liebig, whose research on organic chemistry as applied to physiology (1842) related the calorific values of various nutrients such as fats, sugars and proteins to various organic phenomena involving expenditures of energy; these results were further elaborated and refined by Marcellin Berthelot (1879) and Max Rubner and Wilbur Olin Atwater (1904).

Similarly, technological progress and economic change had subjected human beings to extreme conditions. People had been forced, in war and peace, to endure extremes of temperature, to work at high altitudes, to dive to great depths; others chose to subject themselves voluntarily to extreme conditions, as in sport. To cite just one example, Paul Bert's research on anoxemia at high altitude (1878) paved the way for later studies of phenomena that had to be understood before intercontinental passenger flight could become routine. ["Physiologie," *Encyclopaedia*, p. 1076c–77a]

The Major Problems of Nineteenth-Century Physiology

Bioenergetics

[41] The resolution, through chemistry, of an age-old problem of physiology forced physiologists to confront a problem that physics had yet to resolve: How can energy exist in a variety of forms? In Cartesian mechanics, statics depend on the conservation of work, and dynamics on the conservation of momentum (mv, mass times velocity). Leibniz, in his critique of Cartesian mechanics, considered the quantity mv^2 (mass times the square of velocity, which he called the "live force") to be a substance, that is, an invariant, but he failed to note that in any real mechanical system involving friction, this quantity does not remain constant, due to the generation and loss of heat. The eighteenth century failed to formulate the notion of conservation of energy. At the beginning of the nineteenth century, two forms of energy were recognized: the energy of motion (kinetic or potential) and heat. But observations made by technicians and engineers concerning the operation of the steam engine, the boring of cannon barrels and so on led to study of the relations between work and heat.

The first person to assert the indestructibility and, consequently, the conservation of energy through various transformations was the German physician von Mayer, who based his claims on medical observations made in Indonesia in 1840 having to do with the influence of heat on the oxidation of blood. In 1842, von Liebig published a theoretical paper by Mayer, entitled "Bemerkungen über die Krafte der unbelebten Natur," in the *Annalen der Chemie und Pharmacie*, but it attracted little attention initially. In 1843, James Prescott Joule undertook to determine experimentally the mechanical equivalent of the calorie, and in an 1849 paper read before the Royal Society he claimed responsibility for a discovery – and Mayer then felt compelled to dispute his claim of priority. In 1847, meanwhile, von Helmholtz also published a paper entitled "Über die Erhaltung der Kraft."

Mayer's work actually was more oriented toward biology than Joule's and was therefore more significant for the history of physiology. In 1845, Mayer published the results of his research on dietary energetics under the title "Die organische Bewegung in ihren Zusammenhang mit dem Stoffwechsel." Earlier, in 1842, von Liebig had published his *Organische Chemie und ihre Anwendung auf Physiologie und Pathologie*, in which he demonstrated, through investigation of the caloric content of various nutrients, that all vital phenomena derive their energy from nutrition.

The work of Mayer and von Liebig actually elaborated on studies described even earlier by Théodore de Saussure in his *Recherches chimiques sur la végétation* (1804). Henri Dutrochet, after establishing the laws of osmosis (1826), showed that respiratory phenomena were identical in plants and animals (1837). When the Académie des Sciences sponsored a competition on the origins of animal heat in 1822, two Frenchmen, César Mausuite Despretz, a physicist, and Pierre Louis Dulong, a physician, attempted to reproduce Lavoisier's experiments. Dulong found that the effects of respiration were not enough to account for the full quantity of heat produced. This formed the starting point for further work to determine the amount of energy contributed by nutrition: Henri Victor Regnault and Jules Reiset published their *Recherches chimiques sur la respiration des animaux de diverses classes* in 1849, and their results were later corroborated by Eduard Pflüger's research on the contribution of each nutrient to the total input of nutritional energy, that contribution being measured in each case by the so-called respiratory quotient. In 1879, Berthelot systematized these results in his *Essai de mécanique chimique,* and he also formulated the laws of animal energetics for organisms doing external work and for those simply maintaining themselves. Finally, Rubner, through experiments with dogs carried out between 1883 and 1904, and Atwater, through experiments with human beings conducted between 1891 and 1904, were led to generalize the results of earlier work on the conservation of energy in living organisms.

As for the second law of thermodynamics, concerning the degradation of energy, it was of course first formulated by Nicolas Sadi Carnot in 1824 but little noticed at the time. Benoit Pierre Emile Clapeyron took it up again in 1834, with just as little success; then at mid-century, following further research, it was rediscovered by both Rudolph Julius Emmanuell Clausius and William Thomson (Lord Kelvin). Organisms, like other physicochemical systems, confirm the validity of the second law, which states that transformations of energy — for our purposes, those taking place within living cells — are irreversible, due to an increase in entropy. Organisms, though, are mechanisms capable of reproducing themselves. Like all mechanisms, they are capable of doing work, of accomplishing transformations that are structured and, therefore, less probable than disorganized molecular agitation, or heat, into which all other forms of energy degrade without possibility of reversal. While it is no longer possible to accept Bichat's formulation that "life is the collection of functions that resist death," one can still say that living things are systems whose improbable organization slows a universal process of evolution toward thermal equilibrium — that is, toward a more probable state, death.

To sum up, then, the study of the organism's transformations of the energy it borrows from the environment was the work of chemists as much as of physiologists in the strict sense. Our understanding of the laws of cellular metabolism progressed in parallel with the systematic study of the compounds of carbon, which led to the unification of organic chemistry with inorganic chemistry. Friedrich Wöhler's synthesis of urea in 1828 lent new prestige to the central ideas and methods of von Liebig and his school. But von Liebig's theory of fermentation, which was associated in his mind with the study of the biochemical sources of animal heat (1840), would later be challenged by Louis Pasteur, who was rightly loath to believe that fermentation phenomena were inorganic processes, by nature akin to death, and therefore unrelated to the specific activities of microorganisms. [*Etudes*, pp. 260-62]

Endocrinology

[42] The term "endocrinology," due to Nicholas Pende, was coined only in 1909, yet no one hesitates to use it to refer, retroactively, to any discovery or research related to internal secretions. Work on these secretions in the nineteenth century was not as far-reaching as work on the nervous system, yet the very original nature of that work can nevertheless be seen today as the cause and effect of a veritable mutation in physiological thought. That is why the succinct term "endocrinology" seems preferable to any circumlocution.

Paradoxically, thanks to the work of Claude Bernard, the physiological problem posed by the existence of glands without excretory ducts — organs, originally known as "blood-vessel glands," whose functions could not be deduced from anatomical inspection — was solved by using the same strict methods of chemical investigation that had been applied to the phenomena of nutrition, assimilation through synthesis of specific compounds, disintegration and elimination. [...]

At the beginning of the nineteenth century, nothing was known about the functions of the spleen, thymus, adrenal glands or thyroid. The first glimmer of light came at mid-century in connection with Bernard's research into the digestion and absorption of sugar in the intestine, which revealed the hitherto-inconceivable function of a gland whose affinity with those just mentioned was unsuspected. Moritz Schiff was also working on hepatic glycogenesis and fermentation in Berne in 1859 when he discovered the fatal effects of destroying the thyroid, a result for which he could provide no explanation. It was much later, in Geneva in 1883, that Schiff, revisiting his earlier experiments in the light of Emil Théodore Kocher and Jacques Louis Reverdin's work on the sequellae of surgical excision of goiters (myxedematous cachexia, postoperative myxedema), had the idea of transplanting the thyroid in order to confirm or refute the hypothesis that the gland somehow acted chemically through the blood. Victor Alexander Haden Horsely successfully performed the same experiment on an ape in 1884; Odilon Marc Lannelongue repeated it for therapeutic purposes on a man in 1890. In 1896, Eugen Baumann identified an organic compound of iodine in the thyroid. In 1914, Edward Calum Kendall isolated the active principle in the form of crystallizable thyroxin. Thus, although research into the function of the thyroid began in the physiologist's laboratory, the solution involved the clinician's examining room and the surgeon's operating room.

In the case of the adrenal gland, the point of departure for research lay in clinical observations made between 1849 and 1855 by Thomas Addison and reported in a paper entitled "On the Constitutional and Local Effects of Disease of the Supra-renal

Capsules." In 1856, Charles-Edouard Brown-Séquard read to the Académie des Sciences a series of three papers on "Recherches expérimentales sur la physiologie et la pathologie des glandes surrénales," in which he reported on the lethal effects of removing the capsules as well as of injecting normal animals with blood taken from animals whose "capsules" had been removed. As a result, Brown-Séquard hypothesized that the capsules somehow produced a chemical antitoxic effect on the composition of the blood. That same year, Alfred Vulpian reported his observations in a paper entitled "Sur quelques réactions propres à la substance des capsules surrénales." The cortical cells reacted differently to various dyes than the medullary cells did, from which Vulpian concluded that the latter, which turned green when exposed to iron chloride, secreted a chromogenic substance. This was the first hint of the existence of what would one day be called adrenaline. In 1893, Jean-Emile Abelous and Paul Langlois confirmed Brown-Séquard's experimental results. In 1894, Georges Oliver and Edward Albert Sharpey-Schäfer reported to the London Physiological Society on their observations of the hypertensive effects of injecting aqueous adrenal extract. In 1897, John-Jacob Abel isolated a hypertensive substance from the adrenal medulla, which he called epinephrine. In 1901, Takamine obtained what he called adrenaline in crystallizable form, and Thomas-Bell Aldrich in that same year provided the formula. Adrenaline was thus the first hormone to be discovered. The history of the hormones of the adrenal cortex does not begin until after 1900.

From this brief summary of early experimental work in endocrinology, it is clear that the concept of internal secretion, which Bernard formulated in 1855, did not at first play the heuristic role that one might be tempted to ascribe to it. This was because the concept, which was first applied to the glycogenic function of the liver, initially played a discriminatory role in anatomy rather than an explanatory role in physiology: it distinguished the concept of a gland from that of an excretory organ. But a hormone is a more general concept than an internal secretion: a hormone is a chemical messenger, whereas an internal secretion is simply a distribution or diffusion. Furthermore, the hepatic function, the first-known example of an internal secretion, is special: it places a processed nutriment, a metabolite, into circulation. In this sense, there is a difference between the endocrine secretion of the liver and that of the pancreas: the function of one is supply, of the other, consumption. Insulin, like thyroxin, is the stimulant and regulator of a global mechanism; it is not, strictly speaking, an intermediary, energy-laden compound. Thus, to credit Bernard as the author of the fundamental concept of modern endocrinology is not false, but it is misleading. The concept that proved fruitful was that of the internal environment, which, unlike the concept of internal secretion, was not closely associated with a specific function; rather, from the first it was identified with another concept, that of a physiological constant. When it turned out that living cells depend on a stable organic environment, which Walter Bradford Cannon named "homeostasis" in 1929, the logical possibility arose of transforming the concept of internal secretion into one of chemical regulation. Once the fundamental idea was clear, research on various glands quickly led to the identification and (at least) qualitative description of their functional effects.

It is not surprising, then, that from 1888 on, the work of Moritz Schiff and Brown-Séquard attracted many emulators and stimulated research in endocrinology, usually in conjunction with a desire to correct unsubstantiated pathological etiologies. It was the study of diabetes, for example, which Bernard's work had already clarified, that led Joseph von Mering and Eugene Minkowski to discover the role of the pancreas in the metabolism of glucids (1889), and subsequently to the identification by Frederick Grant Banting and Charles Herbert Best (1922) of the substance that Sharpey-Schäfer had named insulin in 1916. It was the study of acromegaly by Pierre Marie (1886) that led, eventually, to experiments in hypophysectomy by Georges Marinescu (1892) and Giulio Vassale and Ercole Sacchi (1892), and later to work that discriminated between the functions of the anterior and posterior lobes of the pituitary (Sir Henry Dale in 1909, Harvey Cushing in 1910, and Herbert McLean Evans and Crawford Williamson Long in 1921). Brown-Séquard's experiments also spurred work on sex hormones, despite the ironic skepticism of many in the field. The role of the parathyroids, whose anatomical distinctiveness went unnoticed until Ivar Victor Sandström's work of 1880, was elucidated in 1897 through the research of Emile Gley.

Thus, the physiological concept of a chemical regulator, in its current sense, was elaborated in the late nineteenth century, but an expressive term for it had yet to be coined. In 1905, William Bayliss and Ernest Starling, after consulting a philologist colleague, proposed the term "hormone." [*Etudes*, pp. 262-65]

Neurophysiology
[43] Of all the systems whose functions are determined by the need to preserve the integrity of cellular life, the one whose mechanical nature always aroused the fewest objections was the neuromuscular. Mechanistic theories first arose not from the study of plant growth or from viscous and visceral palpation of the mollusk but from observation of the distinctive, sequential locomotion of vertebrates, whose central nervous systems control and coordinate a series of segmentary movements that one can simulate by mechanical means. "An amoeba," Alex von Uexküll maintained, "is less of a machine than a horse." Because some of the earliest concepts of nervous physiology – afferent and effer-

ent pathways, reflexes, localization and centralization – were based in part on analogies with operations or objects that were familiar by dint of the construction and/or use of machines, progress in this branch of physiology, whose discoveries were also incorporated by psychology, earned it widespread recognition. Although terms such as "hormone" and "complex" have entered common parlance, they surely remain more esoteric than a word like "reflex," whose use in connection with sports has made it entirely routine.

If the motor effects of the decapitation of batrachians and reptiles had led eighteenth-century researchers to suspect the role of the spinal cord in the muscular function, and if the experiments of Robert Whytt (1768) and Julien Jean César Legallois (1812) already had a positive character, it was nevertheless impossible to explain what Thomas Willis in 1670 called "reflected movements" in terms of the reflex arc until the Bell-Magendie law had been formulated and verified (1811–22). Marshall Hall's discovery of the "diastaltic" (reflex) function of the spinal cord, simultaneously glimpsed by Johannes Müller, was a necessary consequence of differentiating the various functions of the spinal nerve. That differentiation also led inevitably to identification of functionally specialized bundles of conductors within the spinal cord – by Karl Friedrich Burdach in 1826, Jacob Augustus Lockhart Clarke in 1850, Brown-Séquard in 1850 and Friedrich Goll in 1860. Based initially on experiments involving section and excitation of nerve fibers, this work preceded Friedrich Walter's discovery of spinal degeneration in 1850.

Once the dual significance of conduction along the nervous fiber had been determined, the excitability and conductivity of nervous tissue were studied systematically, along with the contractile properties of muscle. This work was the positive or empirical portion of a large volume of research, some of it magical in character, spurred by the discovery of "animal electricity." The field of electrophysiology began with Luigi Galvani's observations and experiments, his polemics with Alessandro Volta (1794), and Alexander von Humboldt's corroboration of Galvani's results. In 1827, Leopoldo Nobili built an astatic galvanometer sensitive enough to detect very weak currents. Carlo Mateucci established, in 1841, a correlation between muscular contraction and the production of electricity. Du Bois-Reymond virtually invented the entire apparatus and technique of electrophysiology in order to subject Mateucci's work to stringent criticism. He demonstrated the existence of what he called "negative variation," an action potential that generated a current in conjunction with the stimulation of a nerve; he also studied physiological tetanus. Using similar techniques, von Helmholtz in 1850 measured the speed of propagation along the nerve. Although this experiment failed to shed the expected light on the nature of the message transmitted, it did at least refute all theories holding that this message involved the transport of some substance.

After Whytt and George Prochaska identified the spinal cord's sensorimotor coordination function but before Marshall Hall explained its mechanism, Legallois and Pierre Flourens located the center of reflex movement in the medulla oblongata. At around the same time, the ancient concept of a seat of the soul or organ of common sense, whose possible location had been the subject of much speculation in the seventeenth and eighteenth centuries, collapsed. Albrecht von Haller had provided a negative answer to the question, "Do different functions stem from different souls (*An diversae diversarum animae functionum provinciae*)?"[30] In 1808, however, the father of phrenology, Franz Joseph Gall, argued that "the brain is composed of as many distinctive systems as it performs distinct functions," and that it is therefore not an organ but a composite of organs, each corresponding to a faculty or appetite — and, furthermore, that those organs are to be found in the convolutions of the brain's hemispheres, which were reflected in the configuration of the cranial shell.

This is not the place to deal with the allegation that Gall was a charlatan. It is more important to understand why he enjoyed as much influence as he did, and for so long. He provided the physiologists and clinicians of the first two thirds of the nineteenth century with a fundamental idea that one of his critics, Louis François Lelut, called "the polysection of the encephalon." Recall, moreover, that Gall claimed to have come upon his theory through observation of the skulls of certain of his colleagues with a particularly keen memory for words; he located the organ of that memory in the lower posterior portion of the anterior lobe. Now, it happens that the first identification of an anatomical lesion responsible for a clinical diagnosis of aphasia, made by Jean Baptiste Bouillaud in 1825, confirmed Gall's observation. In 1827, Bouillaud published the first experimental findings on the ablation of regions of the cerebral cortex in mammals and birds. From then on, experiments on animals combined with clinical and pathological observation of humans to produce a functional mapping of the cerebral cortex. In 1861, Paul Broca identified the seat of articulate language in the third frontal convolution, which led him to make this declaration of faith: "I believe in the principle of localizations; I cannot believe that the complexity of the cerebral hemispheres is a mere caprice of nature."

In 1870, Gustav Theodore Fritsch and Julius Edward Hitzig provided experimental proof of cerebral localization by employing a revolutionary new technique, electrical stimulation of the cortex. Previously, due to the failure of attempts to stimulate the brain directly during trepanation, direct stimulation had been declared impossible. From experiments with dogs, Fritsch and Hitzig concluded that the anterior and posterior regions of the brain were not equivalent; the anterior region was associated with the motor function, the posterior with the sensory function. Because Hitzig could not apply electrical stimuli to a human brain, in 1874 he instead mapped the motor region in an ape; in 1876, David Ferrier confirmed Hitzig's results. Naming Flourens but aiming his criticism at Friedrich Goltz, Ferrier wrote, "The soul is not, as Flourens and many who came after him believed, some kind of synthetic function of the entire brain, whose manifestations can be suppressed *in toto* but not in part; on the contrary, it is certain that some, and probable that all, psychic functions derive from well-defined centers in the cervical cortex." Similarly, Ferrier's discovery of the role of the occipital lobe in vision led Hermann Munk in 1878 to give the first precise localization of a sensory center. A growing number of experiments, confirmed by clinical observations, provided Carl Wernicke with the material to entitle his 1897 treatise on the anatomy and physiology of the brain the *Atlas des Gehirns*. But it was not until the early twentieth century that Alfred Campbell (1905) and Korbinian Brodmann (1908), drawing on advances in histology from Camillo Golgi to Santiago Ramon y Cajal, were able to lay the foundations for a cytoarchitectonics of the cortex.

In *Leçons sur les localisations* (1876), Jean-Martin Charcot wrote, "The brain is not a homogeneous, unitary organ but an association." The term "localization" was taken literally at the time: it was assumed that the unfolded surface of the cortex could be divided into distinct zones, and that lesions or ablations could explain sensorimotor disturbances described as deficits (a-phasia, a-graphia, a-praxia and so on). Yet Jules Gabriel François Baillarger had pointed out in 1865 that aphasia is not a loss of the memory of words, because some aphasics retain their vocabulary but lose the ability to use words properly – and in anything but an automatic manner. Over the next two decades, Hughlings Jackson, interpreting similar observations in terms of Spencerian evolutionism, introduced the concept of a conservative integration of neurological structures and functions, according to which less complex structures and functions are dominated and controlled at a higher level by more complex and highly differentiated ones, which appear later in the phylogenetic order. Pathological states are not decompositions or diminutions of physiological states; rather, they involve a dissolution or loss of control, the liberation of a dominated function, the return to a more reflexive, although in itself positive, state.

An important event in the history of the localization concept was the International Congress of Medicine held in London in 1881, at which Sherrington, then aged twenty-four, heard the Homeric debate between Ferrier and Friedrich Goltz. Later, when Charles Scott Sherrington visited Goltz in Strasbourg in 1884–85, he learned the technique for taking progressive sections of the spinal cord. His work on the rigidity caused by decerebration (1897) and research on subjects ranging from reciprocal innervation to the concept of an integrative action of the nervous system (1906) enabled him to corroborate and correct Jackson's fundamental ideas without venturing outside the realm of physiology.

Between Marshall Hall and Sherrington, the study of the laws of reflex made little progress apart from Eduard Pflüger's earlier, rather crude statement in 1853 of the rules of irradiation, a concept that implied the existence of an elementary reflex arc. Sherrington showed, to the contrary, that even in the case of the simplest reflex, the spinal cord integrates the limb's entire bundle of nerves. Brain functions merely expand upon this capacity of the spinal cord to integrate various parts of the organism. Following Jackson, Sherrington thus established that the animal organism, seen in terms of its sensorimotor functions, is not a mosaic but a structure. The great physiologist's most original contribution, however, was to explain, with the concept of the cortex, the difference between nervous mechanisms for integrating immediate and deferred movements.

At around the same time, Ivan Pavlov studied another cortical integrating function, which he called "conditioning" (1897). Pavlov showed how the cortical functions could be analyzed by modifying techniques borrowed from reflexology. When an animal (in this case, a dog) was conditioned through the simultaneous application of different stimuli, ablation of more or less extensive regions of the cortex allowed one to measure the degree to which the sensorimotor reflex depended on the integrity of the cortical intermediary. This technique, which Pavlov refined as results accumulated, was taught to large numbers of the great Russian physiologist's disciples. [...]

I will end with a few words about what John Newport Langley, in 1898, called the "autonomic" nervous system, whose functions, because they involve what Bichat called "vegetative" as opposed to "animal" life, were less susceptible of mechanical interpretations than those of the central nervous system. It was Jacob Winslow who in 1732 coined the expression "great sympathetic" nervous system to refer to the ganglionic chain. In 1851, Bernard discovered the effect of the sympathetic system on sensitivity and body temperature; in 1852-54, Brown-Séquard contributed new techniques for exploring the functions of the sympathetic nervous system by sectioning nerves and applying electrical stimuli. Langley was a pioneer in the use of chemical techniques, including the blockage of synapses by nicotine (1889) and the sympathicomimetic property of adrenaline (1901). [*Etudes*, pp. 266-71]

CHAPTER SIX

Epistemology of Medicine

The Limits of Healing

[44] Awareness of the limits of medicine's power accompanies any conception of the living body which attributes to it a spontaneous capacity, in whatever form, to preserve its structure and regulate its functions. If the organism has its own powers of defense, then to trust in those powers, at least temporarily, is a hypothetical imperative, at once prudent and shrewd. A dynamic body deserves an expectant medicine. Medical genius may be a form of patience. Of course, the patient must agree to suffer. Théophile de Bordeu, well aware of this, wrote in his *Recherches sur l'histoire de médecine*: "The method of expectation has something cold or austere about it, which is difficult for the keen sensibilities of patients and onlookers to bear. Thus, very few physicians have practiced it, particularly in nations whose people are naturally ardent, impatient, and fearful."

Not all patients respond to treatment; some recover without it. Hippocrates, who recorded these observations in his treatise *On the Art*, was also, according to legend, responsible for — or, if you will, credited with — introducing the concept of nature into medical thinking: "Natures are the healers of diseases," he wrote in Book Six of *Epidemics*. Here, "healer" refers to an intrinsic activity of the organism that compensates for deficiencies, restores a disrupted equilibrium or quickly corrects a detected deviation. This activity, however, is not the product of innate knowledge: "Nature finds its own ways and means, but not by intelligence: blinking is one such, the various offices of the tongue are another, and so are other actions of this sort. Nature does what is appropriate without instruction and without knowledge."

The analogy between nature as healer and the medical art throws the light of nature on the art, but not vice versa. The medical art must observe, must listen to nature; to observe and to listen in this context is to obey. Galen, who attributed to Hippocrates concepts that one can only call Hippocratic, adopted them in his own right and taught that nature is the primary conservator of health because it is the principal shaper of the organism. However, no Hippocratic text goes so far as to portray nature as infallible or omnipotent. The medical art originated, developed and was perfected as a gauge of the power of nature. Depending on whether nature as healer is stronger or weaker, the physician must either allow nature to take its course, intervene to support it or help it out, or refuse to intervene on the grounds that there are diseases for which nature is no match. Where nature gives in, medicine must give up. Thus, Hippocrates wrote, "To ask art for what art cannot provide and to ask nature for what nature cannot provide is to suffer from an ignorance that is more akin to madness than to lack of education." ["Idée de nature," *Médecine*, pp. 6-7]

[45] To simplify (probably to excess) the difference between ancient (primarily Greek) medicine and the modern medicine inaugurated by Andreas Vesalius and William Harvey and celebrated by Roger Bacon and René Descartes, one might say that the former was contemplative, the latter operational. Ancient medicine was founded upon a supposed isomorphism between the cosmic order and the equilibrium of the organism, reflected in

nature's presumed power to correct disorders on its own. Nature the physician was respected by a therapeutics of watchfulness and support. By contrast, modern medicine was activist in its orientation. Bacon expressed the hope that it would learn from chemistry, and Descartes that it would learn from mechanics. Yet between the Greeks and the Moderns, for all that they were separated by the Copernican revolution and its critical consequences, the difference remained philosophical, without perceptible impact on the health of mankind. The shared project of Bacon and Descartes, to preserve health and to avoid or at least delay the decline of old age – in short, to prolong life – resulted in no notable achievements. Although Nicolas de Malebranche and later Edme Mariotte spoke of "experimental medicine," the phrase remained a signifier in search of a signified. Eighteenth-century medicine remained a symptomatology and nosology, that is, a system of classification explicitly based on that of the naturalists. Medical etiology squandered its energies in the erection of systems, reviving the ancient doctrines of solidism and humorism by introducing new physical concepts such as magnetism and galvanism or by raising metaphysical objections to the procedures of those who would assimilate medicine to mechanics. Therapeutics, guided by pure empiricism, alternated between skeptical eclecticism and obstinate dogmatism. Tragically, medicine could not accomplish its goals. It remained an empty discourse about practices often not very different from magic.

Freud said of ancient medicine that psychic therapy was the only treatment it had to offer, and much the same thing could have been said about medicine in the eighteenth and most of the nineteenth centuries. By this I mean that the presence and personality of the physician were the primary remedies in many afflictions of which anxiety was a major component. [*Ideology and Rationality*, pp. 52–53]

The New Situation of Medicine

A Shift

[46] The gradual elimination from medical understanding of any reference to the patient's living conditions was, in part, an effect of the colonization of medicine by basic and applied science in the early nineteenth century; but it was also a consequence of industrial society's interest (in every sense of the word) in the health of its working populations (or, as some would put it, in the human component of the productive forces). The political authorities, at the behest of, and with advice from, hygienists, took steps to monitor and improve living conditions. Medicine and politics joined forces in a new approach to illness, exemplified by changes in hospital structures and practices. In eighteenth-century France, particularly at the time of the Revolution, steps were taken to replace hospices, which had provided shelter and care to sick patients, many of whom had nowhere else to turn, with hospitals designed to facilitate patient surveillance and classification. By design, the new hospitals operated as, to borrow Jacques René Tenon's phrase, "healing machines." Treating diseases in hospitals, in a regimented social environment, helped strip them of their individuality. Meanwhile, the conditions under which diseases developed were subjected to increasingly abstract

analysis and, as a result, the gap widened between the reality of patients' lives and the clinical representation of that reality. [...] The statistical study of the frequency, social context and spread of disease coincided exactly with the anatomical-clinical revolution in the hospitals of Austria, England and France in the early nineteenth century. ["Maladies," *Univers*, p. 1235a]

[47] Three phenomena altered the situation of European medicine. The first was the institutional and cultural change that Michel Foucault has baptized "the birth of the clinic," which combined hospital reforms in Vienna and Paris with increasingly widespread use of such exploratory practices as percussion (Joseph Leopold Auenbrügger, Jean-Nicholas Corvisart) and mediate auscultation (René-Théophile Hyacinthe Laënnec), and with systematic efforts to relate observed symptoms to anatomical and pathological data. Second, a rational attitude of therapeutic skepticism was fostered and developed in both Austria and France, as Edwin Heinz Ackerknecht has shown.[31] Third, physiology gradually liberated itself from its subservience to classical anatomy and became an independent medical discipline, which at first focused on disease at the tissue level, as yet unaware that eventually it would come to focus even more sharply on the cell. And physiologists looked to physics and chemistry for examples as well as tools.

Hence, a new model of medicine was elaborated. New diseases were identified and distinguished, most notably in pulmonary and cardiac pathology (pulmonary edema, bronchial dilation, endocarditis). Old medications, whose numbers had proliferated with no discernible effect, were discounted. And rival medical theories cast discredit on one another. The new model was one of knowledge without system, based on the collection of facts and, if possible, the elaboration of laws confirmed by experiment. This knowledge, it was hoped, would be capable of conversion into effective therapies, whose use could be guided by critical awareness of their limitations.

In France, elaboration of the new medical model was pursued first by François-Joseph Victor Broussais, then by François Magendie, and finally by Claude Bernard. Despite the traditional claims of medical historians, however, it can be shown that the physiological model remained an ideology. If the goal of the program was eventually achieved, it was reached by routes quite different from those envisioned by the program's authors. [*Ideology and Rationality*, pp. 54–55]

The Physiological Point of View

François-Joseph Victor Broussais
[48] By demolishing the period's most majestic and imposing system, that of Philippe Pinel, Broussais cleared the way for the advent of a new spirit in medicine. "It was Broussais's opinion that pathology was nothing but physiology, since he called it 'physiological medicine.' Therein lay the whole progress in his way of looking at things."[32] To be sure, Broussais's "system of irritation" hindered his understanding unnecessarily, and he discredited himself by overreliance on leeches and bleeding. Yet it should not be forgotten that the publication of his *Examen de la doctrine médicale généralement adoptée* was, in the words of Louis Peisse, "a medical [equivalent of] 1789."[33] In order to refute Pinel's "philosophical nosography" and doctrine of "essential fevers," Broussais borrowed from Bichat's general anatomy the notion that each type of tissue, owing to its specific texture, exhibits certain characteristic alterations. He identified fever with inflammation, distinguished different original sites and paths of propagation for each type of tissue, and thus explained the symptomatic diversity of different fevers. He explained inflammation as the result of an

excessive irritation, which interfered with the movement of a tissue and could in the long run disturb its organization. He stood on its head the basic principle of pathological anatomy by teaching that the dysfunction precedes the lesion. He based medicine on physiology rather than anatomy. All of this is summed up in a well-known passage of the preface to the *Examen* of 1816: "The characteristic traits of diseases must be sought in physiology.... Enlighten me with a scientific analysis of the often confused cries of the suffering organs.... Teach me about their reciprocal influences." Discussing the new age of medicine in his *Essai de philosophie médicale*, Jean Baptiste Bouillaud wrote, "Is not the fall of the system of *Nosographie philosophique* one of the culminating events of our medical era, and is not the overthrow of a system that had governed the medical world a revolution whose memory will not fade?"[34] In a more lapidary fashion, Michel Foucault put it this way in *The Birth of the Clinic*: "Since 1816, the doctor's eye has been able to confront a sick organism."[35] Emile Littré, a man familiar with the concept of "distinguishing" different types of explanation (he refers to "Bichat's great distinction" between occult and irreducible qualities), was thus able to observe in 1865 that "while theory in medicine once was suspect and served only as a target, so to speak, for the facts that demolished it, today, owing to its subordination to physiological laws, it has become an effective instrument of research and a faithful rule of conduct."[36] No doubt Claude Bernard was right to say that Broussais's physiological medicine "was in reality based only on physiological ideas and not on the essential principle of physiology."[37] Yet Broussais's *idea* was well suited to become a *program* and to justify a medical *technique* quite different from the one originally associated with it. François Magendie took Broussais's doctrine and transformed it into a method. That is why Broussais's system brought about a different kind of revolution from other systems. Physiological medicine, even if it mimicked the form of a system, marked a decisive shift from the era of systems to the age of research, from the age of revolution to the epoch of progress, because Broussais's idea looked to techniques within reach of contemporary possibilities. [*Etudes*, pp. 136–38]

François Magendie
[49] What Broussais promised, someone else had already begun to deliver. This man, too, had declared that "medicine is nothing but the physiology of the sick man."[38] Just one year after Broussais's *Histoire de phlegmasies* (1808), this man had published his *Examen de l'action de quelques végétaux sur la moëlle épinière*. He founded the *Journal de physiologie expérimental* a year before Broussais founded the *Annales de la médecine physiologique* and in it in 1822 confirmed Charles Bell's discovery (1811) through his "Expériences sur les fonctions des racines des nerfs rachidiens." From the titles of these works alone we gather the difference between the orientation of Broussais's work and that of this other physician: François Magendie (1783–1855). Whereas Broussais had worked first in military and later in civilian hospitals, Magendie was a man of the laboratory as well as a hospital physician. For him, experimental physiology was the study of the physics of vital phenomena such as absorption. He conducted systematic experiments with animals to test the pharmacodynamic properties of newly isolated classes of chemical compounds such as the alkaloids. As early as 1821, Magendie's *Formulary* carried the subtitle "For the Use and Preparation of Various Medications Such as Nux Vomica, Morphine, Prussic Acid, Strychnine, Veratrine, Iodine, and the Alkalis of Quinquinas" (that is, the quinine of Pelletier and Caventou).

In short, Magendie's experimental medicine differed from Broussais's physiological medicine in three ways: it was centered in the laboratory rather than the hospital; it experimented on animals rather than on men; and instead of Galenic principles it used extracts isolated by pharmaceutical chemistry, for example, replacing opium with morphine and quinquina with quinine. Of these three differences, the second was initially greeted with the greatest incomprehension and criticism. Magendie's vivisections aroused hostile protest and demonstrations, no doubt for reasons more profound than compassion for animal suffering. For to reason from animals to man was to abolish the distance between the two. The practice was held to stem from a materialist philosophy, and success would result in the temptation to extend the experiments to man. When accused of experimenting on humans, Magendie denied the charge. But if administering unproven drugs is experimentation (as Claude Bernard himself was one of the first to admit[39]), then Magendie did experiment on humans, patients in hospitals, which he considered a vast laboratory where patients could be grouped and studied comparatively. [*Ideology and Rationality*, pp. 58–59]

Claude Bernard
[50] A year before his death, Claude Bernard, writing the introduction for a planned *Traité de l'expérience dans les sciences médicales*, took literally a well-known quip of Magendie's. Bernard repeated his predecessor's self-characterization: "He was the ragpicker of physiology. He was merely the initiator of experimentation. Today it is a discipline that has to be created, a method."[40] For Bernard, a self-styled ragpicker was no doubt superior to a dogmatic system-builder who did not even realize that he was building a system, like Broussais. But what are we to make of Bernard's repeated insistence that only he appreciates the true requirements of *the* experimental method?

Insufficient attention has been paid, I think, to two concepts in Bernard's methodological writings that were for him inseparable: theory and progress. Experimental medicine is progressive, he argued, because it elaborates theories and because those theories are themselves progressive, that is, open. Bernard's view is summed up in two *obiter dicta*: "An experimentalist never outlives his work. He is always at the level of progress," and "With theories there are no more scientific *revolutions*. Science grows gradually and steadily."[41] Add to this the two concepts of determinism and action – knowledge of the one being essential for success of the other – and you have the four components of a medical ideology that clearly mirrored the progressive ideology of mid-nineteenth-century European industrial society. In light of more recent concepts, such as Bachelard's epistemological break and Kuhn's structure of scientific revolutions, Bernard's concept of theory without revolution has drawn understandable and legitimate criticism. In Bernard's day, physicists still found in Newton and Pierre-Simon Laplace reasons to believe in principles of conservation. Rudolph Julius Emmanuell Clausius had yet to attract the attention of a large part of the scientific community to Carnot's principle, of which philosophers were *a fortiori* even less aware. Michael Faraday's experiments, André-Marie Ampère's laws and James Clerk Maxwell's calculations had yet to reveal electrical current as a possible substitute for coal as the motor of the industrial machine. In 1872, the German physiologist Emile Du Bois-Reymond (of whom Bernard had on several occasions expressed a rather contemptuous opinion) displayed sufficient confidence in Laplacian determinism to predict when England would burn her last piece of coal (*Über die Grenzen des Naturerkennens*). But in that same year, the Académie des Sciences in Paris, consulted for the second time about the invention of an electrical worker named Zénobe Gramme, finally acknowledged that practice had raced ahead of theory and authenticated a revolution in technology. In short, the concept of a theory without revolution, which Bernard took to be the solid basis of his methodology, was perhaps no more than a sign of internal limitations in his own medical theory: experimental medicine, active and triumphant, which Bernard proposed as a definitive model of what medicine in an industrial society ought to be. He contrasted his model with that of contemplative, watchful medicine, a model appropriate to agricultural societies in which time was governed by quasi-biological rather than industrial norms. The son of a vine grower who maintained a deep attachment to his native soil, Bernard was never able to appreciate fully that science requires not only that the scientist abandon ideas invalidated by facts but also that he give up a personalized style of research, which was the hallmark of his own work. In science, it was the same as in agriculture, where economic progress had uprooted many from the soil.

Paradoxically, the internal limitations of Bernard's theory of disease (etiology and pathogeny) were due to the initial successes of his research as Magendie's successor. For he had discovered the influence of the sympathetic nervous system on animal heat (1852); had generated, in the course of research on glycogenesis, a case of diabetes by a lesion of the pneumogastric nerve at the level of the fourth ventricle (1849–51); and had demonstrated the selective action of curare on the motor nerves. As a result, Bernard conceived an idea that he never repudiated, namely, that all morbid disorders are controlled by the nervous system,[42] that diseases are poisonings, and that infectious viruses are agents of fermentation that alter the internal environment in which cells live.[43] Although these propositions were later adapted to quite different experimental situations, none can be said to have been directly responsible for a positive therapeutic application. What is more, Bernard's stubborn views on the subject of pathogeny prevented him from seeing the practical implications of the work of certain contemporaries whom he held in contempt because they were not physiologists. Convinced of the identity of the normal and the pathological, Bernard was never able to take a sincere interest in cellular pathology or germ pathology. [*Ideology and Rationality*, pp. 60-63]

The Statistical Point of View

René-Théophile Hyacinthe Laënnec
[51] Consider Laënnec. François Magendie mocked him as a mere annotator of signs. The invention of the stethoscope and its use in auscultation as codified in the *De l'auscultation médiate* of 1819 led to the eclipse of the symptom by the sign. A symptom is something presented or offered by the patient; a sign, on the other hand, is something sought and obtained with the aid of medical instruments. The patient, as the bearer and often commentator on symptoms, was "placed in parentheses." A sign could sometimes reveal an illness before a symptom led to its being suspected. In Section 86, Laënnec gives the example of a pectoriloquy as the sign of a symptomless pulmonary phthisis.[44] This was the beginning of the use of man-made instruments to detect alterations, accidents and anomalies, a practice that would gradually expand with the addition of new testing and measuring equipment and the elaboration of subtle test protocols. From the ancient stethoscope to the most modern magnetic resonance imaging equipment, from the X-ray to the computerized tomographic scanner and ultrasound instrument, the scientific side of medical practice is most strikingly symbolized by the shift from the medical office to the testing laboratory. At the same time, the scale on which pathological phenomena are represented has been reduced from the organ to the cell and from the cell to the molecule.

The task of the physician, however, is to interpret information derived from a multiplicity of sources. Though medicine may set aside the individuality of the patient, its goal remains the conquest of disease. Without diagnosis, prognosis and treatment, there is no medicine. Here we find an object suitable for study in terms of logical and epistemological analysis of the construction and testing of hypotheses. We also find ourselves at the dawn of medical mathematics. Doctors were just beginning to become aware of an epistemological limitation already recognized in cosmology and physics: no serious prediction is possible without quantification of data. But what kind of measurement could there be in medicine? One possibility was to measure variations in the physiological functions. This was the purpose of instruments such as Jean Poiseuille's hemodynamometer (1828) and Karl Ludwig's kymograph. Another possibility was to tabulate the occurrence of contagious diseases and chart their propagation; in the absence of confirmed etiologies, these data could be correlated with other natural and social phenomena. It was in this second form that quantification first established a foothold in medicine. ["Statut épistémologique," *Histoire*, pp. 19-20]

Philippe Pinel
[52] The statistical method of evaluating etiological diagnoses and therapeutic choices began with Pierre Louis's *Mémoire* on phthisis (1825), which appeared four years before the publication in London of Francis Bisset Hawkins's *Elements of Medical Statistics* (whose outlook was as social as it was medical). Those who celebrate the first use of statistics in medicine tend to forget Pinel, however. In 1802, in his *Médecine clinique*, he used statistical methods to study the relation between certain diseases and changes in the weather. He also introduced statistical considerations in the revised edition of his *Traité médico-philosophique sur l'aliénation mentale*. Edwin Heinz Ackerknecht says that Pinel was "the veri-

table father of the numerical method." It may be of some interest to recall a little-known judgment concerning him. Henry Ducrotay de Blainville said this in his *Histoire des sciences de l'organisation* of 1845:

> A mathematician, Pinel began by applying mathematics to animal mechanics; a philosopher, he carried on with an in-depth study of mental illness; a naturalist and observer, he made progress in applying the natural method to medicine; and toward the end he lapsed back into his early predilections by embracing the chimerical idea of applying the calculus of probabilities to medicine, or medical statistics, as if the number of diseases could affect the infinite variations of temperament, diet, locale and so on, which influence their incidence and make them so diverse from individual to individual.

This judgment is worth remembering for the light it sheds on the stormy relations between Blainville and Auguste Comte and on the hostility of the positivist philosophers to the calculus of probabilities. The Fortieth Lesson of the *Cours de Philosophie positive* states that medical statistics are "absolute empiricism in frivolous mathematical guise" and that there is no more irrational procedure in therapy than to rely on "the illusory theory of chance." One finds the same hostility in Claude Bernard, despite his skepticism about Comte's philosophy. ["Statut épistémologique," *Histoire*, pp. 20-21]

Pierre-Charles-Alexandre Louis
[53] Louis used statistics in a different spirit from Pinel. His main goals were to substitute a quantitative index for the clinician's personal judgment, to count the number of well-defined signs present or absent in the examination of a patient and to compare the results of one period with those obtained by other physicians in

other periods using the same methods. Experience in medicine is instructive, he insisted, only if numerical records are maintained. But, others argued, tables and charts destroy memory, judgment and intuition. That is why Emile Littré and Charles Robin, both positivists, declared their hostility to "numerics" in the article they published under that rubric in the thirteenth edition of their *Dictionnaire de médecine, chirurgie et pharmacie* (1873). In their view, calculations could never replace "anatomical and physiological knowledge, which alone makes it possible to weigh the value of symptoms." Furthermore, the effect of using the numerical method is that "patients are observed in a sense passively." As with the case of Laënnec, this was a method that set aside the distinctive features of the patient seeking individual attention for his or her pathological situation.

It would be more than a century before "the illusory theory of chance," as Comte called it, would be fully incorporated into diagnosis and therapy through methods elaborated to minimize errors of judgment and risks of treatment, including the computerized processing of biomedical and clinical data. One recent consequence of this technological and epistemological evolution has been the construction of "expert systems" capable of applying various rules of inference to data gleaned from examination and then recommending possible courses of treatment. ["Statut épistémologique," *Histoire*, pp. 21-22]

A Medical Revolution

Bacteriology

[54] The discoveries of Louis Pasteur, Hermann Robert Koch and their students quickly led to a profound epistemological revolution in medicine, so that, strangely enough, these researchers had a greater impact on clinical medicine than did contemporary clinical practitioners. Pasteur, a chemist without medical training, inaugurated a new era in medicine. He freed medical practice from its traditional anthropocentrism: his approach had as much to do with silkworms, sheep and chicken as with human patients. Pasteur discovered an etiology unrelated to organ functions. By revealing the role of bacteria and viruses in disease, he changed not only the focus of medicine but the location of its practice. Traditionally, patients had been cared for at home or in hospitals, but vaccinations could now be administered in dispensaries, barracks and schoolhouses. The object of medicine was no longer so much disease as health. This gave new impetus to a medical discipline that had enjoyed prominence in England and France since the end of the eighteenth century – public health or hygiene. Through public health, which acquired institutional status in Europe in the final third of the nineteenth century, epidemiology took medicine into the realm of the social sciences and eco-

nomics. It became impossible to look upon medicine solely as a science of organic anomalies or changes. The effects of the patient's social and economic situation on the conditions of his or her life now numbered among the factors that the physician had to take into account. The political pressures stemming from public health concerns gradually resulted in changes in medicine's objectives and practices. The accent was shifted from health to prevention to protection. The semantic shift points to a change in the medical act itself. Where medicine had once responded to an appeal, it was now obedient to a demand. Health is the capacity to resist disease; yet those who enjoy good health are nevertheless conscious of the possibility of illness. Protection is the negation of disease, an insistence on never having to think about it. In response to political pressures, medicine has had to take on the appearance of a biological technology. Here, for a third time, the individual patient, who seeks the attention of a clinician, has been set aside. But perhaps individuality is still recognized in the notion of resistance, in the fact that some organisms are more susceptible than others to, say, the cholera bacillus. Is the concept of resistance artificial, serving to cover a gap in the germ theory's determinism? Or is it a hint of some more illuminating concept yet to come, for which microbiology has paved the way?

If medicine has attained the status of a science, it did so in the era of bacteriology. A practice is scientific if it provides a model for the solution of problems and if that model gives rise to effective therapies. Such was the case with the development of serums and vaccines. A second criterion of scientificity is the ability of one theory to give rise to another capable of explaining why its predecessor possessed only limited validity. ["Statut épistémologique," *Histoire*, pp. 22–23]

The German School

[55] Yet it was an extension of microscopic techniques for the study of cell preparations and the use of synthetic aniline stains (manufactured in Germany after 1870) that led, for the first time in the history of medicine, to a therapeutic technique that was both effective and unrelated to any medical theory: chemotherapy, invented by Paul Ehrlich (1854–1915). From Wilhelm von Waldeyer in Strasbourg, Ehrlich had learned how to use stains to examine normal and pathological tissue, and at Breslau he had attended lectures on pathological anatomy given by Julius Cohnheim (1839–1884), a student of Rudolph Ludwig Karl Virchow, who would later show that inflammation was caused by the passage of leucocytes through the capillary wall. Virchow's ideas reached Ehrlich through Julius Cohnheim. Nevertheless, if cellular pathology played an indirect part in the invention of chemotherapy, the role of bacteriology and the discovery of immunity was more direct. The problem that Ehrlich stated and solved can be formulated as follows: Through what chemical compounds with specific affinity for certain infectious agents or cells could one act directly on the cause rather than on the symptoms of disease, in imitation of the antitoxins present in various serums?

This is not the place to delve into the circumstances surrounding the discovery of immunity or to revive a dispute over priority, an exercise useful for reminding us that the constitution of scientific knowledge does not necessarily require the simultaneous existence of all who claim to be its authors.[45] It is of little importance that the Berlin School preceded the Paris School by several months, or that Hermann Robert Koch's pupil Emil Adolf von Behring concluded before Pasteur's pupil Pierre Paul Emile Roux that diphtheria cannot be treated with a vaccine but can only be prevented by injection of serum taken from a convalescent patient – provided one has a convalescent patient, that is, a survivor of the disease. Roux was able to prepare the toxin *in vitro*. Von Behring managed to attenuate its virulence with trichloride of iodine. Roux was more successful than von Behring in increasing the activity of the serum.

Nevertheless, Ehrlich, whom Koch put in contact with von Behring, dreamed that chemistry could one day endow man with powers far beyond those of nature.[46] He hit upon the idea of looking for substances with specific affinities for certain parasites and their toxins on the model of stains with elective histological affinities. For what is a stain but a vector aimed at a particular formation in a healthy or infected organism? When a chemical compound directed at a particular cell penetrates that cell, what happens is analogous to the way in which a key fits into a lock. Ehrlich's first success came in 1904, when in collaboration with Kiyoshi Shiga he discovered that Trypan red destroys the trypanosome that causes sleeping sickness. Later came the discovery in 1910 of Salvarsan, or "606," and Neo-Salvarsan, which proved less effective in combatting syphilis than was believed at first. But Ehrlich's real success lay not so much in the products that he identified himself as in those that would ultimately be discovered in pursuit of his fundamental hypothesis: that the affinities of chemical stains could be used as a systematic technique for developing artificial antigens. Using the same method, in 1935 Gerhardt Domag discovered prontosil red, the first of a glorious series of sulfamides. Its declining efficacy led to the greatest of triumphs to this day, the chemical synthesis of penicillin by Howard Walter Florey and Ernst Chain. This is not to say that therapeutics since the discovery of chemotherapy has been reduced to the automatic and inflexible application of chemical antitoxins or antibiotics, as if it were enough to administer a remedy and let it do its work. Gradually, physicians learned that infectious agents develop resistance to the drugs used against them, and that organisms sometimes defend themselves, paradoxically enough, against their chemical guardians. Hence it was necessary to develop combined treatment regimens.[47] But such flexibility, typical of modern therapies, was made possible only by the rationalist simplification inherent in Ehrlich's program: since cells choose between stains, let us invent stains that will infallibly choose particular cells.

But what does it mean to invent a stain? It means to change the positions of the atoms in a molecule, to alter its chemical structure in such a way that its color can be read out, as it were, from its formula. Ehrlich's project was not simply impossible; it was inconceivable in the time of Magendie. It was not until 1856 that William Perkin, Sr., obtained a mauve dye from aniline as the outcome of research directed toward an entirely different goal. It was not until 1865 that F.A. Kékulé published his paper "The Composition of Aromatic Compounds." After confirming that the carbon atom is tetravalent, Kékulé determined the structure of benzene and gave the name "aromatic" to its derivatives to distinguish them from compounds involving the fatty acids, which, along with the alcohols, were the primary focus of chemical interest in the days of Magendie and Bernard.

The theoretical creation of new chemical substances was confirmed on a vast scale by the chemical industry. Alizarin, the principal component of madder, which Perkin in England and Karl James Peter Graebe and Edme Caro in Germany separately and simultaneously synthesized in 1868, was within ten years' time being produced at the rate of 9,500 tons annually. Finally in 1904, aniline, the most elaborate of the dye compounds, bestowed its prestigious name on the German firms Badische Anilin und Soda Fabrik (BASF) and Anilin Konzern.

Thus, two of the preconditions necessary for the development of chemotherapy as a replacement for the therapies associated with the old medical theories were a new symbolic representation for chemical substances and a new technology fo. producing organic compounds, which supplanted the old extractive processes. These were events with fixed, ascertainable dates; their place in history could not have been deduced in advance. Hence, chemotherapy could not have existed without a certain level of scientific and industrial society. Between Edward Jenner and Ehrlich came the indispensable discovery of aniline, which no one could have foreseen at the beginning of the century. In his study of the "rationalism of color," Gaston Bachelard wrote, "the chemist thinks of color in terms of the very blueprint that guides his creation. Therein lies a communicable, objective reality and a marketable social reality. Anyone who manufactures aniline knows the reality and the rationality of color."[48] [*Ideology and Rationality*, pp. 65-68]

The French School
[56] In considering the precursors of the immunization techniques perfected at the end of the nineteenth century, I shall look at the work of Pasteur rather than at that of Koch, partly because it came first chronologically and partly because Pasteur's work was of more general import, for "it not only modified the relationship between biology and chemistry but changed the representation of the world of living things generally, the relations between beings, and the functions ascribed to chemical reactions."[49]

François Dagognet argues, contrary to a widely held view, that it was not because of technical problems raised by industrialists, artisans and animal breeders ("maladies" of beer, wine, silkworms and sheep) that Pasteur took so long to develop "Pasteurism." Rather, Pasteur encountered technical problems because, from his first encounter with theoretical chemistry, he saw the experimental modification of natural products as a theoretical tool for analyzing reality. For him, the laboratory was a place for reworking substances given by nature or art and a place for freeing dormant or blocked causal mechanisms – in short, a place for revealing reality. Hence, laboratory work was directly affected by what was going on in the world of technology.

The revolution in medical thinking began with the development of two methods for studying the properties of crystals: stereometry and polarimetry. Dissatisfied with Eibhard Mitscherlich's explanations of the effect of polarized light on tartrates and paratartrates, Pasteur discovered the different orientation of the facets of paratartrate crystals. After isolating the two different kinds of crystals, he observed that a solution made with one kind of crystal rotated polarized light to the right, whereas a solution made with the other rotated it to the left. When the two crystals were combined in solution in equal parts, the optical effect was nullified. When a solution of calcium paratartrate was fermented by the effect of a mold, Pasteur noted that only the right-polarizing form of the crystal was altered. He therefore inferred a connection between the properties of microorganisms and molecular asymmetries. Dagognet has shown how microbiology began with this ingenious reversal of a result in biochemistry. A microscopic organism, a mold or a yeast, was shown to be capable of distinguishing between optical isomers. Pasteurism converted chemical separation by bacteria into bacteriological isolation by chemical isomers.[50] Thus confirmed in his belief that there is a structural contrast between the asymmetrical living organism and the mineral, and hence justified in rejecting any explanation receptive to the notion of spontaneous generation, Pasteur linked germ, fermentation and disease in a unified theoretical framework. Since my purpose here is simply to reflect on matters of history and epistemology, there is no need to recall the subsequent progress, doubts, retreats or even temporary errors that

Pasteur made in elaborating this theory. [*Ideology and Rationality*, pp. 68-70]

An Applied Science

[57] Bacteriology provided proof of its militant scientificity by giving rise to the science of immunology, which not only extended and refined Pasteurian medical practices but developed into an autonomous biological science. Immunology replaced the Pasteurian relation of virus to vaccinated organism with the more general relation of antigen to antibody. The antigen is a generalization of the aggressor microbe. The history of immunology has been a search for the true meaning of the prefix *anti-*. Semantically, it means "against," but doesn't it also mean "before"? Perhaps there is a relation, as of key and lock, between these two meanings.

As immunology became aware of its scientific vocation, it confirmed its scientific status through its ability to make unanticipated discoveries and to incorporate new concepts, one very striking example being Karl Landsteiner's discovery in 1907 of the human blood types. Consistency of research findings is another criterion of scientific status. Immunological findings were so consistent, in fact, that immunology's object of research came to be known as the "immune system," where the word "system" connotes a coherent structure of positive and negative responses at the cellular and molecular level. The immune system concept was more effective at "preserving appearances" than the earlier concept of "terrain." In a systemic structure, cyclical effects can appear to impede a causality construed to be linear. The immune system, moreover, has the remarkable property known as idiotypy: an antibody is specific not only to a particular antigen but also to a particular individual. The idiotype is the capacity of the immune system to encode an organic individuality.

However tempting, it would be a mistake to view this phenomenon as betokening a rediscovery of the concrete individual patient set aside by the very medical science whose progress eventually revealed the existence of the idiotype. Although immune identity is sometimes portrayed, through abuse of terminology, as involving an opposition of "self" and "nonself," it is a strictly objective phenomenon. Medicine may sometimes appear to be the application of biological knowledge to concrete individuals, but that appearance is deceiving. The time has now come to consider the epistemological status of medicine as such, leaving historical matters aside. Given what we know about immunology, genetics and molecular biology, or, looking backward in time, about X-rays and cellular staining techniques, in what sense can we say that medicine is an applied science or an evolving synthesis of applied sciences?[...]

It is appropriate to describe medicine as an "evolving synthesis of applied sciences," insofar as the realization of its goals requires the use of scientific discoveries having nothing to do with its intrinsic purposes.[...] In using the term "applied science," the accent, I think, should fall on "science." In saying this, I disagree with those who see the application of knowledge as involving a loss of theoretical dignity, as well as those who think they are defending the uniqueness of medicine by calling it a "healing art." The medical application of scientific knowledge, converted into remedies (that is, into means of restoring a disturbed organic equilibrium), is in no sense inferior in epistemological dignity to the disciplines from which that knowledge is borrowed. The application of knowledge is also an authentic form of experimentation, a critical search for effective therapies based on imported understandings. Medicine is the science of the limits of the powers that the other sciences claim to confer upon it.[...]

If the progress of a science can be measured by the degree to which its beginnings are forgotten, then it is worth noting that when doctors today need to do a blood transfusion, they verify the blood-type compatibility of donor and recipient without knowing that the tests they are ordering are the product of a history that can be traced back through immunology and bacteriology to Lady Montagu and Edward Jenner, indeed to a type of medical practice that doctrinaire physicians once considered heretical. That practice started medicine down a road that brought it into contact with a particular branch of mathematics, the mathematics of uncertainty. Calculated uncertainty, it turned out, is not incompatible with etiological hypotheses and rational diagnosis based on data gathered with the aid of suitable instruments.

What expert is qualified to decide the epistemological status of medicine? Philosophers cannot bestow upon themselves the power to judge nonphilosophical disciplines. The term "epistemology" refers to the legacy, not to say the relics, of the branch of philosophy traditionally known as "theory of knowledge." Because the relation of knowledge to its objects has been progressively revealed by scientific methods, epistemology has broken with philosophical assumptions to give itself a new definition. Rather than deduce criteria of scientificity from *a priori* categories of understanding, as was done in the past, it has chosen to take those criteria from the history of triumphant rationality. Why shouldn't medicine therefore be both judge and party in the case? Why should it feel the need for a consecration of its status within the scientific community? Might it be that medicine has preserved from its origins a sense of the uniqueness of its purpose, so that it is a matter of some interest to determine whether that sense is a tenuous survival or an essential vocation? To put it in somewhat different terms, are what used to be *acts* of diagnosis, decision and treatment about to become *roles* ancillary to some computerized medical program? If medicine cannot shirk the duty to assist individual human beings whose lives are in danger, even if that means violating the requirements of the rational, critical pursuit of knowlege, can it claim to be called a science?

A clever and learned historian of medicine, Karl Rothschuh, has examined this issue in terms borrowed from Thomas Kuhn's historical epistemology. In 1977, he asked whether Kuhn's concepts of "normal science," "paradigm" and "scientific group" could be applied to conceptual advances in clinical medicine; he concluded that Kuhn's framework, while useful for understanding medicine's incorporation of advances in the basic sciences since the early nineteenth century, is inadequate to account for the difficulties encountered by clinical medicine, due to the complexity and variability of its object. He concluded his paper with a quotation from Leibniz: "I wish that medical knowledge were as certain as medical problems are difficult." In the course of his analysis, Rothschuh reports that Kuhn once characterized medicine as a "protoscience," whereas he, Rothschuh, prefers to call it an operational science (*operationale Wissenschaft*). These two appellations are worth pausing over. "Protoscience" is ingenious because it is ambiguous. *Proto-* is polysemic: it suggests "prior" as well as "rudimentary," but it may also refer to hierarchical priority. "Protoscience" is a term that might well be applied to an earlier period in the history of medicine, but it seems somehow ironic to use it when some physicians believe that the time has come to allow computers to guide treatment while others argue that patients ought to be allowed to consult the machines directly. Yet "operational science" seems no more appropriate a term than "applied science," which some nineteenth-century physicians themselves applied to their discipline as they began to treat patients on the basis of their understanding of physical and chemical mechanisms explored by physiologists. For example, the work of Carlo Mateucci, Emile Du Bois-Reymond and Hermann von Helmholtz on animal electricity led Guillaume-Benjamin Duchenne de Boulogne to discover new ways of treating muscular diseases. His major works, published between 1855 and 1867, bear titles incorporating the word "application."

An instructive example is electrotherapy. It suggests that medicine was impelled to become an applied science by the need to discover more effective treatments, as if in obedience to its original imperative. Later, of course, the "science of electricity" led to the development not of therapeutic but of diagnostic devices such as the electrocardiograph (invented by Willem Einthoven in 1903), the electroencephalograph (Johannes Berger, 1924) and endoscopy. By treating the patient as an abstract object of therapy, it was possible to transform medicine into an applied science, with the accent now on science. Like any science, medicine had to evolve through a stage of provisionally eliminating its concrete initial object.

Earlier, I called medicine an "evolving synthesis of applied sciences." Now that I have discussed the sense in which medicine is an applied science, I have only to justify the choice of the words "evolving" and "synthesis." Surely the reader will grant that any science, whether pure or applied, validates its epistemological status by developing new methods and achieving new results. A science evolves because of its interest in new methods for dealing with its problems. For example, the existence of chemical neurotransmitters was acknowledged (not without reservations, particularly in France) when the work of Sir Henry Dale and Otto Loewi filled in blanks in the results obtained by electrical methods a century earlier.

So much for "evolving" – but what about "synthesis?" A synthesis is not a mere addition; it is an operational unity. Physics and chemistry are not syntheses, but medicine *is*, insofar as its object, whose interrogative presence is suspended by methodological choice, nevertheless remains present. That object has a human form, that of a living individual who is neither the author nor the master of his own life and who must, in order to live, sometimes rely on a mediator. However complex or artificial contemporary medicine's mediation may be – whether technical, scientific, economic or social – and however long the dialogue between doctor and patient is suspended, the resolve to provide effective treatment, which legitimates medical practice, is based on a particular modality of life, namely, human individuality. In the physician's epistemological subconscious, medicine is truly a synthesis because, to an ever-increasing degree, it applies science to the task of preserving the fragile unity of the living human individual. When the epistemological status of medicine becomes a matter of conscious questioning, the search for an answer clearly raises questions that fall outside the purview of medical epistemology. ["Statut épistémologique," *Histoire*, pp. 23-29]

Part Three

History

Chapter Seven

Cell Theory

Theories Never Proceed from Facts

[58] Is biology a theoretical or an experimental science? Cell theory is an ideal test case. We can see light waves only with reason's eyes, but we appear to view the cells in a plant section with the same eyes we use to look at everyday objects. Is cell theory anything more than a set of observational protocols? With the aid of a microscope, we can see that macroscopic organisms consist of cells, just as we can see with the naked eye that the same organisms are elements of the biosphere. Yet the microscope extends the powers of intelligence more than it does the powers of sight. Furthermore, the first premise of cell theory is not that living things are composed of cells but that *all* living things consist of *nothing but* cells; every cell, moreover, is assumed to come from a preexisting cell. Such an assertion cannot be proven with a microscope. At best, the microscope can serve as a tool in the task of verification. But where did the idea of the cell come from in the first place?[1]

Robert Hooke is generally given too much credit for the formulation of cell theory. True, he was the first to discover the cell, somewhat by accident, as he pursued a curiosity awakened by microscopy's earliest revelations. After making a thin slice in a piece of cork, Hooke observed its compartmentalized structure.[2] He also coined the word "cell" while under the spell of an image: the section of cork reminded him of a honeycomb, the work of an animal, which then further reminded him of a work of man, the honeycomb being like a building made up of many *cells*, or small rooms. But Hooke's discovery led nowhere: it failed to open up a new avenue of research. The word disappeared, only to be rediscovered a century later.

The discovery of the cell concept and the coining of the word are worth dwelling on for a moment. As a biological concept, the cell is surely overdetermined to a considerable degree. The psychoanalysis of knowledge has been sufficiently successful in the past that it now constitutes a distinct genre, to which additional contributions may be added as they arise, even without systematic intention. Biology classes have familiarized all of us with what is now a fairly standard image of the cell: schematically, epithelial tissue resembles a honeycomb.[3] The word "cell" calls to mind not the prisoner or the monk but the bee. Ernst Heinrich Haeckel pointed out that cells of wax filled with honey are in every way analogous to cells of plants filled with sap.[4] I do not think that this analogy explains the appeal of the notion of the cell. Yet who can say whether or not the human mind, in consciously borrowing from the beehive this term for a part of an organism, did not unconsciously borrow as well the notion of the cooperative labor that produces the honeycomb? Just as the alveola is part of a structure, bees are, in Maeterlinck's phrase, individuals wholly absorbed by the republic. In fact, the cell is both an anatomical and a functional notion, referring both to a fundamental building block and to an individual labor subsumed by, and contributing to, a larger process. What is certain is that affective and social values of cooperation and association lurk more or less discreetly in the background of the developing cell theory.

A few years after Hooke, in 1671, Marcello Malpighi and Nehemiah Grew simultaneously but independently published their work on the microscopic anatomy of plants. Although they did not mention Hooke, what they discovered was the same thing he had discovered, even if the word was different. Both men found that living things contain what we now call cells, but neither claimed that living things are nothing but cells. According to Marc Klein, moreover, Grew subscribed to the theory that cells are preceded by and grow out of a so-called vital fluid. The history of this biological theory is worth exploring in greater detail for what it can teach us about scientific reasoning in general.

As long as people have been interested in biological morphology, their thinking has been dominated by two contradictory images — continuity versus discontinuity. Some thinkers imagine living things growing out of a primary substance that is continuous and plastic; others think of organisms as composites of discrete parts, of "organic atoms" or "seeds of life." Continuity versus discontinuity, continuum versus particle: the mind imposes its forms in biology just as it does in optics.

The term "protoplasm" now refers to a constituent of the cell considered as an atomic element of a composite organism. Originally, however, the word referred to the vital fluid out of which all life presumably arose. The botanist Hugo von Mohl, one of the first to observe the birth of new cells by division of existing ones, proposed the term in 1843: in his mind, it referred to a fluid present prior to the emergence of any solid cells. In 1835, Felix Dujardin had suggested the term "sarcode" for the same thing, namely, a living jelly capable of subsequent organization. Even Theodor Schwann, the man regarded as the founder of cell theory, was influenced by both images: he believed that a structureless substance (the cytoblasteme) gives rise to the nuclei around which cells form. In tissues, cells form wherever the nutrient liquid penetrates. This theoretical ambivalence on the part of the authors who did most to establish cell theory led Marc Klein to make a remark that has considerable bearing on what I wish to argue here: "What we find, then, is that a small number of basic ideas recur insistently in the work of authors concerned with a wide variety of objects from a number of different points of view. They certainly did not take these ideas from one another. These fundamental hypotheses appear to represent persistent modes implicit in the nature of scientific explanation." Translating this epistemological observation into philosophical terms, it follows that *theories never proceed from facts*, a finding that conflicts with the empiricist point of view that scientists often adopt uncritically when they try to philosophize about their experimental findings. Theories arise only out of earlier theories, in some cases very old ones. The facts are merely the path – and it is rarely a straight path – by which one theory leads to another. Auguste Comte shrewdly called attention to this relation of theory to theory when he remarked that since an empirical observation presupposes a theory to focus the attention, it is logically inevitable that false theories precede true ones. [...]

Thus, if we wish to find the true origins of cell theory, we must not look to the discovery of certain microscopic structures in living things. [*Connaissance*, pp. 47-50]

Comte Buffon, or the Discontinuous Imagination
[59] In the work of Buffon, who, as Marc Klein points out, made little use of the microscope, we find a theory of the composition of living things – indeed, a system, in the eighteenth-century sense of the term. Buffon proposed a series of axioms to explain certain facts having to do chiefly with reproduction and heredity. In Chapter Two of the *Histoire naturelle des animaux* (1748), he set forth his "theory of organic molecules." In Buffon's words,

"animals and plants that can multiply and reproduce in all their parts are organized bodies composed of other, similar organic bodies, whose accumulated quantity we can discern with the eye but whose primitive parts we can perceive only with the aid of reason." From this, Buffon deduced that there are infinitely many organic parts, each composed of the same substance as "organized beings." These organic parts, common to animals and plants, are primitive and incorruptible. What is called "generation" in biology is merely the conjunction of some number of primitive organic parts; similarly, death is merely the dispersion of those parts.

The hypothesis that organized beings consist of primitive organic parts is the only one, Buffon argues, capable of avoiding the difficulties encountered by two earlier theories that claimed to explain the phenomena of reproduction, namely, ovism and animalculism. Both of these theories assumed that heredity is unilateral: ovists, following Regner de Graaf, claimed that it was maternal, whereas animalculists, following Anthonie van Leeuwenhoeck, argued that it was paternal. Buffon, alert to phenomena of hybridization, believed that heredity must be bilateral, as is clear from Chapter Five of his work. The facts reinforced this belief: a child could resemble either his father or his mother. Thus, he writes in Chapter Ten, "The formation of the fetus occurs through combination of organic molecules in the mixture composed of the seminal fluids of two individuals."[...]

In Buffon's view, Newtonian mechanics explicitly had jurisdiction over the organization of living things:

> It is obvious that neither the circulation of the blood nor the movement of the muscles nor the animal functions can be explained in terms of impulse or any of the laws of ordinary mechanics. It is just as obvious that nutrition, development and reproduction obey other laws. Why not acknowledge, then, that there are forces penetrat-

ing and acting upon the masses of bodies, since we have examples of forces in the substance of bodies in magnetic attractions and chemical affinities?[5]

Organic molecules attract one another in obedience to a law of morphological constancy, constituting an aggregate that Buffon called the "internal mold." Without the hypotheses of internal mold and organic molecule, nutrition, development and reproduction would be unintelligible. [...]

There can be no doubt that Buffon hoped to be the Newton of the organic world, much as David Hume at around the same time hoped to become the Newton of psychology. Newton had demonstrated that the forces that move the stars are the same as those that move objects on the surface of the earth. Gravitational attraction explained how simple masses could form more complex systems of matter. Without such a force of attraction, reality would be not a universe but just so much dust.

For Buffon, the hypothesis that "matter lost its force of attraction" was equivalent to the hypothesis that "objects lost their coherence."[6] A good Newtonian, Buffon believed that light was a corpuscular substance:

> The smallest molecules of matter, the smallest atoms we know, are those of light.... Light, though seemingly blessed with a quality the exact opposite of weightiness, with a volatility that might be thought essential to its nature, is nevertheless as heavy as any other matter, since it bends when it passes near other bodies and finds itself within reach of their sphere of attraction.... And just as any form of matter can convert itself into light through extreme subdivision and dispersion through impact of its infinitesimal parts, so, too, can light be converted into any other form of matter if, through the attraction of other bodies, its component parts are made to coalesce.[7]

Light, heat and fire are different modes of existence of the same common material. To do science was to try to find out how, "with this single source of energy and single subject, nature can vary its works ad infinitum."[8] If, moreover, one assumes that living matter is nothing but ordinary matter plus heat, a corpuscular conception of matter and light inevitably leads to a corpuscular conception of living things:

> All the effects of crude matter can be related to attraction alone, all of the phenomena of living matter can be related to that same force of attraction coupled with the force of heat. By living matter I mean not only all things that live or vegetate but all living organic molecules dispersed and spread about in the detritus or residue of organized bodies. Under the head of living matter I also include light, fire and heat, in a word, all matter that appears to us to be active by itself.[9]

This, I believe, is the logic behind the theory of organic molecules, a biological theory that owed its existence to the prestige of a physical theory. The theory of organic molecules is an example of the analytic method in conjunction with the discontinuous imagination, that is, a penchant for imagining objects by analogy with discrete rather than continuous models. The discontinuous imagination reduces the diversity of nature to uniformity, to "a single source of energy and a single subject." That one element, the basis of all things, then forms compounds with itself that produce the appearance of diversity: nature varies its works ad infinitum. The life of an individual, whether an animal or a plant, is therefore an effect rather than a cause, a product rather than an essence. An organism is a mechanism whose global effect is the necessary consequence of the arrangement of its parts. True, living individuality is molecular, monadic.

The life of an animal or plant, it seems, is merely the result of all the actions, of all the little, individual lives (if I may put it that way) of each of its active molecules, whose life is primitive and apparently cannot be destroyed. We have found these living molecules in all living or vegetating things: we are certain that all these organic molecules are equally essential to the nutrition and therefore to the reproduction of animals and plants. It is not difficult to imagine, therefore, that when a certain number of these molecules are joined together, they form a living thing: since there is life in each of its parts, life can also be found in the whole, that is, in any assemblage of those parts.

[*Connaissance*, pp. 52–56]

Lorenz Oken, or the Continuous Imagination

[60] Charles Singer and Marc Klein, as well as Emile Guyénot, though to a lesser degree, did not fail to note the credit due to Oken for the formulation of cell theory. Oken belonged to the Romantic school of nature philosophers founded by Schelling.[10] The speculations of this school had as much influence on early-nineteenth-century German physicians and biologists as on men of letters. There is no rupture of continuity between Oken and the first biologists that would offer deliberate empirical support for cell theory. Matthias Jacob Schleiden, who first formulated cell theory for plants in his *Beiträge zur phytogenesis* (1838), taught at the University of Jena, where memories of Oken's teaching were still fresh. Theodor Schwann, who between 1839 and 1842 generalized cell theory to all living things, had seen a good deal of Schleiden and his teacher, Johannes Müller, who had been a nature philosopher in his youth.[11] Singer is thus fully justified in remarking that Oken "in a sense *sowed* the ideas of the authors regarded in his stead as the *founders* of cell theory."[...]

Here, the idea that organisms are composites of elementary life forms is merely a logical consequence of a more basic notion, which is that the elements of life are released when the larger forms to which they belong disintegrate. The whole takes precedence over the parts. Klein states this explicitly:

> The association of primitive animals in the guise of living flesh should not be thought of as a mechanical coupling of one animal to another, as in a pile of sand where the only relation among the grains of which it is composed is one of proximity. Just as oxygen and hydrogen disappear in water, just as mercury and sulfur disappear in cinnabar, what takes place here is a true interpenetration, an intertwining and unification of all the animalcules. From that moment on, they have no life of their own. All are placed at the service of a higher organism and work toward a unique and common function, or perform that function in pursuing their own ends. Here, no individual is spared; all are sacrificed. But the language is misleading, for the combination of individualities forms another individuality. The former are destroyed, and the latter appears only as a result of that destruction.[12]

We are a long way from Buffon. The organism is not a sum of elementary biological entities; it is, rather, a higher entity whose elements are subsumed. With exemplary precision, Oken anticipated the theory of degrees of individuality. This was more than just a presentiment, though it did anticipate what techniques of cell and tissue cultures would teach contemporary biologists about differences between what Hans Petersen called the "individual life" and the "professional life" of cells. Oken thought of the organism as a kind of society, but that society was not an association of individuals as conceived by the political philosophy of the Enlightenment but, rather, a community as conceived by the political philosophy of Romanticism. [...]

Comparison is inevitable between Oken's biological theories and the political philosophy of the German Romantics whom Novalis influenced so deeply. Novalis's *Glaube und Liebe: der König und die Königin* appeared in 1798; his *Europa oder die Christenheit* was published in 1800; Oken's *Die Zeugung* came out in 1805. The first two works are vehemently critical of revolutionary thinking. Novalis alleged that universal suffrage pulverized the popular will and failed to give due weight to social, or, more precisely, communal, continuity. Anticipating Hegel, Novalis (like Adam Heinrich Müller a few years later) considered the state to be a reality willed by God, a fact surpassing individual reason to which the individual must sacrifice himself. If there is an analogy between these sociological views and biological theory, it is, as has often been remarked, because the Romantics interpreted political experience in terms of a "vitalist" conception of life. Even as French political thinkers were offering the ideas of the social contract and universal suffrage to the European mind, the vitalist school of French medicine was proposing an image of life as transcending analytical understanding. Vitalists denied that organisms could be understood as mechanisms; life, they argued, is a form that cannot be reduced to its material components. Vitalist biology provided a totalitarian political philosophy with the means to propose certain theories of biological individuality, though philosophy was under no compulsion to do so. How true it is that the problem of individuality is indivisible.[13] [*Connaissance*, pp. 58-63]

Enduring Themes
[61] Did the concepts of individuality that inspired these speculations about the composition of organisms disappear altogether among biologists truly worthy of being called scientists? Apparently not.

Claude Bernard, in his *Leçons sur les phénomènes de la vie com-*

muns aux animaux et aux végétaux, published after his death by Dastre in 1878-79, described the organism as "an aggregate of elementary cells or organisms," thereby affirming the principle of autonomy for anatomical constituents. This is tantamount to asserting that cells behave in association just as they would behave in isolation if the milieu were the same as that created for them within the organism by the action of nearby cells. In other words, cells *would live in liberty exactly as they do in society.* Note, though, in passing, that if the regulative substances that control the life of the cell through stimulation and inhibition are the same in a culture of free cells as in the internal environment of the organism, one cannot say that the cells live in liberty. Nevertheless, Bernard, hoping to clarify his meaning by means of a comparison, asks us to consider a complex living thing "as a city with its own special stamp," in which individuals all enjoy the same identical food and the same general capacities yet contribute to social life in different ways through their specialized labor and skills.

In 1899, Ernst Heinrich Haeckel wrote, "The cells are truly independent citizens, billions of which compose our body, the cellular state."[14] Perhaps images such as the "assembly of independent citizens" constituting a "state" were more than just metaphors. Political philosophy seems to dominate biological theory. What man could say that he was republican because he believed in cell theory or a believer in cell theory because he was a republican?

To be sure, Bernard and Haeckel were not altogether immune to philosophical temptation or exempt from philosophical sin. The second chapter of Marcel Prenant, Paul André Bouin and Louis-Camille Maillard's 1904 *Traité d'histologie*, which Marc Klein credits, along with Felix Henneguy's *Leçons sur la cellule* (1896), with being the first classical work to introduce cell theory in the teaching of histology in France,[15] was written by Prenant. The author's sympathies for cell theory did not blind him to facts that might limit the scope of its validity. With admirable clarity he wrote, "*What is dominant in the notion of cell is the character of individuality;* this might even suffice as a definition." But, then, any experiment showing that seemingly hermetic cells are in reality, in Wilhelm His's words, "open cells" in communication with one another, tends to devalue cell theory, which leads Prenant to this conclusion:

> The individual units may vary as to their degree of individuality. A living thing is born as an individual cell. Later, the individuality of the cell disappears in the individuality of the individual or person composed of many cells. The individuality of the person may in turn be effaced in a society of individuals by a social individuality. What happens when one examines the individual and society, those ascending series of multiples of the cell, can also be found in cellular submultiples: the parts of the cell in turn possess a certain degree of individuality partially subsumed by the higher and more powerful individuality of the cell. Individuality exists from top to bottom. Life is not possible without individuation of the living.

How far is this from Lorenz Oken's view? Once again, it seems, the problem of individuality is indivisible. Perhaps insufficient attention has been paid to the fact that, etymologically speaking, individuality is a negative concept: the individual is that which cannot be divided without losing its characteristic properties, hence a being at the limit of nonbeing. This is a minimum criterion for existence; but no being in itself is a minimum. The existence of an individual implies a relation to a larger being; it calls for, it requires (in the sense that Octave Hamelin gives these terms in his theory of the opposition of concepts) a background of continuity against which its discontinuity stands out. In this regard, there is no reason to confine the power of individuality within the limits of the cell. When Prenant recognized in 1904 that the parts of the cell possess a certain degree of individuality subsumed within the individuality of the cell, was he not looking forward to later discoveries concerning the submicroscopic structure and physiology of the protoplasm? "Are protein viruses living or nonliving?" biologists ask themselves. This is tantamount to asking whether crystals of nuclear protein are or are not "individuals." "If they are living," Jean Rostand argues, "they represent life in the simplest conceivable form. If they are not, they represent a state of chemical complexity that prefigures life."[16] But what is the point of saying that protein viruses are simple living things if one concedes that it is their complexity that is a prefiguration of life? Individuality, in other words, is not an endpoint but a term in a relation. It is misleading to interpret the results of research intended to shed light on that relation as revealing some ultimate truth. [*Connaissance*, pp. 69–71]

Toward a Fusion of Representations and Principles
[62] Consider now works of three authors published between the two world wars. These three men exhibit not only different casts of mind but different scientific specialties. In 1929, Rémy Collin published an article entitled "Théorie cellulaire et la vie." In 1935, Hans Petersen published *Histologie und Mikroskopische Anatomie*, whose first few chapters I shall focus on here. And in 1939, Dr. Jules Duboscq lectured on the place of cell theory in protistology (the study of unicellular organisms). Using different arguments and emphases, all three works converge toward a similar solution, which Duboscq expressed thus: "It is a mistake to take the cell to be a necessary constituent of living things." In the first place, it is difficult to regard metazoa (pluricellular organisms) as republics of cells or composites of individualized cellular building blocks, given the role of such essential systems

as the muscular system or of such formations as plasmodia or syncitia consisting of continuous masses of cytoplasm with scattered nuclei. In the human body, only the epithelia are clearly cellularized. Between a free cell such as a leucocyte and a syncytium such as the cardiac muscle or the surface of the chorial villosities of the fetal placenta, there are intermediate forms, such as the giant multinuclear cells (polycaryocytes), and it is difficult to say whether syncitia develop through fusion of once-independent cells or vice versa. Both mechanisms can, in fact, be observed. Even in the development of an egg, it is not certain that every cell comes from the division of a preexisting cell. Emile Rhode was able to show in 1923 that individual cells, in plants as well as animals, frequently result from the subdivision of a primitive plasmodium (multinucleate mass).

But the anatomical and ontogenetic aspects of the problem are not the whole story. Even authors who, like Hans Petersen, acknowledge that the real basis of cell theory is the development of metazoa, and who see the production of chimeras – living things created by artificially combining egg cells from different species – as supporting the "additive" composition of living things are obliged to admit that *the explanation of the functions of these organisms contradicts the explanation of their genesis*. If the body is really a collection of independent cells, how does one explain the harmonious functioning of the larger unit? If the cells are closed systems, how can the organism live and act as a whole? One way to resolve the difficulty is to look for a coordinating mechanism: the nervous system, say, or hormonal secretions. But the connection of most cells to the nervous system is unilateral and nonreciprocal; and many vital phenomena, especially those associated with regeneration, are rather difficult to explain in terms of hormonal regulation, no matter how complex. Petersen therefore remarked:

> Perhaps one can say in a general way that all the processes in which the body participates as a whole (and in pathology there are few processes where this is not the case) are difficult to understand in terms of *the cellular state or the theory of cells as independent organisms.* [...] Given the way in which the cellular organism behaves, lives, works, maintains itself against the attacks of its environment and regains its equilibrium, the cells are organs of a uniform body.

Here the problem of individuality comes up again: a totality, initially resistant to division of any kind, takes priority over the atomistic view derived from an attempt to subdivide the whole. Petersen quite pertinently quotes a remark made by Julius Sachs in 1887 concerning multicellular plants: "Whether cells seem to be elementary independent organisms or simply parts of a whole depends entirely on *how we look at things.*"

In recent years, increasing doubts and criticisms have been voiced about cell theory in its classical form, that is, in the fixed, dogmatic form in which it is presented in textbooks, even those intended for advanced students.[17] There is far less objection today to noncellular components of organisms and to mechanisms by which cells can be formed out of continuous masses of protoplasm than there was when Rudolph Virchow, in Germany, criticized Theodor Schwann's idea of a cytoblasteme and Charles Robin, in France, was looked upon as a cantankerous, old-fashioned iconoclast. In 1941, Tividar Huzella showed in his *Zwischen Zellen Organisation* that intercellular relations and extracellular substances (such as the interstitial lymph and noncellular elements of connective tissue) are just as important biologically as the cells themselves. The intercellular void that one can see in those preparations made to be viewed through a microscope is by no means devoid of histological function. In 1946, P. Busse Grawitz concluded on the basis of his research that cells can appear in basically acellular substances.[18] According to cell theory, fundamental substances (such as the collagen of the tendons) must be secreted by the cells, even if it is not possible to say precisely how the secretion takes place. Here, however, the order is reversed. Of course, the experimental argument in such a theory is negative in nature: the researcher trusts that sufficient precautions have been taken to prevent the migration of cells into the acellular substance in which cells are seen to emerge. In France, Jean Nageotte had observed, in the development of a rabbit embryo, that the cornea of the eye first appears to be a homogeneous substance containing no cells during the first three days of growth — yet, in light of Virchow's law, he believed that those cells that appeared subsequently must have arrived there through migration. Yet no such migration was ever observed. [*Connaissance*, pp. 73-76]

[63] It is not absurd to conclude that biology is proceeding toward a synthetic view of organic structure not unlike the synthesis that wave mechanics brought about between concepts as seemingly contradictory as wave and particle. Cell and plasmodium are among the last incarnations of the contradictory demands of discontinuity and continuity which theorists have faced ever since human beings began to think. Perhaps it is true that scientific theories attach their fundamental concepts to ancient images — I would even be tempted to say myths, if the word had not been so devalued by its recent use in philosophies obviously created for purposes of propaganda and mystification. For what, in the end, is this continuous initial plasma, this plasma that biologists have used in one form or another ever since the problem of identifying a structure common to all living things was first posed in order to deal with the perceived inadequacies of the corpuscular explanation? Was it anything other than a logical avatar of the mythological fluid from which all life is supposed to

arise, of the frothy wave that bore Venus on its foam? Charles Naudin, a French biologist who came close to discovering the mathematical laws of heredity before Gregor Mendel, thought that the primordial blasteme was the "clay" mentioned in the Bible.[19] This is why I have argued that theories do not arise from the facts they order – or, to put it more precisely, facts *do* act as a stimulus to theory, but they neither engender the concepts that provide theories with their internal coherence nor initiate the intellectual ambitions that theories pursue. Such ambitions come to us from long ago, and the number of unifying concepts is small. That is why theoretical themes survive even after critics are pleased to think that the theories associated with them have been refuted. [*Connaissance*, p. 79]

CHAPTER EIGHT

The Concept of Reflex

Epistemological Prejudices

[64] Broadly speaking, the various histories of research into reflex movement have failed to discriminate sufficiently among description of automatic neuromuscular responses, experimental study of anatomical structures and their functional interactions, and formulation of the reflex concept and its generalization in the form of a theory. This failure accounts for the surprising discrepancies, when it comes to awarding credit for an original discovery or anticipation to a particular individual, among historians as well as biologists engaged in backing the claims of certain of their colleagues.

Here I propose to distinguish points of view that are all too often confounded. My purpose is not to right wrongs, like some scholarly avenger, but to draw conclusions of potential value to epistemology and the history of science. Indeed, the ultimate reason for the existence of divergent histories has to do with two rather widespread prejudices. One of these involves all the sciences: people are disposed to believe that a concept can originate only within the framework of a theory — or, at any rate, a heuristic — homogeneous with the theory or heuristic in terms of which the observed facts will later be interpreted. The other involves biology in particular: it is widely believed that, in this science, the only theories that have led to fruitful applications and positive advances in knowledge have been mechanistic in style. [...]

In the nineteenth century, the mechanist theory, based on the generalization of a concept whose basic outline was clear by 1850, produced a retroactive effect on the way in which its origins were conceived. It seemed only logical that a phenomenon which, along with many others, provided justification for a mechanical explanation of animal life could have been discovered and studied only by a mechanist biologist. If the logic of history thus pointed toward a mechanist, the history of physiology provided a name – Descartes. This coincidence seemed to foreclose further discussion, though no one knew or cared to know whether the logic confirmed the history or the history inspired the logic. From the incontestable fact that Descartes had proposed a mechanical theory of involuntary movement and even provided an excellent description of certain instances of what would later, in the nineteenth century, be called "reflexes," it was deduced, in surreptitious anticipation of what was to come, that Descartes had described, named and formulated the concept of the reflex because the general theory of the reflex was elaborated in order to explain the class of phenomena that he had explained in his own fashion.

My own view is that, in the history of science, logic per se ought to take precedence over the logic of history. Before we relate theories in terms of logical content and origin, we must ask how contemporaries interpreted the concepts of which those theories were composed – for if we do not insist on internal consistency, we risk falling into the paradox that logic is ubiquitous except in scientific thought. There may be a logic, moreover, in the succession of doctrines in themselves illogical. Even if one holds that the principle of noncontradiction is obsolete, and even if one substitutes for logic some currently more prestigious term, the essence of the case remains unchanged. Indeed, even if theories engender one another dialectically, the norms of scientific theory are not those of myth, dream or fairy tale. Even if virtually none of the principles of a theory remain intact, the theory can be called false only in terms of a judgment based on those principles and their consequences. Thus, the elements of a doctrine are supposed to fit together in a way that is not haphazard; its concepts are supposed to combine in some way that is not mere juxtaposition or addition.

We must, accordingly, look in some new direction for conceptual filiations. Rather than ask who the author of a theory of involuntary movement that prefigured the nineteenth-century theory of the reflex was, we ask what a theory of muscular movement and nerve action must incorporate in order for a notion like reflex movement, involving as it does a comparison between a biological phenomenon and an optical one (reflection), to make sense (where "making sense" means that the notion of reflex movement must be logically consistent with some set of concepts). If a concept outlined or formulated in such a context is subsequently captured by a theory that uses it in a different context or with a different meaning, it does not follow that the concept as used in the original theory is nothing but a meaningless word. Some concepts, such as the reflection and refraction of light, are theoretically polyvalent, that is, capable of being incorporated into both particle theory and wave theory. Furthermore, the fact that a concept plays a strong role in a certain theoretical domain is by no means sufficient grounds for limiting research into the origins of that concept to similarly constituted domains.

By adhering to these methodological precepts, I came not to discover Thomas Willis – for some nineteenth-century physiologists aware of the history of the reflex concept *had* mentioned his name — but to confirm his legitimate right to a title that had previously been open to doubt or challenge. [*Formation du réflexe*, pp. 3-6]

René Descartes Did Not Formulate the Reflex Concept

[65] When Descartes proposed his general theory of involuntary movement, he, like many others before him, associated such movements with phenomena that we today refer to as reflexes. Does it follow, then, that he belongs among the naturalists and physicians who helped to delineate and define the *concept* of reflex? The answer to this historical and epistemological question must, I think, be deferred until detailed, critical study of the Cartesian anatomy and physiology of the nerve and muscle enables us to decide whether or not Descartes could have anticipated, however confusedly, the essential elements of the concept.

Descartes, of course, believed that all physiological functions could be explained in purely mechanical terms. Hence, he saw only a limited number of possible interactions among an organism's parts: contact, impulse, pressure and traction. The importance of this fact cannot be overemphasized. Descartes's whole conception of animal movement derives from this principle together with what he considered a sufficient set of anatomical observations. [*Formation du réflexe*, p. 30]

[66] In Article 10 of *The Passions of the Soul*, Descartes claims that the animal spirits, born in the heart[20] and initially carried by the blood, build up in the brain as pressure builds in an air chamber. When released by the brain, these spirits are transmitted through the nerves to the muscles (other than the heart), where they determine the animal's movements. Descartes says that muscles are balloons filled with spirits, which, as a result of their transversal expansion, contract longitudinally, thus moving the articulated bone structures or organs such as the eye in which

they are inserted.[21] Morphologically, this tells us little, but that little suffices for Descartes's physiology of movement. Every nerve is a bundle of fibers contained within a tube, a marrow consisting of fine threads extending from the cerebral marrow and rather loosely sheathed in an arterylike tubular skin.[22] One might say, borrowing an image from modern technology, that Descartes envisioned the nerve as a sort of electrical cable run through a conduit. As a bundle of wires, the nerve served as a sensory organ,[23] while as a conduit it served as a motor organ.[24] Thus Descartes, unlike Galen and his followers, did not distinguish sensory nerves from motor nerves. Every nerve was both sensory and motor, but by virtue of different aspects of its structure and by way of different mechanisms.[25] The centripetal sensory excitation was not something that propagated along the nerve but, rather, an immediate and integral traction of the nervous fiber. When the animal sees, feels, touches, hears or tastes, the surface of its body shakes the brain by way of the nerve fiber. The centrifugal motor reaction, on the other hand, is a propagation, a transport. The spirits flow out through the pores of the brain, opened up in response to the pulling on the fibers, and into the empty space between the fibers and the conduit through which they run. If pressed, they press; if pushed, they push. Hence the muscle swells, that is, contracts.[26] Involuntary movement is thus different from action in all of its elements and phases. [*Formation du réflexe*, pp. 34-35]

[67] Basically, the concept of reflex consists of more than just a rudimentary mechanical explanation of muscular movement. It also contains the idea that some kind of stimulus stemming from the periphery of the organism is transmitted to the center and then reflected back to the periphery. What distinguishes reflex motion is the fact that it does not proceed directly from a center or central repository of immaterial power of any kind. Therein

lies, within the genus "movement," the specific difference between involuntary and voluntary. Now, according to Cartesian theory, movement that manifests itself at the periphery, in the muscles or viscera, originates in a center, the center of all organic centers, namely, the cardiac vessel. This is a material center of action, to be sure, not a spiritual one. The Cartesian theory is thus certainly mechanical, but it is not the theory of the reflex. The very image that suggested the word "reflex," that of a light ray's reflection by a mirror, requires homogeneity between the incident movement and the reflected movement. In Descartes's theory, though, the opposite is true: the excitation of the senses and the contraction of the muscles are not at all similar movements with respect to either the nature of the thing moved or the mode of motion. What does pulling on a bell cord have in common with blowing air into the pipe of an organ? Both are mechanical phenomena. [*Formation du réflexe*, p. 41]

[68] To sum up, while it is true that Descartes's work contains the theoretical equivalent of certain nineteenth-century attempts to formulate a general reflexology, rigorous examination turns up neither the term nor the concept of reflex. The downfall of Cartesian physiology, one cannot overemphasize, lay in the explanation of the movements of the heart. Descartes failed to see William Harvey's theory as an indivisible whole. To be sure, he was well aware that the explanation of the heart's movements was, for the seventeenth century, the key to the problem of movement generally.[27] This would continue to be the case in the eighteenth century. One fact turned out to be crucial in the Baconian sense for any theory purporting to explain the neuromuscular causes and regulations of movement — namely, the movement of excised organs, especially the heart. If the brain did not cause spirits to flow into these organs, what caused them to contract? Descartes did not have to confront this question. Removed from the body,

the heart retained its heat, and traces of blood remaining in it could vaporize and cause it to expand.[28] But for those who held that the heart was a muscle, it became difficult to argue that the brain was the essential central controller of all organ movements. Thus, it became necessary to look to places other than the brain, if not for the cause then at least for factors governing certain movements. [*Formation du réflexe*, p. 52]

Thomas Willis Deserves Credit for the Reflex Concept

[69] What distinguished Willis from Descartes were his conceptions of the motion of the heart and the circulation of the blood, which he took wholesale from William Harvey; namely, his conceptions of the nature of animal spirits and their movement through the nerves; of the structure of nerves; and of muscular contraction.

According to Willis (and Harvey), the heart is a muscle and nothing more. If it is the *primum movens* of the other muscles, it is so only by virtue of the rhythm of its function; its structure is identical. "It is not a noble organ, first in the hierarchy, but a mere muscle."[29] The only possible cause of the circulatory movement of the blood was the action of the spirits on the heart, as on any other muscle: that action made the heart into a hydraulic machine.[30] Willis distinguished between the circulation of the blood, a mechanical phenomenon, and its fermentation, a chemical one. Fermentation heated the blood, which then imparted its heat to the heart – not vice versa.[31] In Willis's mind, this distinction was sharp: circulation exists in all animals, whereas fermentation, he believed, is found only in the higher animals.[32] Willis deserves credit first of all for not feeling obliged, as Descartes did, to correct Harvey on a fundamental point of cardiac anatomy and physiology, as well as for not granting the heart a privileged role and preeminent nature in comparison to other muscles. For

Willis as for Harvey, the heart was simply a hollow muscle.

As for the animal spirits, Willis looked upon them as distilled, purified, sublimated, spiritualized blood. All four terms, listed in order of increasing dignity, are found in his writing. The brain and cerebellum functioned as stills to separate the animal spirits from the blood, a separation that occurred nowhere else in the body.[33] Functionally, the spirits flowed along nerves and fibers from the brain to the periphery — membranes, muscles, parenchyma — and from the periphery back to the brain. On the whole, however, if the flow of blood was a circulation, the flow of animal spirits was more in the nature of an irrigation: emanating from the brain, they were dispersed at the periphery. In this respect, there was no difference between Willis and Descartes. Willis, however, distinguished between the cause of the blood's circulation and that of the flow of animal spirits, and he acknowledged that the spirits flowed through the nerve in both directions. Above all, he saw the animal spirits quite differently from Descartes. [...]

According to Willis, the animal spirit was a potentiality in need of actualization. It was full of surprises. Though it seemed to be merely a ray of light, it could be explosive, and when it exploded its effects were magnified in accordance with rules that were not those of either arithmetic or geometry.[34] Descartes held that the spirits were expelled from the heart and sped toward the muscles in the manner of a current of air or stream of water, whereas Willis argued that they were propagated from the brain to the muscle in much the same way as heat or light. Slowed and transported by a liquid juice filling the interstices of the nervous structure, the spirits, upon reaching the peripheral organs, drew energy and heightened motor potential from the arterial blood bathing them. This energy came from the addition of nitrosulfurous particles to their own salt spirits, igniting the mixture and setting off an explosion, as of gunpowder in a cannon. This intramuscular explosion caused the muscle to contract and thus produced movement.[35] [*Formation du réflexe*, pp. 60-63]

[70] What distinguishes Willis from Descartes, however, is not simply his greater fidelity to Harvey's physiology or his notion, more chemical than mechanical, of the animal spirits. Unlike Descartes, Willis does not assume that the structure of the nerves allows them to play different roles in the sensory and motor functions. The nerves, he argues, have a single structure, fibrous and porous. They are neither conduits enclosing thin strands nor solid rods. They contain gaps, empty spaces into which animal spirits may enter. They are prolonged by fibers, which are not their only capillary extensions; some of these originate outside, and independent of, the nerves, through epigenesis. Just as animal spirits flow through, or reside in, the nerves, so too do they flow through, or reside in, the fibers. They may flow in either direction, and in wavelike motions. They flow first one way, then the other, in paths radiating from a center, the brain.[36]

These anatomical and physiological concepts were necessary conditions for Willis to do what Descartes was precluded from doing as regards the problem we are addressing. Though necessary, however, they were not yet sufficient. Willis's originality is more apparent in the powers of imagination that caused him to pursue the ultimate consequences of the explanatory comparisons he employed. Because he conceived of the anatomical structure of the nervous system as radiant rather than ramified, with the brain emitting nerves as the sun emits rays, Willis thought of the propagation of spirits in terms of radiation.[37] Now, the essence of the animal spirit itself could not be explained entirely in terms of any known chemical substance. Since it originated in the "flame" of the blood, it was comparable to a ray of light.[38] This analogy is pursued to the end: the nervous discharge was instantaneous, just like the transmission of light. Even the final stage of transmission, the excitation of the muscle by the nerve, supported the comparison. Just as light corpuscles produced light only if they encountered ethereal particles disseminated in the air, the animal spirits released the power in them only if they met sulfurous or nitrous particles disseminated in the interstitial blood. The resulting spasmodic intramuscular explosion caused the muscle to contract. Thus, the animal spirit was light only until it became fire. Its transport was analogous to illumination, whereas its effect was analogous to an explosive detonation. In this physiology the nerves are not strings or conduits but fuses (*funis ignarius*).[39] [*Formation du réflexe*, pp. 65–66]

[71] We know that we have encountered a concept because we have hit upon its definition – a definition at once nominal and real. The term *motus reflexus* is applied to a certain class of movements, of which a familiar example is provided: the automatic reaction of scratching. In addition to the object being defined, we have a defining proposition, which fixes its meaning. We have a word that establishes the adequacy of the defining proposition to the object defined (*scilicet*). The definition itself requires few words: it is not a full-blown theory but a précis. It is a definition that works by division, for it is associated with the prior definition of direct movement, the two together covering the entire range of possible causes of movement. Given the clearly stated principle (*quoad motus originem seu principium*), the division is exhaustive: every movement originates either at the center or at the periphery. This biological definition relies on a physical and, indeed, a geometric one. In sum, we find in Willis the thing, the word and the notion. The thing, in the form of an original observation, a cutaneous reflex of the cerebrospinal system, the scratch reflex; the word, reflex, which has improperly entered the language both as an adjective and a noun;[40] and the notion, that is, the possibility of a judgment, initially in the form

of an identification or classification and subsequently in the form of a principle of empirical interpretation. [*Formation du réflexe*, pp. 68-69]

The Logical and Experimental Consequences

[72] Thomas Willis assumed that all muscular motions are caused by a centrifugal flux of animal spirits from the brain, but he distinguished between voluntary motions governed by the cerebrum, such as locomotion, and natural or involuntary motions governed by the cerebellum and medulla oblongata, such as respiration and heartbeat. Hence, he also distinguished between two souls – one sensitive and reasonable, found in man alone, the other sensitive and vital, found in both man and animals.[41]

In man both souls were situated within the striated bodies, the seat of the *sensorium commune* of the reasonable soul. This was the stage at which a discrimination was made between those sensory impressions that were reflected into motions without reference to consciousness and those explicitly perceived as such by the soul. [...]

It should come as no surprise, therefore, that Jean Astruc (1684-1766) of Montpellier located the seat of common sense in the white matter of the brain. This localization enabled Astruc to propose an explanation of sympathetic phenomena that contained, for the first time since Willis, the notion of reflex motion (*An sympathia partium a certa nervorum positura in interno sensorio?*, 1736). How was it that a stimulus or injury to one part of an organism gave rise to a reaction in another part? Astruc rejected the explanation, common at the time, that certain fibers of communication connected the nerves. He argued that all nerve fibers are separate and independent from the brain to the periphery of the organism. Astruc explained sympathetic reaction in terms of a physical reflection of impressions that he believed took place in the medulla. When animal spirits, stirred by some stimulus, were carried to the brain by the nerve, they encountered fibers in the texture of the medulla, so that, "being reflected with an angle of reflection equal to the angle of incidence," they might enter the orifice of a motor nerve situated at that precise location. [...]

Like Astruc, Robert Whytt of Edinburgh rejected the explanation of sympathies in terms of extracerebral communication between nerves, yet he could not accept Astruc's mechanistic ideas, nor could he envision, as Haller did, a muscular irritability distinct from sensibility. He was therefore forced to propose a truly novel conception of the functions of the spinal cord. In his *Essay on the Vital and Other Involuntary Motions of Animals* (1751), Whytt attempted to prove by observation and experiment that all motions are caused by the soul, in response sometimes to an explicit perception, sometimes to a confused sensation of a stimulus applied to the organism. The central idea of his theory of involuntary motion is that every involuntary motion has a manifest purpose, namely, to eliminate the causes of disagreeable impressions. For example, when the pupil of the eye contracts in response to light, it is not the effect of a direct action of the light on the iris but rather of an importunate bedazzlement transmitted to the retina and the optic nerve. "The general and wise intention of all involuntary motions is the removal of everything that irritates, disturbs or hurts the body." It is this vital sense of all motions (which Whytt does not hesitate to compare to an immediate, prelogical moral sense) that precludes understanding them in terms of purely mechanical causes. Whytt nevertheless denies that he is a "Stahlian," one of those "who hold that one cannot explain these motions in terms of the soul without accepting the whole of the Stahlian view." The "sensitive principle" is not the "rational and calculating" soul. Or, rather, it is the same soul – for there is only one – insofar as it eschews calculation and reasoning and confines itself to immediate, hence unconscious, sensibility. Physiologically, this means that muscles contract only if innervated and sensorially stimulated, which means that they must be connected to the seat of the soul. Of course Whytt was not unmindful of the arguments that Haller, with the aid of his theories, drew from the observation of muscular motions in decapitated animals and separated organs. This led him to suspect the role of the spinal column as a sensory cause of motion, "because the spinal column does not appear to be exclusively an extension of the brain and cerebellum. It is probable that it prepares a nervous fluid of its own, and this is the reason why vital and other movements persist for several months in a tortoise whose head has been severed."[...]

Johann August Unzer (1727-1799) was critical of Whytt on the grounds that nervous sensation is distinct from sensibility per se and that movement in living things is not necessarily caused by the soul, even if it cannot be explained in terms of a mechanical phenomenon. The animal organism is indeed a system of machines, but those machines are natural or organic, that is, they are machines even in their very tiniest parts, as Leibniz had explained. An animal-machine need not have a brain and a soul. It does not follow from this that the nervous force in a brainless organism is merely a mechanical action. The nervous force is a force of coordination and subordination of organic machines. For this function to operate, it is enough for ganglia, plexi or junctions of other sorts to make it possible for a nervous impression from an external source to be reflected in the form of an internally originated excitation destined for one organ or another. The movements of the brainless polyp, for instance, can be explained in this way. The explanation also explains movement in a decapitated vertebrate. "Such a nervous action, due to an internal sense impression, not accompanied by a representation, stemming from the reflection of an external sense impression, is what takes place, for example, when a decapitated frog jumps in response to a pinch of its digit."[42] Unzer's originality should now be apparent: he refused to identify antimechanism with animism, and he decentralized the phenomenon of reflection of stimuli, which Willis and Astruc had been able to conceive only in terms of a cerebral seat.

George Prochaska, professor of anatomy and ophthalmology at Prague and Vienna, would succeed in combining Whytt's observations on the functions of the spinal cord with Unzer's hypotheses about extending the reflex function outside the brain. In *De functionibus systematis nervosi commentatio* (1784), Prochaska argued that the physiology of the nervous system had confined itself too narrowly to the brain, ignored comparative anatomy, and therefore, until Unzer, failed to recognize that the *vis nervosa*, or nervous force (no more talk of animal spirits), required only one thing: an intact connection of the nerve fiber to the *sensorium commune*, distinct from the brain. Even without a connection to the brain, a sensory nerve can link up, through the *sensorium commune*, to a motor nerve inserted into muscle, and thus transform an impression into a movement. Even if Prochaska did not definitively reject the opinion that the spinal cord is a bundle of nerves, he made the radical assertion that it, together with the medulla oblongata, is the seat of the *sensorium commune*, the necessary and sufficient condition of the nerve function. In dividing, moreover, one divided the nervous force without abolishing it, thereby explaining the persistence of excitability and movement in the frog whose medulla had been sectioned. It was at the level of the medulla, Prochaska argued, that impression was reflected into movement. Unlike Astruc, Prochaska did not believe that this reflection was a purely physical phenomenon governed by a law similar to the law of optical reflection; in the same spirit as

Whytt, rather, he argued that medullary reflection of nervous impressions was governed by a biological law of the conservation of living things. The examples cited by Prochaska were the same ones that Descartes and Astruc had described: occlusion of the eyelids and sneezing. Prochaska defined the relation of reflex motion to consciousness better than any of his predecessors: he explicitly distinguished the aspect of obligatory automatism from the aspect of optional, intermittent unconsciousness, and he supported this distinction with arguments from comparative anatomy. As one ascends from lower to higher animals, a brain is added to the *sensorium commune*. In man, soul and body have been joined by God. Nevertheless, the soul "produces absolutely no action that depends wholly and uniquely on it. All its actions are produced, rather, through the instrument of the nervous system." Thus Prochaska ends where Descartes began: in the case of involuntary motions, the soul uses an apparatus that can also function without its cooperation and permission. But the anatomo-physiological context of this assertion is quite different, since Prochaska conceives of the nervous system not "in general," like Descartes, but as an increasingly complicated hierarchical series, of which the human brain is the highest development though not the characteristic type. ["Physiologie animale," *Histoire générale*, vol. 2, pp. 613-16]

[73] In the eighteenth century, Astruc used the notion of a reflection of the nervous influx, based on the physical law of reflection of light, in a mechanistic theory of sympathies that assumed the brain to be the unique center of reflection. Whytt described the reflex phenomenon without using the word or notion, but the laws governing that phenomenon were assumed not to be purely physical, due to the connection between the reflex reaction and the instinct of self-preservation. Whytt argued that the relation between the sensory and motor functions was

not centralized but diffuse and not mechanical but psychic, and he therefore saw no reason to ascribe it to any specific anatomical structure. Unzer also believed that the law governing the phenomenon was not strictly mechanical, but he systematically used the term and the notion of reflection in a decentralized theory of the sensorimotor relationship, which he ascribed to a number of anatomical structures (the nervous ganglia and plexus as well as the brain). Prochaska, finally, retained both the word and the notion of reflection but treated its physical mechanism as subordinate to the organic entity's sense of self-preservation, decentralized the reflex function by locating its explicit anatomical support in the medulla oblongata and spinal cord (and also, probably, in the sympathetic ganglia), and was apparently the first to note that not all automatic reactions were unconscious. Legallois then went on to prove something that Prochaska never did, namely, that the spinal column does not have the structure of a nerve. Without using the term reflex or the notion, he located the reflex function in the medulla, whose metameric division he established experimentally.

Thus, by 1800 the definition of the reflex concept was in place, a definition ideal when considered as a whole but historical in each of its parts. It can be summarized as follows (with the names of the authors who first formulated or incorporated certain basic notions indicated in parentheses): a reflex movement (Willis) is one whose immediate cause is an antecedent sensation (Willis), the effect of which is determined by physical laws (Willis, Astruc, Unzer, Prochaska) – in conjunction with the instincts (Whytt, Prochaska) – by reflection (Willis, Astruc, Unzer, Prochaska) in the spinal cord (Whytt, Prochaska, Legallois), with or without concomitant consciousness (Prochaska).[43] [*Formation du réflexe*, pp. 130–31]

Corrections

[74] Taking this definition as our starting point, we can see precisely what elements stood in need of correction. One of the best reference texts is Johannes Müller's *Handbuch der Physiologie des Menschen*, where the illustrious German physiologist compares his ideas on reflex movement with those of Marshall Hall.[44] Müller makes it clear that in 1833, when both Hall's paper and the first edition of the *Handbuch* were published, the reflex concept was a principle of explanation, a theoretical instrument for interpreting phenomena defined as "movements following sensations." The theoretical content of this concept consisted of two elements, one positive, the other negative: negatively, the concept rejected the theory of anastomoses between sensory and motor fibers; positively, the concept required a central intermediary between the sensory impression and the determination of the motor reaction. It was for the express purpose of denoting the true function of the *medulla spinalis*, or spinal medulla (rather than spinal cord), that Marshall Hall coined the term "diastaltic" to indicate that the medulla could provide a functional connection between sensory and motor nerves only if situated between them as an authentic anatomical structure distinct from the brain. The diastaltic (reflex) function of the spinal medulla determined its relation to the esodic, or anastaltic, function of the sensory nerve and the exodic, or catastaltic, function of the motor nerve.

On this fundamental point Müller and Hall agreed. In Müller's words, "the phenomena I have described thus far on the basis first of my own observations and then those of Marshall Hall's have one thing in common, namely, that the spinal medulla is the intermediary between the sensory and the motor action of the nervous principle." Bear in mind that the two physiologists' agreement about the specific central function of the spinal cord was the result of twenty years of research and controversy concerning the validity and interpretation of the Bell-Magendie law (1811–22). [...]

The Bell-Magendie law was a necessary ingredient for the formulation of the reflex concept, insofar as that concept includes the specific function of the spinal cord. What Hall called the diastaltic (or diacentric) function was conceivable only in conjunction with two mutually independent properties of the nerve. Only if those two properties existed was a nervous center required to divert the nervous impulse to a new destination. [...] The course that Müller followed from 1824 to 1833 shows that it took Bell's idea and Magendie's experiments to relate the reflex concept to the physiological function of the spinal cord.

The second respect in which the nineteenth century rectified the eighteenth-century concept had to do with the relation of reflex movement to consciousness, that is, with psychological matters. It was expressly on this point that Müller disagreed with Hall. In describing a reflex as a movement that follows a sensation, Müller, like Willis, Whytt, Unzer and Prochaska before him, was in a sense obliging himself to unravel a mystery: how could a movement depend on a sensation when the nervous circuit had been broken by decapitation, thus removing the interconnecting sensory organ, the brain? Although Müller disagreed with Whytt, who believed that reflex movements involved both conscious sensations and spontaneous reactions, and although he praised Prochaska for having pointed out that a reflex might or might not be accompanied by a conscious sensation, he regarded the reflex as the effect of a centripetal action propagated toward the spinal cord by the sensory nerve, which then might or might not continue on to the common sensorium and, thus, might or might not become conscious. Reflex movement was therefore one species within a genus comprising all movements conditioned on the action of the sensory nerves. Hall, on the other hand, felt that one ought to consider the centripetal (anastaltic) impression without reference to the brain or to consciousness, and that the concepts of sensation and even sensitivity ought not to enter into the concept of a reflex. The reflex function did not even depend on sensory or motor nerves but, rather, on specific nervous fibers that Hall called "excito-motor" and "reflecto-motor" fibers. Hall's 1833 Royal Society paper on "The Reflex Function of the Medulla Oblongata and the Medulla Spinalis" explicitly distinguishes reflex movement not only from voluntary movement directly controlled by the brain but also from the respiratory movement controlled by the medulla oblongata, as well as from involuntary movement initiated by direct stimulus of nerve or muscle fiber. A reflex movement is not a spontaneous, direct response emanating from a central source; it presumes a stimulus applied at some distance from the reacting muscle being transmitted to the spinal cord and from there reflected back to the periphery. Hall oriented the reflex concept toward a segmental and explicitly mechanistic conception of the functions of the nervous system.

This was difficult for Müller to accept. To be sure, he was open about his disagreement with Prochaska, and he ascribed all reflex movements to a teleological principle of instinctive organic self-preservation. But as Fearing has pointed out, Müller's interest in the phenomena of associated movements and radiant sensations and his elaborate attempts to explain the latter in terms of a reflex function of the brain and spinal cord show that he was a long way from conceiving of reflexes as segmental and local mechanisms. In fact, Müller's observations of associated movements in narcotized animals and general reflex convulsions led him to two simultaneous conclusions: reflex movements can involve the entire body in response to the most insignificant local sensation, and the more extensive a reflex movement is, the less it is synchronized.

Müller's concept of reflex, which maintained a connection

with sensation — that is, with the brain — as well as the possibility that a local sensation might produce reflected effects throughout the organism, sidestepped most of the objections that had been raised against Hall's ideas. Hall had scandalized many physiologists by attributing to the spinal cord a power to regulate movement still widely believed to be an exclusive province of the brain. [...]

It was in 1853, four years before Hall's death, that Eduard Pflüger published *Die sensorischen Functionen des Rückenmarks der Wirbeltiere*. The well-known laws of reflex activity (homolateral conduction, symmetry, medullary and cerebral irradiation, generalization) essentially recast, in apparently more experimental form, Müller's notion of the association of movements and the radiation of sensations. In fact, Pflüger followed Müller in using the reflex concept to explain so-called sympathetic or consensual phenomena, whose interpretation had previously divided proponents of the principle of anastomosis of the peripheral nerves (Thomas Willis, Raymond Vieussens, Paul-Joseph Barthez) from believers in the principle of a confluence of impressions in the *sensorium commune* (Jean Astruc, Robert Whytt, Johann August Unzer, George Prochaska). According to Prochaska, the reflex concept preserved the explanation of sympathies in terms of the *sensorium commune* but located the latter outside the brain in the medulla oblongata and spinal cord. Unlike Whytt, Prochaska distinguished the *sensorium commune* from the soul but continued to credit it with a teleological function, according to which the reflex action was a form of self-preserving instinct (*nostri conservatio*). So it is hardly surprising that Pflüger in 1853 felt that Prochaska had had a better understanding of the nature of the reflex process in 1784 than Hall had managed in 1832-33. For the same reasons that had persuaded Prochaska to hold on to the concept of a *sensorium commune* Pflüger believed in the existence of a medullary soul (*Rückenmarksseele*), which enabled him to explain the purpose of reflex actions. Hall, on the other hand, drew a sharp distinction between adaptive or intentional movement, deliberate and stemming from the brain, and reflex movement, which he characterized as "aimless." Less mechanistic than Hall, Müller had raised the rigidity caused by certain generalized reflexes as an objection against Prochaska's view, though it is true that Müller was careful to note that this occurred only "in a suitably prepared animal." Pflüger's concept of the reflex must be regarded as a misleading dialectical synthesis: its experimental basis was as old as Marshall Hall, whereas the philosophical context that made it meaningful was as old as Prochaska would have been, had he not died in 1820.

In fact, Pflüger did not succeed in 1853 in finding a strictly physiological solution to a problem that Hall, rather than really facing, had sidestepped by attributing what he called "excito-motor powers" to nerve fibers. The problem lay in the terms "sensation" or "sensibility" as they were used in the earliest definitions of the reflex. Willis had said that "reflex motions immediately follow sensation" (*motus reflexus est qui a sensione praevia immediatus dependens, illico retorquetur*), whereas Prochaska had said that "one of the common sensory functions is to reflect sense impressions as motor impulses" (*praecipua functio sensorii communis consistat in reflexione impressionum sensoriarum in motorias*). Müller began his chapter on reflex movements by saying, "Movements that follow sensations have always been known." As long as people continued to speak of "sensation," they remained on the terrain of psychology. It was logical to look for a seat of the psyche, and why not suspect the spinal cord? In 1837, Richard Dugard Grainger correctly noted that contemporary physiologists appeared to believe in the existence of two kinds of sensation, one conscious, the other unconscious. Edward George Tandy

Liddel points out that when Charles Todd coined the term "afferent" in 1839, a major step was taken toward distinguishing between the two kinds of sensation. Yet it may be that the truly major step came only later, when the subjective concept of sensibility (*le sens de l'influx*) was replaced by a purely objective one defined in terms of the histology of receptors.

What is interesting about the history of the reflex concept between Pflüger's work and Charles Scott Sherrington's first publications is its importation from physiology into clinical work, which began with Hall. The latter was the first to use the disruption or disappearance of certain reflexes as diagnostic symptoms. The concept of the reflex arc gradually took on meaning beyond that associated with the schematic structure introduced by Rudolph Wagner in 1844; incorporated thus into symptomatology and clinical examination, it influenced therapeutic decision-making. But as the reflex concept passed from the laboratory into the hospital, it did not go unchanged. While most physiologists tended to look upon reflexes as fundamental, unvarying mechanisms, a few clinicians, among them Emil Jendrassik, who followed up the work of Wilhelm Heinrich Erb and Carl Friedrich Otto Westphal (1875) by looking systematically for tendon reflexes, were surprised to discover that such reflexes were neither constant nor uniform, and that their absence was not necessarily a pathological symptom. It would not be long before physiologists would be obliged to abandon the idea of a reflex as a simple arc establishing a one-to-one relationship between stimulus and muscular response.

The generalization of cell theory, the identification of neurons under the microscope and technological advances in histology demonstrated, of course, that nerves could be decomposed analytically into smaller – in some sense atomic – structures. The concept of a segmental reflex was thereby corroborated. New clinical observations thus forced physiologists to consider segments in the context of the organism as a whole.

When Sherrington discovered that the scratch reflex was not inextricably associated with a strictly defined reflexogenic zone, he laid the groundwork for a new rectification of the concept. The reflex was now seen not so much as the reaction of a specific organ in response to a stimulus as an already coordinated movement determined in part by stimuli in a certain part of the organism and in part by the organism's global state. Reflex movement, even in its simplest, most analytical form, was a form of behavior, the reaction of an organic whole to a change in its relation to the environment.

Although the word "integration" did not appear in Sherrington's vocabulary until after the nineteenth century had ended, the concept of integration was the crowning achievement of nineteenth-century neurophysiology. Sherrington's work on rigidity due to decerebration (1898), reciprocal innervation and synapses converged on a demonstration of the fact that a basic reflex involves medullary integration of a muscle bundle into an entire member through convergence of afferent influxes and combination of antagonistic reactions. The functions of the brain are an extension of the medullary integration of the parts to the entire organism. In adapting Hughlings Jackson's concept of integration, Sherrington was interested not in its evolutionary implications but only in its structural ones.

It seems reasonable to say that Sherrington achieved, in the field of physiology, the dialectical synthesis of the reflex concept with the concept of organic totality that first Prochaska and then Müller had sought and that Pflüger had misleadingly achieved by interpreting the results of his physiological experiments in metaphysical terms.

By the end of the nineteenth century, the reflex concept had thus been purged of any teleological implications, while it had also ceased to be seen – as Hall had seen it – as nothing more than a simple mechanical reaction. Through a series of corrections, it had become an authentically physiological concept. [*Etudes*, pp. 296-304]

CHAPTER NINE

Biological Objects

A Principle of Thematic Conservation

[75] The history of a science would surely fail of its goal if it did not succeed in representing the succession of attempts, impasses and repetitions that resulted in the constitution of what the science today takes to be its object of interest. Unlike geometry and astronomy, terms that are more than two thousand years old, the term biology is not yet two hundred years old. When it was first proposed, geometry had long since ceased to be the science of figures that can be drawn with a straightedge and compass, while astronomy had only recently expanded its scope of interest beyond the solar system. In both cases, the signifier of the scientific discipline remained the same, but the discipline in question had broken with its past. By contrast, the concept of biology was invented to characterize, in retrospect, a discipline that had not yet broken with its past.

The word "biology" occurs for the first time in Jean-Baptiste Lamarck's *Hydrogéologie* (1802). When he mentioned the word again, in the preface to his *Philosophie zoologique* (1809), it was in allusion to a treatise to be entitled *Biologie*, which he never actually wrote. Strikingly, this preface is concerned with general problems of animal organization "as one traverses their entire series from the most perfect to the most imperfect." The idea of a hierarchical series of animals, a chain of being, indicates that the object of the new biology was the same as that of Aristotle's *Historia animalium* and *De partibus animalium*. Hence, Lamarck's own invention — modification in the organs through force of habit and under the influence of changing environmental conditions — was explicitly intended to reestablish "the very order of nature" beyond the lacunae and discontinuities in the system of classification proposed by naturalists — in other words, to establish a clear progression and gradation in organization that could not be overlooked despite any "anomalies."

As for the other inventor of the term and concept of biology, Gottfried Reinhold Treviranus, the very title of the book he published in 1802, *Biologie oder Philosophie der Lebenden Natur für Naturforscher und Arzte* (volume 2 in a six-volume series, the last of which was published in 1822), indicates that he had no wish to separate or distinguish the naturalist from the physician as to their philosophical or general conception of the phenomena of life. Thus, at the turn of the nineteenth century, a new way of looking at the study of living things, which entailed a new logic, was in fact limited by the traditional association of the standpoint of the naturalist with that of the physician, that of the investigator with that of the healer. [...]

Since the turn of the nineteenth century, however, definitions of biology's specific object have been purged of value-laden concepts such as perfection or imperfection, normality or abnormality. Therapeutic intentions, which once informed or, more accurately, deformed, the biologist's view of laboratory work, have since been limited to the applications of biological knowledge. Hence, it would seem that the question of "normality" in the history of biology ought to be classed as a matter of historical rather than current interest. I shall attempt to prove the contrary. To that end, I direct the reader's attention to the end of the historical process. For contemporary biochemists, the functions of self-preservation, self-reproduction and self-regulation are characteristic properties of microorganisms such as bacteria. The model often proposed by scientists themselves and not just by popularizers of their work is that of the "fully automated chemical factory."[45] The organic functions are acknowledged to be superior to their technological counterparts in reliability, if not infallibility, and in the existence of mechanisms for detecting and correcting reproductive errors or flaws. These facts make it reasonable to ask whether there is not some principle of thematic conservation at work in the historical constitution of biology. On this view, which contrasts with an idea of science elaborated by historians and philosophers in the era when physics dealt with macroscopic objects, biology is different from the other sciences, and the history of biology ought to reflect that fact in the questions it asks and the way in which it answers them. For the alleged principle of thematic conservation in the history of biology is perhaps only a reflection of the biologist's acceptance in one way or another of the indisputable fact that life, whatever form it may take, involves self-preservation by means of self-regulation. [*Ideology and Rationality*, pp. 125-28]

Various Manifestations of the Biological Object

In antiquity
[76] The fundamental concepts in Aristotle's definition of life are those of soul and organ. A living body is an animate and organized body. It is animate because it is organized. Its soul is in fact act, form and end. "Suppose that the eye were an animal – sight would have been its soul.... We must now extend our consideration from the 'parts' to the whole living body; for what the

departmental sense is to the bodily part which is its organ, that the whole faculty of sense is to the whole sensitive body as such."[46] The organs are the instruments of the soul's ends. "The body too must somehow or other be made for the soul, and each part of it for some subordinate function, to which it is adapted."[47] It is impossible to overstate the influence of Aristotle's use of the term *organon* to designate a functional part (*morion*) of an animal or vegetal body such as a hand, beak, wing, root or what have you. Until at least the end of the eighteenth century, anatomy and physiology preserved, with all its ambiguities, a term that Aristotle borrowed from the lexicon of artisans and musicians, whose use indicates implicit or explicit acceptance of some sort of analogy between nature and art, life and technics.

As is well known, Aristotle conceived of nature and life as the art of arts, by which he meant a process teleological by its very nature, immanent, unpremeditated and undeliberated – a process that every technique tends to imitate, and that the art of medicine approaches most closely when it heals by applying to itself rules inspired by the idea of health, the telos and form of the living organism. Aristotle, a physician's son, thus subscribed to a biological naturalism that had affinities with the naturalism of Hippocrates.

Life's teleological process is not perfectly efficient and infallible, however. The existence of monsters shows that nature does make mistakes,[48] which can be explained in terms of matter's resistance to form. Forms or ends are not necessarily and universally exemplary; a certain deviation is tolerated. The form of an organism is expressed through a rough constancy; it is what the organism appears to be most of the time. Hence, we can consider a form to be a norm, compared to which the exceptional can be characterized as abnormal. [*Ideology and Rationality*, pp. 128-29]

In the seventeenth and eighteenth centuries
[77] Descartes contradicted Aristotle's propositions point by point. For him, nature was identical with the laws of motion and conservation. Every art, including medicine, was a kind of machine-building. Descartes preserved the anatomical and physiological concept of an organ but eliminated any distinction between organization and fabrication. A living body could serve as the model for an automaton or vice versa. Yet there was an ambiguity in this reversibility. The intention behind the construction of an automaton was to *copy* nature, but in the Cartesian theory of life the automaton serves as an intelligible *equivalent* of nature. There is no room in Cartesian physics for an ontological difference between nature and art. "[S]o it is no less natural for a clock constructed with this or that set of wheels to tell the time than it is for a tree which grew from this or that seed to produce the appropriate fruit."[49][...]

To begin with, the Cartesian watch is no less subject to the laws of mechanics if it tells the time incorrectly than if it tells the time correctly.[50] Similarly, it is no less natural for a man to be sick than to be healthy, and sickness is not a corruption of nature.[51] Yet the thirst that drives the victim of dropsy to drink is a "veritable error of nature," even though it is an effect of the substantial union of soul and body, whose sensations, such as thirst or pain, are statistically valid indicators of things or situations favorable or harmful "to the conservation of the human body when it is fully healthy."[52] This idea is confirmed at the end of the "Conversations with Burman" (1648), in which the medicine of the physicians, not based on sound Cartesian mechanics, is denigrated and ridiculed in favor of a course of conduct amenable, as animals are, to the silent lessons of nature concerning "self-restitution." "Every man is capable of being his own physician."[53] Even for Descartes, self-preservation remains the primary distinctive characteristic of the living body. [...]

Undoubtedly it was Georg Ernst Stahl who most stubbornly defended, in his *De diversitate organismi et mecanismi* (1706), the irreducibility of the organism, that is, the idea that a certain *order* obtains in the relations of the parts of a mechanism to the whole. A living body is both instrumented and instrumental. Its efficient structure (*structura, constructio, ordinatio, distributio* are all used in paragraph 19) reveals cooperation on the part of mediate or immediate agents. The material constitution of the body is subject to rapid corruption. Stahl observes, however, that disease is an exceptional condition. Hence, there must be some power of conservation, some immaterial power offering active resistance to decomposition, permanently at work in the bodies of living things. Self-preservation of the organism is achieved as a result not of some mechanical but of natural "autocracy."[54][...]

It is not only the history of anatomy and physiology that begins with Aristotle but also the history of what was long called "natural history," including the classification of living things, their orderly arrangement in a table of similarities and differences, study of their kinship through morphological comparison and, finally, study of the compatibility of different modes of existence. Natural history sought to explain the diversity of life forms able to coexist in a given environment. In 1749 Carolus Linnaeus referred to this coexistence as the *oeconomia naturae*. [...]

In the eighteenth century, the status of species was the foremost problem of the naturalists, as can be seen most clearly of all in the work of Comte Buffon and Linnaeus. The latter did not experience as much difficulty as the former in holding that the species were fixed at creation and perpetuated from generation to generation. Buffon attempted to resolve the problem with his theory of "internal molds" and "organic molecules." Organic molecules, he maintained, were indestructible; they survived the process of reproduction from generation to generation, accumulating in the bodies of living things in specific forms shaped by internal molds. The latter, determined by the form of the organism, dictated the way in which the parts had to be arranged in order to form a whole.

Consider for a moment the internal mold metaphor. Molds are used in smelting and masonry to impose a certain three-dimensional shape. Etymologically, the word is related to *"modulus"* and "model." In common usage, it indicates a structural norm. In living organisms, however, the structural norm can accommodate irregularities, to which Buffon refers on more than one occasion as anomalies (*êtres anomaux*). An organic anomaly is not the same as a physical irregularity, however. Initially, Buffon conceived of generation as analogous to crystallization, but ultimately he came to think of crystallization as a form of organization. He was unable to avoid associating anomalies with degeneration, hence with the problem of the mutability of species. On this point, Buffon was never able to achieve certainty. He did not regard the idea of derivative species as absurd on its face, but he believed or professed to believe that observation confirmed the teachings of the Bible.[55]

Pierre Louis Moreau de Maupertuis was bolder in theorizing, perhaps because he possessed less extensive empirical information. For him, structural variation was the rule of organic progression. In paragraph 31 of the *Système de la nature* (1751), he set forth a theory of generation based on the existence of elementary particles of matter endowed with appetite and memory, whose "arrangement" reproduces the possibly miraculous structure of the first individuals. The phenomena of resemblance, miscegenation and monstrosity could be explained, he argued, in terms of the compatibility or incompatibility of "arrangements" in seeds mingled through copulation. Thus, later, in paragraph 45, he asks,

Can we not explain in this way how from just two individuals the most dissimilar species could have multiplied? Originally they may have stemmed from fortuitous productions in which the elementary parts did not retain the order they occupied in the father and mother animals. Each degree of error could have produced a new species, and repeated errors could have given rise to the infinite diversity of animals that we see today.

It is tempting to read this text with spectacles provided by contemporary biochemical and genetic theory. *Order* and *error* occur both here and in contemporary accounts of hereditary biochemical defects as ground and cause of both normality and abnormality. But today biochemistry and genetics offer us a way of interpreting organic abnormalities that was worked out in cooperation with the Darwinian explanation of the origin of species and the adaptation of organisms. Hence Maupertuis's propositions should be regarded more as fictions than as anticipations of scientific theories to come. He was unable to overcome the difficulty posed by the natural mechanism for normalizing differences. Both he and Buffon believed that human intervention – through techniques of husbandry or agronomy – was the only way to stabilize variations within species. [*Ideology and Rationality*, pp. 129-35]

In the nineteenth century

[78] The publication of *On the Origin of Species by Means of Natural Selection; or the Preservation of Favoured Races in the Struggle for Life* in 1859 occasioned doubts in the minds of some early readers because of the traditional meaning of certain concepts mentioned in the title and frequently alluded to in the body of the work. The theory of natural selection states that certain deviations from the norm can be seen *a posteriori* to provide a tenuous

advantage for survival in novel ecological situations. Darwin thus substituted a random fit for a preordained adaptation. Natural selection is eliminative. Disadvantaged organisms die; the survivors are all different in one degree or another. The reader who takes literally such Darwinian terms as "selection," "advantage," "adaptation," "favor" and "disfavor" may partially overlook the fact that teleology has been excluded from Darwin's theory. Does this mean that all value-laden terms have been excluded from the idea of life? Life and death, success or failure in the struggle for survival — are these value-neutral concepts, even if success is reduced to nothing more than continued existence? Does Darwin's language reveal his thought or does it suggest that even for Darwin a causal explanation of adaptation could not abolish the "vital meaning" of adaptation, a meaning determined by comparison of the living with the dead? As Darwin observed, variations in nature would have remained without effect had it not been for natural selection. What could limit the ability of this law, operating over a long period of time and rigorously scrutinizing the structure, overall organization and habits of every creature, to promote good and reject evil?[56]

And Darwin's work ends with a contrast: "while this planet has gone circling on according to the fixed law of gravity, from so simple a beginning endless forms most beautiful and most wonderful have been and are being evolved."

In suggesting that individual variations, deviations in structure or instinct, are useful because they yield a survival advantage in a world in which relations of organism to organism are the most important of all causes of change in living beings, Darwin introduced a new criterion of normality into biology, a criterion based on the living creature's relation to life and death. By no means did he eliminate morality from consideration in determining the object of biology. Before Darwin, death was considered to be the regulator of the quantity of life (Buffon) or the sanction imposed for infractions of nature's order, the instrument of her equilibrium (Linnaeus). According to Darwin, death is a blind sculptor of living forms, forms elaborated without preconceived idea, as deviations from normality are converted into chances for survival in a changed environment. Darwin purged from the concept of adaptation any reference to a preordained purpose, but he did not separate it completely from the concept of normality. In the spirit of Darwinism, however, a norm is not a fixed rule but a transitive capacity. The normality of a living thing is that quality of its relation to the environment which enables it to generate descendants exhibiting a range of variations and standing in a new relation to their respective environments, and so on. Normality is not a quality of the living thing itself but an aspect of the all-encompassing relation between life and death as it affects the individual life form at a given point in time.

Thus, the environment decides, in a nonteleological way, which variations will survive, but this does not necessarily mean that evolution does not tend to create an organic order firm in its orientation if precarious in its incarnations. Heredity is an uninterrupted delegation of ordinal power. What difference does it make if, in Salvador Luria's words, "evolution operates with threats, not promises"?[57][...]

The physiologists took their inspiration from a distinction first made by Xavier Bichat:

> There are two kinds of life phenomena: (1) the state of health, and (2) the state of sickness. Hence, there are two distinct sciences: physiology, which is concerned with phenomena of the first state, and pathology, which is concerned with those of the second. The history of phenomena in which the vital forces have their natural type leads us to that of phenomena in which those forces are distorted. Now,

in the physical sciences, only the first history exists; the second is nowhere to be found. Physiology is to the motion of living bodies what astronomy, dynamics, hydraulics, hydrostatics and so forth... are to the motions of inert bodies. The latter have no science that corresponds to them as pathology corresponds to the former.[58]

But not all physiologists agreed with Bichat that there exist vital forces not subject to the laws of physics. Here I must cite Claude Bernard once more, because his position is so up to date. He admitted, first of all, that vital phenomena are subject only to physical and chemical causes, but he also held that the organism develops from the egg according to an immanent design, a plan, a regularity, which is responsible for its ultimate organization, for its harmony, persistence and, if need be, restoration.

What Bernard described in images is today explained by the theorems of macromolecular biochemistry. Like the metaphor of the "internal mold," the images of "design," "plan," "guiding idea" and "order" are given retroactive legitimacy by the concept of a program encoded in sequences of nucleotides.[59] For the first time in the history of biology, all the properties of living things – growth, organization, reproduction, hereditary continuity – can be explained in terms of molecular structure, chemical reactions, enzymes and genes. [*Ideology and Rationality*, pp. 136-39]

In the twentieth century
[79] The level of objectivity at which the opposition between normal and abnormal was legitimate was shifted from the surface to the depths, from the developed organism to its germ, from the macroscopic to the ultramicroscopic. Now it is the transmission of the hereditary message, the production of the genetic program, that determines what is normal and what is a deviation from the normal. Some human chromosomal anomalies such as mongolism

can be observed directly in the clinic. Others, such as Klinefelter's syndrome, are tolerated without apparent ill effect and manifest themselves only in special ecological circumstances. As for genetic anomalies, I shall mention only "innate errors of metabolism" – to use the phrase that Archibald Edward Garrod coined in 1909 – that is, specific biochemical lesions that result from the presence of a mutant gene, which is called "abnormal" not so much because of its statistical rarity as because of its pathological or even fatal effects (for example, hemophilia, Huntington's chorea and so on). A new nomenclature of disease is thus established, referring disease not to the individual considered in its totality but to its morphological and functional constituents: diseases of the hemoglobin, hormonal diseases (such as hyperthyroidism), muscle diseases and so forth. Gene mutations that block chemical syntheses by altering their enzyme catalysts are no longer interpreted as deviations in Maupertuis's sense but as errors in reading the genetic "message," errors in the reproduction or copying of a text.

The term "error" does not imply that science has returned to the Aristotelian and medieval notion that monsters are errors of nature, for the failure here is not some lack of skill on the part of the artisan or architect but a mere copyist's slip. Still, the new science of living things has not only not eliminated the contrast between normal and abnormal but it has actually grounded that contrast in the structure of living things themselves.[60] [*Ideology and Rationality*, pp. 140-41]

A New Historical Crux

[80] Perhaps the epistemologist may now be allowed to remain skeptical about dogmatic reductionist views, given what can be learned if we look at the history of biology, without any simplifying *a priori* assumptions, in light of the various manifesta-

tions of what I have proposed calling the principle of thematic conservation.

I anticipate one possible objection, however. In looking for a distinctive concept of normality in biology, have I not confused the issue by considering different orders of biological objects? Astronomers from Sir William Herschel to Edwin Hubble revolutionized their discipline by magnifying their object to an unimaginable degree, revealing galaxies beyond the solar system and metagalaxies beyond the galaxies. By contrast, biologists have discovered the nature of life by making their objects smaller and smaller: bacterium, gene, enzyme. In the preceding discussion, am I dealing with observations at one level and explanations at another? Normality appears to be a property of the organism, but it disappears when we look at the elements that make up that organism.

At all levels, however, biologists have identified ordering structures that, while generally reliable, sometimes fail. The concept of normality is intended to refer to these ordering structures. No such concept is needed in the epistemology of physics. By introducing it as I have done here, I in no way intend to deny that biology is based on physics and chemistry. I do intend to prevent the coalescing of two properly distinct approaches to history. In the history of biology, the pseudotheoretical content of prescientific conceptualizations of structural and functional normality was abandoned, but the conceptualizations themselves have been preserved, in "displaced" form, as indices of the objective uniqueness of the living organism. Dmitry Mendeleyev's periodic table does not justify Democritus's intuitions *a posteriori*, but the decoding of the genetic program does provide *a posteriori* justification of Claude Bernard's metaphors. Even within the terms of a monist – indeed a materialist – epistemology, physics remains radically different from biology. Physics was produced, sometimes at risk of life and limb, by living things subject to sickness and death, but sickness and death are not problems of physics. They are problems of biology.

Between the bacteria in a laboratory culture and the biologists who observe them, there is a whole range of living things permitted to exist by the filter of natural selection. Their lives are governed by certain norms of behavior and adaptation. Questions about the vital meaning of those norms, though not directly matters of chemistry and physics, are questions of biology. As Marjorie Greene points out, alongside the biochemists there is room in biology for a Frederik Jacobus Buytendijk or a Kurt Goldstein.[61] History shows that she is right. [*Ideology and Rationality*, pp. 142–44]

Part Four

Interpretations

Chapter Ten

René Descartes

Relations between Theory and Technology

[81] What did Descartes know about technology, and what did he hope to learn from it? His correspondence, reread with this question in mind, gives a strong impression of a man with a wide curiosity about practical techniques and keen to discover principles or laws that might account for their efficacy. The subjects that recur most frequently in his meditations are, of course, the grinding of lenses for optical instruments, the construction of machines and the art of medicine. Yet he also found in the routines of peasants and soldiers and the lore of travelers material for comparison and opportunities to put his theories to the test. The influence of the soil on the growth of transplanted plants, the maturation of fruits, the separation of materials of different density in the manufacture of butter, the way a child's legs flail while mounting a horse, the ringing of bells in order to cause thunder clouds to burst — these commonplaces of rural life provided Descartes with occasions for reflection. As a soldier, he rubbed the tip of his pike with oil and noticed sparks. And as a resident of Amsterdam, he was aware of all that a great port had to offer in the way of practical and luxury goods, and of all that a population that each day welcomed travelers from the antipodes

could teach about human diversity. With astonishment and admiration we watch Descartes apply his scrupulously methodical intelligence to the most diverse and specialized technical problems: smoking chimneys, water pumps and marsh drainage, medical diagnosis, drugs, allegedly miraculous fountains, automata, the trajectory of cannonballs, the velocity of bullets, the strength of a sword thrust, the sound of bells. Descartes's interest in artillery, medicine and automata was, of course, shared by many of his contemporaries in France and Italy; but underlying his attention to the most minute details and problems was a comprehensive doctrine incorporating the smallest details and difficulties of physics and metaphysics. Yet his ambition to achieve mastery of the natural world seems almost modest in comparison with his dreams: to restore sight to the blind, to view the animals on the moon (if any), to make men wise and happy through medicine, to fly like a bird. Medical observations are scattered throughout his work. He confessed to the Marquess of Newcastle that the primary purpose of his studies had always been to preserve health,[1] and he probably believed, as Constantijn Huygens reported, that "that vexing custom, death, will one day disappear" (II, 550). His technical preoccupations with optics can be found in his correspondence with David Ferrier (1629-1638) as well as in the *Optics*. As for his research and experiments with machinery, apart from the brief treatise on lifting engines written for Huygens in 1637, we have only Adrien Baillet's account of Descartes's relations with Villebressieu, the king's engineer (I, 209, 214, 218). Baillet lists those of Villebressieu's inventions allegedly due to suggestions from Descartes: a water pump, a wheeled bridge for use in attacking fortresses, a portable folding boat for crossing rivers and a wagon chair for the transport of wounded soldiers. This brief résumé of Descartes's technological interests, insignificant though it may seem, is nevertheless worth remembering because it was

Descartes's willingness to "lower his thought to the least of the mechanics' inventions" (I, 185) that enabled him to conceive of the relation between theory and practice in a way that is important not only for understanding his thought but for grasping the nature of philosophical reflection in general.

How did Descartes conceive of the relation between theory and technology? To answer this question, let us turn to the texts. In any number of passages he deplored the failure of artisans to learn from what was known about the materials and phenomena they used in daily practice. All purposive action, he maintained, should be subordinate to its associated science. He had contempt for technique without understanding (I, 195) and inventors without method (X, 380) and was extremely wary of artisans who refused to take his directions (I, 501, 506). The most significant passages in this regard are to be found in the *Rules for the Direction of the Mind*. From the outset, Descartes contrasts the diversity of technological skills with the unity of theoretical understanding and proposes using theory to achieve total knowledge. As each acquisition of truth becomes a rule of method, thought moves from truth to truth, and it thereby acquires the ability to act reliably and efficiently. This ability is the result of a sustained attentiveness that the specialized artisan, limited and partial in his views, seeks in vain to achieve. In Rule Five, Descartes mentions among the illusions that his method tends to eliminate that of people who "study mechanics apart from physics and, without any proper plan, construct new instruments...."[2] Countering such presumption is this admirable affirmation of *Principles of Philosophy*: "All the rules of Mechanics belong to Physics, so that all things that are artificial are also natural" (IX, 321). That is why "one must first explain what the laws of nature are and how nature usually behaves before one can teach how those laws can be made to produce unusual effects" (II, 50). To do without understanding the why of what he is doing is the lot of the mere technician. To promise without performing is the definition of the charlatan. To obtain effects at will through understanding of their causes is the ambition of Descartes. We learn what is technologically possible by studying what is theoretically necessary. Thus far, there is nothing in Descartes's philosophy concerning technology that does not seem obvious, if by obvious we mean something that has become familiar owing to modern philosophy's long-standing interest in a theme that, from da Vinci to the Encyclopedists on to Auguste Comte and Karl Marx, became a standard topos.

In Descartes's thinking, however, there were important restrictions on the conversion of knowledge into action. Descartes saw obvious "difficulties" in moving from theory to practice which not even perfect intelligence could resolve by itself. Even given perfect knowledge, the technological embodiment of that knowledge would in some cases contain inevitable imperfections. No Archimedean mirror, even one polished by an angel, could burn an object a league away unless it were made extremely large (I, 109). Even if an angel were to give instructions, based on theory, for building a steelyard balance capable of weighing objects up to two hundred pounds, "it is almost impossible to be so precise in all aspects of construction that there should be no fault in the scale, and thus practice will discredit theory" (II, 469). The instrument must therefore be calibrated empirically, Descartes recommends. Five years after formulating a theory of lenses, he wrote Marin Mersenne that in lensmaking the gap between theory and practice was so great that theoretical perfection could never be achieved (III, 585). Note that the three examples discussed thus far — mirrors, lenses and scales — involve the theories of optics and levers, which were among the earliest successes of Cartesian science. Even more explicitly, Descartes held that if men could not fly, the problem was not theoretical but practical: "One can indeed make a machine that sustains itself in the air like a bird, *metaphysice loquendo*, for birds themselves, in my opinion at any rate, are such machines, but not *physice* or *moraliter loquendo*, because it would require springs so light yet so powerful that humans could not manufacture them" (III, 163).

Descartes never explained his thinking about the difference between science and construction, two human activities that his philosophy seems to suggest not only stem from a common source but are convertible, in the sense that knowledge can be converted into construction. Hence, it is up to us to clarify his meaning by looking at the texts and comparing various strands of his thinking. Descartes maintained that one should be able to deduce empirical results from intuitive principles that he called "seeds of truth" or "simple [or occasionally, absolute] natures." An effect had not been explained, he held, until one could say how by an act of God it might have been made different but no less intelligible. The celebrated passage of Part Six of the *Discourse on the Method* in which the impossibility of completely deducing effects from causes leads to acknowledgment of the need to "judge the causes in terms of their effects" clearly indicates that technologically useful "forms or types of bodies" may be impediments to analytic deduction. From first causes, the scientist can deduce "the heavens, the stars, an earth, and even, on earth, water, air, fire and minerals," that is, "ordinary effects," "common and simple things." But whereas science treats matter as homogeneous and without distinctive identity, the technician, who relates matter to "our use of it," treats it as particular and diverse — hence the need for experimental trial and error. The passage in the *Discourse* in which Descartes proceeds from theory to technology is greatly elucidated, I think, by another passage, this one from the *Principles of Philosophy*, which proceeds in the other direction, from technology to theory:

Medicine, Mechanics and, in general, all the arts in which knowledge of physics may be useful have but one goal: to apply sensitive bodies to one another in such a way that, owing to natural causes, sensible effects are produced. In this we do just as well if the series of causes thus imagined is false as if it were true, since the series is supposed to be similar insofar as its sensible effects are concerned (IX, 322).

If, in many cases, practice "discredits theory," it is because "any application of sensitive bodies to one another," or, in other words, any technical synthesis, will normally include unpredictable and unanticipated effects, given that we are working with substances about which not everything can be deduced.

Descartes also believed that knowledge and construction were related in another way, however: things could be built without knowledge of the theory of how they worked, and this in turn could provide theoretical opportunities. This, I think, is the lesson of the *Optics* when reread in the light of the problem that concerns us here. Optical theory began with the invention of the magnifying glass, which initially was the fruit of trial and error, and luck. That initial success was later blindly copied. Yet the new invention still suffered from many deficiencies, and Descartes believed in the need for scientific study of what constituted a good lens. He proposed to deduce the proper shape of lenses from the laws of light. Thus, a purely fortuitous technological discovery provided the occasion for "many good minds to find out a number of things about optics" (VII, 82). In particular, it provided Descartes with the "opportunity to write this treatise" (VII, 82, 159).

According to the *Optics*, knowledge of nature depends on technology in two ways. First, technology provides instruments, in this case the magnifying glass, that lead to the discovery of new phenomena (VII, 81, 226). Second and more important, the imperfections of technology provide "the occasion" for theoretical research aimed at resolving "difficulties." Science, therefore, grows out of technology, not in the sense that the true is a codification of the useful, a record of success, but, rather, in the sense that technological obstacles, setbacks and failures lead to questions about the nature of the resistance encountered by human art. Obstacles to progress are seen to be independent of human desires, and this leads to a search for true knowledge. Science may later claim to impose discipline on technologies born without permission from any theorist. But where do such technologies originate? Not in the faculty of understanding, although that faculty might enable learned men to surpass "the ordinary artisan" (VII, 227), but, rather, in the exigencies of *life*. Thus Descartes, who had long dreamed of an infallible medical science, felt an urgent need for it when his hair began to turn white (I, 435) and he sensed that death might deprive him of the "more than century-old hope" that justified his concern with his body (I, 507). Before he could write the *Optics*, moreover, his failing eyes which were easily deceived had interfered with his ability to perceive useful things. Since "we cannot make ourselves a new body" (VII, 148), we must augment our internal organs with external organs (VII, 148) and supplement our natural ones with artificial ones (VII, 165). The impetus to create new technologies stems from man's needs, appetites and will (IX, *Principles of Philosophy*, 123). In his theory of the union of soul and body, Descartes was careful to emphasize the irreducibility of the emotions, and in his theory of error he stressed the primal importance of the will. These emphases suggest that he believed that life, whose philosophy consists in the desire to live well, cannot be apprehended in terms of pure understanding alone, that is, within a system of purely intellectual judgments. Thus, the conviction that technology cannot be reduced to science and construction cannot be reduced to understanding, together with the converse belief that the whole edifice of science cannot simply be converted into action, comes down to a belief in the existence of a unique "power." Liberty and will are not subject to the same limitations as intelligence, not only in the human mind but also in God. For Descartes, technology was always to some degree a synthetic and, as such, unanalyzable form of action, but I do not believe that he viewed it consequently as unimportant; rather, he saw it as a form of creation, though admittedly an inferior one.

If the foregoing analysis is correct, one question remains unanswered: Why is there no theory of creation in Descartes's philosophy? Or, to put it another way, why is there no aesthetics? Of course, it is difficult to draw any conclusion whatsoever from an absence – but there are grounds for asking whether Descartes might not have felt an obscure sense that admitting the possibility of a general aesthetics might have contradicted his general theory. For Descartes, the intelligibility of reality derived from mechanics and mathematical physics. For him, movement, along with extension and number, was a fundamental, intuitive concept of which it was safe to neglect all qualitative and synthetic aspects. And although he saw movement as the source of all material variety, he simultaneously precluded himself from raising the issue of diversification, which is one aspect of the problem of creation. As we know from the *Discourse on the Method*, he candidly admitted that geometric analysis had its limits, but he may not have wished to acknowledge, or admit to himself, that the impossibility of a "definitive" morality (since action normally involves desire and risk) also implied the impossibility of a "definitive" analytic science (as he wished his own to be). ["Descartes," *Travaux*, pp. 79–85]

The Theory of the Animal-Machine

[82] Descartes's theory of the animal-machine is inseparable from his famous dictum, "I think, therefore I am." The radical distinction between soul and body, thought and extension, implies the substantial unity of matter, whatever its form, and thought, whatever its function.[3] Since judgment is the soul's only function, there is no reason to believe in the existence of an "animal soul," since animals, bereft of language and invention, show no sign of being capable of judgment.[4]

The denial that animals possess souls (or the faculty of reason) does not imply they are devoid of life (defined as warmth in the heart) or sensibility (insofar as the sensory faculties depend on the disposition of the organs).[5]

The same letter I cited above reveals one of the moral underpinnings of the theory of the animal-machine. Descartes does for animals what Aristotle did for slaves: he devalues them in order to justify using them as instruments. "My opinion is no more cruel to animals than it is overly pious toward men, freed from the superstitions of the Pythagoreans, because it absolves them of the hint of crime whenever they eat or kill animals." Surprisingly, we find the same argument stood on its head in a letter from Leibniz to Conring: if we must look upon animals as something more than machines, then we should become Pythagoreans and give up our dominion over the beasts.[6] This attitude is typical of Western man. The theoretical mechanization of life is inseparable from the technological utilization of the animal. Man can claim possession of and mastery over nature only by denying that nature has any purpose in itself, and then only by regarding all of nature other than himself – even that which appears to be animate – as a means to an end.

Such an attitude justified the construction of a mechanical model of the living body, including the human body – for Descartes, the human body, if not man himself, was a machine. Descartes found the mechanical model he was looking for in automata, or moving machines.[7]

In order to bring out the full significance of Descartes's theory, let us turn now to the beginning of the *Treatise on Man*, a work first published in Leyden in 1662 in the form of a Latin copy and only later published in the original French, in 1664. He wrote there:

> These men will be composed, as we are, of a soul and a body. First I must describe the body on its own, then the soul, again on its own; and finally I must show how these two natures would have to be joined and united in order to constitute men who resemble us.
>
> I suppose the body to be nothing but a statue or machine made of earth, which God forms with the explicit intention of making it as much as possible like us. Thus God not only gives it externally the colors and shapes of all the parts of our bodies, but also places inside it all the parts required to make it walk, eat, breathe, enabling it to imitate all those functions which seem to proceed from matter and to depend solely on the interacting movements of the organs.
>
> We see clocks, artificial fountains, mills and other such machines which, although man-made, seem to move of their own accord in various ways; but I am supposing this machine to be made by the hands of God, and so I think you may reasonably think it capable of a greater variety of movements than I could possibly imagine in it, and of exhibiting more artistry than I could possibly ascribe to it.[8]

Reading this text as naively as I possibly can, I come to the conclusion that the theory of the animal-machine makes sense only by virtue of two hypotheses that often receive less emphasis than they are due. The first is that God the fabricator exists, and the second is that the existence of living things must precede the construction of the animal-machine that models their behavior. In other words, in order to understand the animal-machine, one must think of it as preceded, in the logical as well as the chronological sense, by God, as efficient cause, and by a preexisting living thing, as formal and final cause. In short, I propose to read the theory of the animal-machine, which is generally interpreted as involving a break with the Aristotelian concept of causality, as one in which all the types of causality that Aristotle invokes can be found, but not simultaneously and not where Aristotle would have placed them.

The text explicitly states that the construction of the living machine is to mimic that of a preexisting organism. The mechanical model assumes a live original. Hence, Descartes in this text may be closer to Aristotle than to Plato. The Platonic demiurge copies Ideas. The Idea is a model of which the natural object is a copy. The Cartesian God, *Artifex maximus*, tries to equal the living thing itself. The living machine is modeled on the living thing. Think of approximating a circle by means of a series of inscribed polygons, each with one more vertex than the preceding one: in order to conceive of the passage from polygon to circle, one has to imagine extending this series to infinity. Mechanical artifice is inscribed in life in the same way: in order to imagine the passage from one to the other, one has to imagine an extrapolation to infinity, that is, to God. This is what Descartes appears to mean by the final remarks of the above quotation. Hence, the theory of the animal-machine is to life as a set of postulates is to geometry, that is, a mere rational reconstruction that only pretends to ignore the existence of what it is supposed to represent and the priority of production over rational justification.

This feature of Descartes's theory was clearly perceived, moreover, by a contemporary anatomist, the celebrated Nicolaus Steno, who delivered a *Dissertation on the Anatomy of the Brain* in Paris in 1665, one year after the publication of the *Treatise on Man* appeared. His homage to Descartes is all the more surprising in that anatomists did not always approve of the philosopher's way of looking at their subject. Steno noted, in fact, that Descartes saw man not as anatomists saw him but as reconstructed by the philosopher from a divine vantage point. In substituting mechanism for organism, Descartes might appear to be banishing theology from the philosophy of life, but that appearance is misleading: the theology was reinstated in the very premise of the argument. An anatomical form had been substituted for a dynamic one, but since that anatomical form was a product of technology, the only possible teleology was implicit in the technique of production. In fact, mechanism and anthropomorphism are not diametrically opposed, for while the operation of a machine can be *explained* in purely causal terms, the construction of that same machine cannot be *understood* without a telos or, for that matter, without man. A machine is made by man and for man in order to produce certain effects and move toward some goal.[9]

What is positive in Descartes, therefore, and in his project to understand life in terms of mechanics, is the elimination of all anthropomorphic aspects of finality. In carrying out this project, however, one kind of anthropomorphism appears to have been substituted for another – technological anthropomorphism for political anthropomorphism.

In the *Description of the Human Body and All of Its Functions*, a short treatise written in 1648, Descartes took up the question of voluntary movement. With a clarity that dominated all theories of automatic and reflex movement down to the nineteenth century, he offered the observation that the body obeys the soul only if it is mechanically disposed to do so. The decision of the soul is not a sufficient condition for the movement of the body: "The soul cannot produce any movement in the body without the appropriate disposition of the bodily organs which are required for making the movement. On the contrary, when all the bodily organs are appropriately disposed for some movement, the body has no need of the soul to produce that movement."[10] In other words, the soul's relation to the body is not that of a king or a general issuing orders to his subjects or troops. Envisioning the body in terms of a clockwork mechanism, Descartes saw the various organs as controlling one another in much the same way as the gears of a clock. Descartes thus replaced a political image of command and a magical type of causality (involving words or signs) with a technological image of "control" and a positive type of causality involving a mechanical meshing or linkage. [*Connaissance*, pp. 110-14]

[83] In short, it may seem that, appearances to the contrary notwithstanding, the Cartesian explanation fails to get us beyond teleology. This is because mechanism can explain everything once we assume the existence of machines constructed in a certain way, but it cannot explain why the machines are built that way in the first place. There is no machine for building machines, so that, in a sense, to explain the workings of an organ or organism in terms of a mechanical model is to explain it in terms of itself. In the end, we are dealing with a tautology. Tools and machines are kinds of organs, and organs are kinds of tools or machines.[11] Hence, it is hard to detect any incompatibility between mechanism and teleology. No one doubts the need for tools to achieve certain goals. And conversely, every mechanism must have a purpose, because a mechanism is a determinate as opposed to a random set of motions. The real contrast is between mechanisms whose purpose is obvious and those whose purpose is obscure. The purpose of a lock or a clock is obvious, whereas the purpose of a crab's pressure sensor, which is often cited as a marvel of adaptation, is obscure.

It is undeniably true that certain biological mechanisms serve certain purposes. To take an example that mechanistic biologists often cite, consider the broadening of the female pelvis prior to birth. Given that the fetus is 1–1.5 centimeters larger than the pelvic opening, birth would be impossible if a relaxation of the pubic symphyses and a posterior movement of the sacrococcygian bone did not increase the diameter of the aperture. Given a phenomenon whose biological purpose is so clear, one can legitimately refuse to believe that the mechanism that makes it possible (and that is essential for it to occur) has no biological purpose. A mechanism is a necessary sequence of operations, and to verify the presence of a mechanism, one must determine what effect those operations produce. In other words, one must find out what the intended purpose of the mechanism is. The shape and structure of a machine tell us about its use only if we already know how machines of similar shape and structure are used. Hence, we must find out how a machine works in order to deduce its function from its structure. [*Connaissance*, p. 115]

The Distinctiveness of the Animal-Machine

[84] Descartes began by attempting to formulate what he himself called a "theory of medicine,"[12] that is, a purely speculative anatomical and physiological science as rigorous and exact as mathematical physics and just as receptive to conversion into practical applications, or therapies. But what was to be deduced from the physics of the human body, namely, a determination of "vital utility," was in fact present from the beginning in the subject's principles. [*Formation du réflexe*, p. 53]

[85] According to Descartes, the distribution of the spirits upon leaving the brain depends on several factors. First are the effects of objects that stimulate the senses, or excitations, which Descartes compares to the fingers of an organist touching the keys

of the instrument.[13] Second is the position of the pineal gland in relation to the brain, together with the state of the gland's outer surface; Descartes devoted a great deal of time to the effects of the will, memory, imagination and common sense on these variables.[14] The instincts are the third and last factor affecting the animal spirits. To understand what Descartes meant by instincts, recall his distinction between external movements and internal movements, or passions.[15] External movements can be further broken down: they are either expressive (laughing or crying, for example), and thus purely circumstantial, or adaptive, that is, "useful for pursuing desirable things or avoiding harmful ones,"[16] which is to follow "the *instincts* of our nature."[17] Thus, for Descartes, the physiological mechanism that determined the distribution of animal spirits emanating from the brain depended on what can only be called a biological teleology (the pursuit of desirable things and the avoidance of harmful ones). This was not a lapse. Other, similar examples can be found elsewhere in the *Treatise on Man*. In the *Primae cogitationes circa generationem animalium*, Descartes invokes the *commoda* and *incommoda naturae* as causes of various animal movements, the mechanism of which is explained in terms of animal spirits.[18] Martial Gueroult, moreover, has pointed out the remarkable significance of the *Sixth Meditation*.

To sum up, Descartes distinguished three types of factors influencing the flow of animal spirits: external and contingent factors (sensory excitations), acquired and individual factors (memory) and natural and specific factors (instincts). In this he showed a remarkable alertness to the biological phenomenon of interaction between organism and environment. [*Formation du réflexe*, pp. 31–32]

[86] A vitalist principle of sorts thus remained part of the explanation of movements that, according to the original proj-

ect, were to be explained exclusively in terms of material laws. Gueroult is correct, then, when he says that Descartes began with a conception of medicine as pure physics which he later rejected, and, further, that "one of his chief reasons for confessing the failure of his medical project was his growing conviction that mechanical concepts alone would never suffice to create a medical science because the human body is not pure extension but in part a psychophysical substance."[19] Following Gueroult, perhaps, but more boldly, I would ask whether the attempt to reduce animal biology to mechanics did not reveal the resistance of vital phenomena to full expression in mechanical terms. I earlier alluded to the passage in the *Primae cogitationes* in which *commoda* and *incommoda naturae* were seen to influence the movements of organic parts and even entire organisms.[20] True, Descartes, who prided himself on explaining what we would call the natural appetites or inclinations of animals "solely in terms of the rules of mechanics alone,"[21] pointed out that "brutes have no knowledge of what is advantageous or harmful" – meaning that they are not conscious of such things or able to articulate such knowledge, so that what we observe is simply an association between certain movements and certain events that enable animals to grow.[22] Here, however, we touch on what is probably the limit of mechanistic explanation, for the three aspects of animal life and development – conservation, individuation and reproduction[23] – point to a distinctive difference between animal-machines and mechanical ones. To be sure, Descartes continually insisted on the identity of the two types of machine: "Since art copies nature, and people can make various automatons which move without thought, it seems reasonable that nature should even produce its own automatons, which are much more splendid than artificial ones – namely, the animals."[24] Another passage expresses the same idea: "It is no less natural for a clock constructed with this or that set of wheels to tell the time than it is for a tree which grew from this or that seed to produce the appropriate fruit."[25] But may we not reverse the order of this relation and say that whatever is natural, that is, mechanical, in the animal organism is also artificial, given that animal-machines are automatons constructed, as it were, by God? And in constructing these machines, did God not provide for their conservation, individuation and reproduction by mechanical means? In other words, were not certain teleological ends incorporated into the assemblage of mechanical parts? Since those ends surpass our understanding, however, cannot and should not the science of living things leave them out of its account? Thus, in positing mechanical equivalents for living things, Descartes banished teleology from the realm of human knowledge only to reinstate it in the (immediately forgotten) realm of divine knowledge.

If, moreover, a poorly made clock obeys the same laws of mechanics as a well-made one, so that the only way to distinguish between the two is to invoke "the maker's desire" and "the use for which the maker intended"[26] his creations, it follows that any working machine is an assemblage of parts embodying a purpose. What defines the machine is not the laws of mechanics that dictate how it works but the purpose for which it was built. If an animal that lives in this world is also a machine, it must be the embodiment of some purpose. The fact that the purpose eludes both the animal's awareness and human understanding does not alter this state of affairs in any fundamental way, for otherwise there would be no difference between the living animal and the dead animal, between *accretio viventium* and *accretio mortuorum*. Gueroult, I think, was clearly aware that if, in considering organisms, we abstract from all teleological considerations, organisms cease to be indivisible entities: "If we remove one hoof from a horse, does it become less 'horselike' than other horses?"[27] And if, in the special case of man, there is no way to avoid recourse to

"God's transcendent purpose, namely, that the laws of mechanism alone should suffice to engender and preserve machines whose parts are arranged so as to fulfill the requisite conditions for a union of body and soul, that is, a relation of means to end"[28] – does this not imply, then, as Gueroult suggests, that if we assume that machines lack this "same organization and interdependence of parts and whole,"[29] we must accept an "incomprehensible division" between men and animals? Indeed, without such interdependence, which allows a mechanical relation of structure to be transformed into a teleological relation of fitness for purpose, the indivisible functional unity of the organism becomes inconceivable. The incomprehensible division is tolerable only when presented as an "unfathomable mystery" that situates man in relation to God's wisdom.[30]

In short, only a metaphysician could have set forth the principles of a mechanistic biology without falling at once into contradiction (contradiction that must in any case emerge in the end). Few historians of biology have noticed this, and even fewer historically minded biologists. It is more regrettable that philosophers have made the same mistake. [*Formation du réflexe*, pp. 54–56]

CHAPTER ELEVEN

Auguste Comte

The Montpellier School

[87] After being banished to Montpellier for his role in the closing of the Ecole Polytechnique, Auguste Comte took courses at the Faculty of Medicine, where Paul-Joseph Barthez had taught until his death ten years prior to Comte's arrival. The man who actually introduced the father of positivism to biology was Henry Ducrotay de Blainville, a former professor at the Muséum and the Sorbonne. Having met him at Claude Henri de Saint-Simon's, Comte attended Blainville's course in general and comparative physiology from 1829 to 1832. He admired his teacher's encyclopedic knowledge and systematic mind. The *Cours de philosophie positive* was in fact dedicated to Blainville and Charles Fourier, and its fortieth lesson is full of praise for Comte's erstwhile teacher. [...]

In portraying the eras that preceded the advent of the positive spirit in philosophy, Comte liked to sketch the history of biology in broad strokes, drawing on a keen awareness of the interrelatedness of biological discoveries that he took from Blainville's lectures. A striking example can be found in the fifty-sixth lesson of the *Cours*, which concerns the naturalists of the eighteenth century.[31] Comte excelled at giving summary descriptions of the contributions of various scientists and at weighing their relative importance. Among those whom he singled out as precursors of positivism were Hippocrates, Barthez, Bichat, Johann Friedrich Meckel, Lamarck and, of course, Claude Bernard. The range of the citations proves that Comte was genuinely learned in the subject, whence the ease with which he attained a lofty vantage from which he was able to conceive of the history of science as a *critical* history, that is, a history not only oriented toward the present but judged against the norms of the present. Thus, in the forty-third lesson Comte's account of the controversy between mechanists and vitalists was planned to reveal the "obviously progressive intent" of the Montpellier vitalists, especially Barthez and Bichat, whose work was so unjustly decried at the time in Paris. [*Etudes*, pp. 62-63]

[88] In a note in the twenty-eighth lesson of the *Cours*, Comte hailed the illustrious Barthez as "a far more influential philosopher" than Condillac, and in his preface to the *Nouveaux éléments de la science de l'homme* he praised it as a text "of eminent philosophical power" and an "excellent logical theory," far superior to the "metaphysician" Condillac's *Traité des systèmes*. In the forty-third lesson, Barthez is praised for having established "the essential characteristics of sound philosophical method, after having so triumphantly demonstrated the inanity of any attempt to discover the primordial causes and intimate nature of phenomena of any order, as well as having reduced all true science to the discovery of the actual laws governing phenomena." There can be no doubt that it was from a medical treatise published in 1778 that Comte took the fundamental tenets of his positive philosophy, which he believed were confirmed by Pierre-Simon Laplace's 1796 *Exposition du système du monde* and Fourier's 1822 *Théorie analytique de la chaleur*.

It should now be clear why Comte, who characterized Georg Ernst Stahl's doctrine as "the most scientific formulation of the metaphysical state of physiology," maintained that Barthez's "vital principle" pointed to "a metaphysical state of physiology farther removed from the theological state than the formulation used by Stahl assumed." Unlike so many of his own contemporaries and so many of Barthez's, Comte refused to be misled by a mere change of terminology. He did not believe that Barthez had merely substituted a new name for what Stahl had called "the soul." On this point, he made a profound and pertinent remark: "For so chimerical an order of ideas, such a change in terminology always indicates an authentic modification of the central idea."

Barthez's invaluable historian, his friend Jacques Lordat, points out that Albrecht von Haller was primarily responsible for the misinterpretation that Comte avoided. It was von Haller who wrote in the second volume of his *Anatomical Library* that Barthez believed that what he called the "vital principle" was the ultimate source of the life force.[32] But in thanking Barthez for sending a copy of his 1772 inaugural address to the Montpellier Faculty of Medicine, "De Principio vitali hominis," von Haller indicated that he himself was not so bold as to "accept a principle of a novel and unknown nature."

Note, moreover, that while Barthez's work was certainly one source of Comte's philosophy, it is at least plausible that Barthez's *Exposition de la doctrine médicale*, which Lordat published in 1818, influenced Comte's judgment of that work. Jacques Lordat was a professor of anatomy and physiology at Montpellier when Comte, who was banished to Montpellier in 1816, attended courses there. When Comte characterized Barthez's expression "vital principle" as a mere "formula," he was actually using the same term that Lordat had used in criticizing von Haller's failure to understand that the phrase implied no belief in a special substance or entity distinct from body and soul. Comte encountered the teachings of the Montpellier School in Montpellier itself, and that, coupled with his outspoken animosity toward certain leading figures of the Paris School, may have had something to do with the admiration that enabled him to form a clear picture of Montpellier's doctrine. [*Etudes*, pp. 75-77]

[89] Comte was able to perceive the direct, authentic insight into biological realities that lay hidden behind the abstract concept of the vital principle. From Barthez as well as Bichat, he learned of the intimate relations among the concepts of organization, life and consensus. This debt to Barthez may explain Comte's tendency to present him as the sole representative of the Montpellier School. He overlooked, or pretended to overlook, Théophile de Bordeu. The idea that the life of an organism is a synthesis of elementary lives, an idea that delighted Diderot in *D'Alembert's Dream*, would no doubt have seemed as unsatisfactory to Comte as did the theory of organic molecules – and he would have raised against it the same objections that he leveled, in the forty-first lesson of the *Cours*, at the first formulations of cell theory. If Bichat dissuaded Comte from following Lorenz Oken, Barthez overshadowed de Bordeu in his mind. The concept of complex living things composed of organic molecules or animalcules suggested a misleading analogy between chemistry and biology. Life is necessarily a property of the whole organism: "The elementary animalcules would obviously be even more incomprehensible than the composite animal, even apart from the insoluble difficulty that one would thereby gratuitously create concerning the effective mode of so monstrous an association." Very much in the spirit of Barthez, Comte held that "every organism is by its very nature an indivisible whole, which we divide into component parts by mere intellectual artifice only in order to learn more about it and always with the intention of subsequently reconstituting the whole." The statement reveals as many taboos as it does scruples. [*Etudes*, pp. 78-79]

Biological Philosophy

[90] The invention of the term "biology" reflected a growing awareness on the part of physicians and physiologists that their subject matter was fundamentally different from that of the physical sciences. The coining of the word suggests an assertion of the discipline's autonomy, if not of its independence. Comte's biological philosophy provided systematic justification for that assertion: it connoted full acceptance of, as well as a need to consolidate, "the great scientific revolution which, under Bichat's leadership, transferred overall priority in natural philosophy from astronomy to biology."[33] Comte was not entirely wrong to see the disappointments he had suffered in his career as consequences of the fact that he, a mathematician, had taken up cudgels on behalf of the biological school in the struggle to maintain, "against the irrational ascendancy of the mathematical school, the independence and dignity of organic studies."[34]

Comte's conception of the milieu justified his belief that biology could not be a separate science. And his conception of the organism justified his belief that biology must be an autonomous science. The originality and force of his position lies in the correlation – or, some would say, dialectical relation – between these two concepts.

Comte took the Aristotelian term "milieu" from Lamarck via Blainville. Although it was in common use in seventeenth- and eighteenth-century mechanics and the physics of fluids, it was Comte who, by reverting to the word's primary sense, transformed it into a comprehensive, synthetic concept that would prove useful to later biologists and philosophers. When he suggested, in the forty-third lesson of his *Cours* in 1837, that the first duty of biology is to provide a general theory of milieus, Comte, who may not have known the work of William Edwards (1824) or Etienne Geoffroy Saint-Hilaire (1831) in this area, thought he was proclaiming Lamarck's superiority over Bichat. Bichat's distaste for the methods of the eighteenth-century iatromathematicians had led him to insist not only that the distinction between living and inert was legitimate but also that the living and the inert were fundamentally antagonistic. Against this, Comte argued that "if all that surrounds living bodies really tended to destroy them, their existence would be fundamentally unintelligible."[35]

Comte's successive judgments of Lamarck are revealing, however, of the deeper meaning of his biological views. [...] Beyond the first consequence of the Lamarckian theory of the milieu – namely, the variability of species and the gradual inception of new varieties – Comte perceived a possibly monist, and ultimately mechanist, tendency. If the organism is conceived of as being passively shaped by the pressure of the environment, if the living thing is denied all intrinsic spontaneity, then there is no reason not to hope that the organic might someday be explained in terms of the inert. But here the spirit of Bichat rose up in Comte against the threat of "cosmological usurpation,"[36] against the shouldering aside of Larmarck's insights in favor of an uncompromising mathematical approach.

Similarly, Comte held, like Bichat and following his lead, that the tissue was the lowest possible level of anatomical analysis; he therefore denied that the cell, which he called the "organic monad," could be the basic component of all complex organisms. It was not simply that he was suspicious of microscopy, whose techniques were still relatively primitive; Comte's opposition to cell theory was primarily logical. For him, an organism was an *indivisible* structure of *individual* parts. Actual living things were not "individuals" in any simple sense. Neither his superficial knowledge of German nature philosophy, especially that of Oken, nor his reading of Henri Dutrochet (at around the time he was preparing the *Cours*), nor even his reading of Theodor Schwann, to whom he alludes, enabled him to see, in the earliest formulations of cell theory, the first glimmerings of a theory of "degrees of individuality." For Comte, the very concept of the cell implied a misleading analogy between organic bodies and inorganic compounds composed of indivisible molecules.[37] [*Etudes*, pp. 63–66]

[91] Clearly, the idea underlying all of Comte's positions on biology was the necessary duality of life and matter. In biological philosophy, the eighteenth century bequeathed two temptations to the nineteenth: materialism and hylozoism, that is, the doctrine that matter is animated or that matter and life are inseparable. Comte, like Descartes, battled on two fronts, and his tactics were, if nothing else, Cartesian. The matter/life dualism was the positivist equivalent of the Cartesian metaphysical dualism of extension and thought. For Comte, dualism was a prerequisite of universal progress, which to him meant nothing other than the subjugation and control of inert matter by the universe of the living under the guidance of humankind. "We are, at bottom, even less capable of conceiving of all bodies as living," wrote Comte,

> than as inert, because the mere notion of life implies the existence of things not endowed with it.... Ultimately, living beings can exist only in inert milieus, which provide them with both a substrate and a direct or indirect source of nourishment.... If everything were alive, no natural law would be possible, for the variability that is always inherent in vital spontaneity is really limited only by the preponderance of the inert milieu.[38]

Even in beings where the only manifestation of life is vegetative, one finds a "radical contrast between life and death." Between plants and animals there is simply a "real distinction," whereas between plants and inert substances there is a "radical separation." The traditional division of nature into three kingdoms (animal, mineral, vegetable) allowed one to imagine a gradual transition from one species to another along a chain of being; Comte therefore proposed replacing that tripartite scheme with a new one consisting of two "empires" (living and inert). He was convinced that "vital science cannot exist without this irreducible dualism."[39]

In essence, Comte saw, between Lamarck and Descartes, a parallel that no one would think of disputing today. Perhaps more perspicacious with respect to the future than accurate in his perception of the present, Comte anticipated the consequences of the idea that animals can be conditioned by their environments – that is, he foresaw the possibility of behaviorism. The assumption of a direct muscular reaction to external impressions is incompatible, Comte argued, with the idea of "animal spontaneity, which at the very least implies that inner motives are decisive."[40] This would lead to a "restoration of Cartesian automatism, which, though incompatible with the facts, continues in one form or another to mar our leading zoological theories."[41]

Now we can see why Comte ascribed such importance to the theories of Franz Joseph Gall, who argued that the fundamental inclinations and drives of human and animal behavior are innate. His cranioscopic method, so easy – all too easy – to celebrate or ridicule, actually stemmed from his principled hostility to sensualism. If it could be shown that certain areas of the brain were by their very nature associated with certain psychic faculties, then one must ascribe primordial existence to those faculties. Hence, nothing could have been more alien to Gall's (or Comte's) thinking than the Lamarckian idea that the biological functions are independent of the organs that embody them (and may even influence the development of those organs). True, Gall did map cerebral topography by studying the mental functions of his patients, but in doing so his intention was to refute, not to corroborate,

Lamarck's doctrine. Gall provided Comte with an argument in favor of innate aptitudes and, more generally, of innate functions – an argument that Comte elaborated into a guarantee of continued progress through development of a preexisting order.

Comte claimed to have achieved comprehensive, critical insight into the biology of his time. If I have correctly identified the grounds of his self-confidence, it should now be possible to state his most important conclusions in a systematic fashion. First, Comte believed that he, following Georges Cuvier, had eliminated teleology from biology: the "conditions of existence" replaced the dogma of final causes, and the only relation assumed to exist between an organism and its environment, or between an organ and its functions, was one of compatibility or fitness, implying nothing more than viability. "Within certain limits," Comte states in the *Cours*, "everything is necessarily arranged in such a way that existence is possible."[42] The harmony between function and organ "does not go beyond what actual life requires."[43] Since, moreover, organisms depend on their environments, living things are subject to cosmic influences. Biology is therefore related to cosmology; hence, the principle that nature's laws are invariable, first formulated in astronomy and eventually extended to chemistry, could now be extended to biology, thereby invalidating the belief that variability and instability are essential to organic processes. Finally, generalizing a principle borrowed from François Joseph Victor Broussais, Comte held that all pathological phenomena could be explained by the laws of physiology. Thus, he argued that the difference between health and disease was a matter of degree rather than of kind – hence medicine should base its actions on the analytic laws of anatomophysiology.

Yet, as even the *Cours* made clear, the very organic structure of living things constituted an obstacle to further progress in positive, experimental physiology. An organism, Comte argued, is a *consensus* of organs and functions. The harmony that exists among the functions of the organism is "intimate in a very different sense from the harmony that exists between the organism and the milieu."[44] An organism, Comte maintained, is a unified whole; to dissect it, to divide it into component parts, was "mere intellectual artifice."[45] The biologist, then, must work from the general to the specific, from the whole to the parts: "How can anyone conceive of the whole in terms of its parts once cooperation attains the point of strict indivisibility?"[46] Between Immanuel Kant and Claude Bernard, Comte once again made finality, in the guise of totality, an essential element of the definition of an organism.

This was not the only place where the positivist method violated the principle of working from the simple to the complex and the known to the unknown. In celebrating the promotion of anatomy to the quasi-philosophical dignity of comparative anatomy, a system that provided a basis for classifying the multitude of specific forms, Comte was led to reject Cuvier's fond notion that the animal kingdom consists of a number of distinct branches and to accept instead Lamarck's and Blainville's theory of a unique series. Once again, his grounds for making this choice involved a subordination of the simple to the complex, of the beginning to the end: "The study of man must always dominate the complete system of biological science, either as point of departure or as goal."[47] This is because the general notion of man is "the only immediate" datum we have.[48] Comte thus claimed to be keeping faith with his general program, "which consists in always reasoning from the better known to the lesser known," even though he insisted on arranging the animal series in order of decreasing complexity – this in order to read the series as "revealing a devolution from man rather than a perfection from the sponge." It would strain credulity to draw a parallel between Comte's approach here and that of Kurt Goldstein, to find in the former a phenomenological biology *avant la lettre* and in the latter a hitherto-neglected positivist inspiration. In fact, Comte had an idea, albeit a confused one, of where he was going. The intellectual function was the distinguishing feature of animal life. To interpret all life as a series devolving from man, the perfect embodiment of that function, was to treat biology as subordinate to sociology, for the true theory of intelligence was to be found, Comte believed, in sociology and not in psychology. [*Etudes*, pp. 67–71]

[92] Comte's biological philosophy, that edifice of erudition and learning, hid an intuitive conviction whose implications were far-reaching. The impetus behind that conviction no doubt stemmed from the fact that a utopian spirit breathed life not only into the bold assertions of a brand-new science but also into the time-tested truths of a philosophy almost as old as life itself. Simply put, this was the conviction that life takes place but does not originate in the world of the inert, where it abandons to death individual organisms that stem from elsewhere. "The collection of natural bodies does not form an absolute whole." This belief, combined with the idea of a continuous, linear series of living things culminating, logically as well as teleologically, in man, was eventually transformed into the idea of Biocracy as the necessary condition of Sociocracy. This was the positivist equivalent of the old metaphysical idea of a Realm of Ends. [*Etudes*, p. 73]

Positive Politics

[93] The superiority of positive politics "results from the fact that it *discovers* what others *invent*." The discovery that the inventor of positive politics claimed as his own was that "the natural laws that govern the march of civilization" are derived from the laws of human organization. To the extent that "the state of social organization is essentially dependent on the state of civilization,"

social organization is nothing other or more than an aspect of human organization "not subject to major change" (so far as we can see). What we know of human organization, moreover, is the result of a methodological decision "to envisage man as a term in the animal series, indeed, from a still more general point of view, as one of a collection of organized bodies or substances." Seemingly faithful to Claude Henri de Saint-Simon's terminology, Comte gave the name "physiology" to the "general science of organized bodies." But a difference between his use of the term and Saint-Simon's is already evident. For Comte, physiology was not just a discipline recently instituted for the study of man as living being, one whose method could serve as a model for the study of man in society; more than that, the content of physiology was to become the nucleus of a new science. Physiology owed its content to medicine, and medicine taught this lesson: "Long having hoped that he might learn to repair any disturbance to his organization and even to resist any destructive force, [man] finally realized that his efforts were futile as long as they did not cooperate with those of his organization, and still more futile when the two were opposed." And further: "The fact that many illnesses were cured in spite of defective treatments taught physicians that every living body spontaneously takes powerful steps to repair accidental disturbances to its organization." Hence, politics is like medicine in that both are disciplines in which perfection requires observation. And just as there were two schools of medical thought, so, too, were there two schools of political thought: the "politics of imagination" involved "strenuous efforts to discover remedies without sufficient consideration of the nature of the disease"; the "politics of observation," on the other hand, knowing "that the principal cause of healing is the patient's vital strength [*force vitale*]," is content, "through observation, to remove the obstacles that empirical methods place in the way of a natural resolution of the crisis." The linking of the terms "vital force" and "crisis" alerts us to what is going on here: this was Hippocratic medicine reinterpreted in the light of the Montpellier School's doctrine.

In Comte's text, the term "crisis" took on a pathological and therapeutic significance that it lacked in Saint-Simon. It was a term freighted with all the weight and decked out with all the majesty of a medical tradition. Thus, "nature" was continually invoked as the ultimate reason why unfavorable political circumstances failed to prevent "the advance of civilization," which in fact "nearly always profits from mistakes rather than being delayed by them." This recourse to nature is so basic that it enables Comte to naturalize, as it were, the most distinctive feature of human history, namely, the labor or industry whereby society pursues its ends: this Comte described as "action on nature to modify it for man's benefit." This teleological end was "determined by man's rank in the natural system as indicated by the facts, something not susceptible of explanation."[...]

This limitation of man's power to knowledge of nature's laws and prediction of their effects, hence to harnessing natural forces to human designs, has more in common with the prudence of Hippocratic diagnostics than with the demiurgic dream of denaturing nature through history.

But reading between the lines of the text is not enough. What of the sources that Comte drew on? The text quoted above contains such phrases as "the political impetus peculiar to the human race" and "the progress of civilization," which "does not march in one straight line" but, rather, proceeds by "a series of oscillations not unlike the oscillations we see in the mechanism of locomotion." And Comte refers to "one of the essential laws of organized bodies," which can be applied "equally well to the human race acting collectively or to an isolated individual" – a law that linked the development of strength to the presence of resistance. From this I venture to conclude that well before he added Anthelme Richerand's *Eléments de physiologie* and Barthez's *Nouveaux éléments de la science de l'homme* to the annals of positivism, Comte had read what both authors had to say about animal movement. Richerand wrote of "zigzag movement in the space between two parallel lines." And Barthez, in his *Nouvelle mécanique des mouvements de l'homme et des animaux*, discussed waves and reciprocating motions. Comte also used the word oscillation. And when he spoke of the peculiar impetus leading to improvements in the social order, he again referred to Barthez, to the Barthez who, in his *Nouvelle mécanique*, tried to refute the idea that animals move for no other reason than that the ground repels their feet. And again, it was Barthez – specifically, Chapter Four of *Nouveaux éléments* – from whom Comte borrowed the law relating strength to resistance.

To sum up, then, Comte left the Ecole Polytechnique and pursued the study of biology, as he indicates in the preface to the sixth and final volume of the *Cours de philosophie positive*. At that time, he discovered and made his own an idea of the organism that became the key concept of his theory of social organization. When Saint-Simon published *De la physiologie appliquée à l'amélioration des institutions sociales* in 1813, he did not attempt to impose a biological model on social structure. His conception of an "organized body" required no such analogy, and his conception of "crises" implied no necessary relation to pathology. Comte, on the other hand, found in Barthez, still in a metaphysical form, the idea that organized systems are to some extent self-regulating or autonomous. And from a lecture that Barthez gave in 1801, entitled "Discours sur le génie d'Hippocrate," Comte drew the Hippocratic conclusion: all organisms (or organizations, as Comte liked to call them) have a spontaneous capacity to preserve and perfect themselves. By interpreting this capacity as an inherent property of the nature of organization, he was able to keep faith with the precepts of positivism.

On December 25, 1824, Comte wrote Jacques-Pierre Fanny Valat: "The state in which we find society today is a long way from normal.... It is, rather, a very violent state of crisis." Because he viewed organization as a normative property of organisms, he could on three different occasions characterize political projects or practices as "monstrosities" or "monstrous" and on four occasions characterize conduct or behavior as "defective." These terms were borrowed from teratology, a science intimately associated with the emerging field of embryology: Etienne Geoffroy Saint-Hilaire's treatise on *Les Monstruosités humaines* had been published in 1822. Comte's philosophy clearly implies a concept of normal as opposed to pathological development. In fact, if Comte, in the *Plan des travaux scientifiques*, invokes the nature of things as frequently as he does, it is because by "things" he means life and by "life" he means a distinct capacity to persist in a "normal" direction. To borrow an expression frequently employed by François Perroux, Comte's conceptualization is "implicitly normative." By reintegrating the human into the organic, the history of man into the history of things, Comte bestowed a guarantee of necessity on the moral destination of the species. He was able to do so without contradiction only because, under cover of the positive term "nature," he superimposed an order of meaning on an order of law. ["A. Comte," *Etudes philosophiques*, pp. 294-97]

The Positivist Disciples
[94] In the *Système de politique positive* (1851), Comte described two young physicians, Dr. Louis-Auguste Segond and Dr. Charles Robin, as his disciples. In 1848, the two men founded the Société de Biologie, an organization whose reports and journals give the

most comprehensive and vivid image we have of biological research in France over the past century or more. The Société's first governing board was chaired by Dr. Rayer, who later became dean of the Faculty of Medicine; Claude Bernard and Charles Robin served as vice chairmen; and Charles-Edouard Brown-Séquard and Robin were the secretaries. The group's first charter was drafted by Robin, and its first article stated that "the Société de Biologie is instituted for the study of the science of organized beings in the normal state and in the pathological state." The spirit that animated the founders of the group was that of positive philosophy. On June 7, 1848, Robin read a paper "On the Direction That the Founding Members of the Société de Biologie Have Proposed to Answer to the Title They Have Chosen." In it, he discussed Comte's classification of the sciences, examined biology's mission in much the same spirit as Comte had done in the *Cours de philosophie positive*, and noted that one of the most urgent tasks facing the discipline was to investigate the milieus in which life existed. Robin even had a name for this proposed subdiscipline – "mesology." When the Société celebrated its fiftieth anniversary in 1899, the physiologist Emile Gley read a report on the evolution of the biological sciences in France, in which the impetus that positivism gave to the subject is frequently alluded to. Gley's report still makes interesting reading.[49]

In 1862, Charles Robin became the first person to hold the chair in histology at the Faculty of Medicine in Paris.[50] From that position he remained faithful to one tenet of Comte's biological philosophy in his refusal to teach cell theory in the dogmatic form in which it had been expressed by Rudolph Virchow. Robin taught instead that the cell was one of many anatomical components rather than the fundamental component of living organisms. In 1865, a student in Robin's school defended a thesis on "The Generation of Anatomical Elements." Its author, who would later translate John Stuart Mill's book *Auguste Comte and Positivist Philosophy* into French, subsequently achieved a fame that has tended to overshadow his early interest in biology. His name was Georges Clémenceau.

Robin was also, along with Emile Littré, the author of the *Dictionnaire de médecine*, which in 1873 supplanted the series of revised editions of Pierre Hubert Nysten's *Dictionnaire*. This reminds us that Comte's biological philosophy also left its mark on the development of lexicography in France as well as on the production of critical editions of medical texts and on the history of medical science. [*Etudes*, pp. 71–72]

[95] With an author as careful about the meaning of words as was Littré, one must take literally what he said about his personal relations with Comte. On at least two occasions he stated, "I subscribe to the positive philosophy."[51] He also said that he had chosen Comte's great book as a "model," adding, "There, happily, I feel that I am a disciple."[52] He described his allegiance to positivism as a kind of conversion: "Having been a mere freethinker, I became a positivist philosopher."[53] When Littré died, his journal, *La Philosophie positive,* sought to counter rumors that he had converted to Catholicism by publishing its late editor-in-chief's final editorial under the title "For the Last Time": "The positive philosophy that kept me from being a mere negator continues to accompany me through this final ordeal."[54]

If there was one principle of the positive philosophy set forth in the *Cours* about which Littré never expressed the slightest reservation, and which he tirelessly defended, it was the hierarchy of the six fundamental sciences, expressing the historical progression of human knowledge. What interested him, of course, was the relation of biology to its predecessors, physics and above all chemistry, but he may have been even more interested in the relation of sociology to biology. This was the source of his disagreement with Herbert Spencer, who argued in the "Classification of the Sciences" that hierarchy ought to be replaced by interdependence. Littré held that no change in the relative ranking of the sciences was possible,[55] and he was able to persuade Mill on this point.[56] An immediate consequence of the hierarchical principle was that importing a method valid for the study of a lower level or stage of phenomena into a discipline at a higher level was "the greatest theoretical mistake one could make."[57] Littré's philosophy of biology, hence of medicine as well, can be summed up in one brief passage: "Biological facts must first obey the laws of chemistry. Any correct interpretation must respect this principle. But the reverse is not true: chemical facts need not obey the laws of biology, for which they lack one thing, namely, the characteristic of life."[58] That "one thing" would persist to the end of Littré's life: for him it was an incontrovertible obstacle, "the crucial difference between mechanism and organism."[59] Littré was, to use a modern term, an implacable enemy of "reductionism." In 1846, for example, in a study of Johannes Müller's *Handbuch der Physiologie des Menschen*, Littré came to the defense of the "irreducible": "It is important to determine the irreducible properties of things.... Irreducible means that which one cannot effectively reduce. In chemistry, for instance, effectively indecomposable compounds are called irreducible."[60] In 1856, in a major article on François Magendie, Littré found that Magendie had been more an opponent than a disciple of Xavier Bichat. In essence, Magendie had failed to distinguish between the occult and the irreducible, the immanent properties of living matter, whereas Bichat had recognized the irreducible while exorcising the occult. Magendie had been unable or unwilling to state a clear position on the reducibility of biological phenomena to the laws of physics and chemistry or on the irreducibility of vital organization. Littré was also critical of Léon Rostan, the author of the medical theory known as "organicism," for neglecting the irreducibility of the properties of living matter. Note, by the way, that both Littré's *Dictionnaire de la langue française* and his *Dictionnaire de médecine* contain articles on the word *irréductible*. ["Littré," *Actes du Colloque Emile Littré, 1801–1881*, pp. 271-73]

[96] In what respects did Xavier Bichat influence Emile Littré and other positivist physicians such as Charles Robin (not only directly but also through Auguste Comte)? To begin with, there was his celebrated distinction between two forms of life, vegetative and sensitive (or animal), the latter being subordinate to the former. Littré alluded to this distinction in his article on François Magendie, where he criticized his subject for not having respected the order in which the functions ought to be studied: in *Physiologie*, Magendie had taken up the sensory functions before considering nutrition.[61] But the main thing that the positivists took from Bichat was his contention that the tissues were the ultimate elements of anatomical analysis, a view that tended to push the new science of histology in one direction rather than another. Bichat's views, repeated by Comte in the forty-first lesson of the *Cours*, explain the persistent skepticism of French physicians in the first half of the nineteenth century with respect to cell theory and microscopic techniques, which were disparaged in favor of such histological methods as dissection, desiccation, maceration and treatment with acids. True, the microscopes available at the time were mediocre, and Louis Ranvier noted in his 1876 inaugural lecture at the Collège de France that Bichat had been right to be wary of them. Nevertheless, positivist physicians displayed persistent hostility to microscopy, partly in obedience to Henry Ducrotay de Blainville's authoritative *Cours de physiologie générale* (1829). René-Théophile Hyacinthe Laënnec also numbered among the instrument's detractors. Thanks to Marc Klein's work on the history of cell theory, there is no need to belabor Robin's oppo-

sition to any form of research that claimed to go beyond what he took to be the basic constituent of the anatomy (tissue). Even as late as 1869, ten years after the publication of Rudolph Virchow's celebrated work, Robin wrote in *La Philosophie positive* that the cell was a metaphysical construct and commented ironically on "the allegedly typical or primordial organic cell."[62] When Littré reviewed Robin's *Anatomie et physiologie cellulaires* in the same journal in 1874, he accepted his friend's doubts as fact. Yet in an 1870 article on the "Origine de l'idée de justice," Littré had discussed two kinds of brain cells, affective and intellectual: Was this a theoretical concession or a mere stylistic convenience?[63]

In what respects, moreover, did François-Joseph-Victor Broussais influence Littré (either directly or through Comte)? Surely, Littré inherited Broussais's stubbornness in defending the theories of physiological medicine, which were based on a belief in the identity of the normal and the pathological, as well as on a refusal to view disease as introducing any new functional process in the organism (a case made even before Broussais by John Hunter). Littré thus accepted and championed what Comte called "Broussais's Principle." In the preface to the second edition of *Médecine et médecins*, Littré stressed the need for medicine to revise its theories in light of physiology's having attained the positive stage of development. Pathology had thus become "physiology of the disturbed state," and this, Littré argued, was an "essential notion." This Broussaisist dogma would later prove to be one of the obstacles to understanding microbiology. But for the moment, let us ask ourselves what the practical effects of this revolution were. In an 1846 article containing a new translation of Celsus, Littré was not afraid to write that "so long as physiology was not fully constituted as a science, there remained gaps in which hypotheses could emerge. But now that it has become, almost before our eyes, a science, every medical system is discredited in advance."[64] ["Littré," *Actes du Colloque Emile Littré, 1801-1881*, pp. 274-75]

[97] Littré set forth his views on hygiene in a commentary on the *Traité d'hygiène publique et privée* by Michel Lévy, the former chief physician of the Armées d'Orient and director of the Val-de-Grâce hospital, whom Littré described as an "eminent author," although Jean-Michel Guardia saw him as more of a rhetorician than a scholar. Public health had been a lively medical subspecialty in France since the work of Jean-Noël Hallé and François Emmanuel Fodéré early in the nineteenth century; it had profited from the experience of such military physicians as Villermé, who had served as surgeon-major in Napoleon's army. This medical subspecialty had no doubt lent credence to the notion of milieu, first put forward in the works of Blainville and Lamarck. Hygiene, according to Littré, is the science of actions and reactions between milieus and organisms, humans included. As for milieu, Littré noted in 1858 that the term had a technical meaning, and he gave a detailed definition in many respects reminiscent of the table of physical agents that Blainville had called "external modifiers." The scientific elaboration of the word "milieu" in the nineteenth century required the participation of a number of sciences that had achieved the stage of "positivity" – physics, chemistry and biology. The term also served in part as an ideological substitute for the notion of "climate," which had been used extensively by eighteenth-century authors, particularly Montesquieu. According to Littré, however, the study of man's own milieu was the province of sociology as much as of physics or biology, so that the prescriptions of "private hygiene" could claim only a historical or empirical rather than a theoretical basis.[65] Having written several articles on the cholera of 1832, the contagiousness of equine glanders and the transmission of the plague, Littré could hardly fail to comment on Lévy's

observations on endemic and epidemic diseases. Not a word was said about etiologies involving microorganisms, although the article on "Leptothrix" in the *Dictionnaire de médecine* reported on Casimir Davaine and Pierre François Olive Rayer's research on the anthrax bacillus, and Littré surely knew about this work owing to his relations with Rayer. It was not until 1880, in an article entitled "Transrationalisme," that Littré mentioned "the circulation of infinitesimally small [creatures] that cause infectious diseases";[66] but by then it was no longer possible to ignore Louis Pasteur's work. Nevertheless, Littré's remarks on public hygiene in the third article are worthy of attention. For Littré, history and sociology serve as instruments of analysis. Littré seems to have been particularly alert to sociomedical issues associated with the rise of industrial society. He strikes off a fine phrase, reminiscent of Saint-Simon: "Civilized man...has assumed responsibility for administering the earth, and as civilization advances, that arduous administration demands ever more ingenuity and industry."[67] Human life, though, suffers from the unanticipated yet inevitable effects of the conflict between work and nature. "Having become so complex, industries cannot do without the oversight of a higher agency that appreciates the dangers, preserves the environment, and does not leave such important issues to the self-interest of private individuals."[68] Thus, Littré had some pertinent remarks to offer on the subject of ecology, years before the word was coined.[69] And no one can deny the clarity or courage with which he expressed his astonishment that no civilized nation had yet seen fit to establish a ministry of public health.[70] ["Littré," *Actes du Colloque Emile Littré, 1801–1881*, pp. 276–77]

[98] We can no longer avoid a brief survey of the reactions of Littré's contemporaries to this biological philosophy. Little attention need be paid to the inevitably biased judgments of official spiritualists such as Paul Janet and Edme Caro; however, greater

importance must be attached to various articles that appeared in Charles Renouvier's journal, *La Critique philosophique*, the very title of which was antipositivist. In 1878, the journal published three articles by Pillon on biology and positivism, two of which set Claude Bernard up as Comte's judge. In the same year Renouvier posed, and answered in the negative, this question: "Is the *Cours de philosophie positive* still abreast of science?" Claude Bernard was proposed as an ideological antidote to Comte. No brief account can do justice to the relations between Littré's biological positivism and Bernard's guiding philosophy. On the one hand, Bernard was a founder of and participant in the Société de Biologie along with Rayer, Louis-Auguste Segond and Charles Robin, who drafted its charter in a frankly positivist spirit and tone. Littré's journal *La Philosophie positive* showed great interest in Bernard's work, publishing, in the year of his death, a very balanced article by Mathias Duval and an article by Littré on determinism. These facts may muddy the waters, but they do not justify any blurring of the lines, for as attentive readers of Bernard already know, he scarcely concealed his hostility to Comte's dogmatism. That hostility is expressed openly at the end of Bernard's most widely read work, the *Introduction à l'étude de la médecine expérimentale*: "Positivism, which in the name of science rejects philosophical systems, errs as they do by being such a system." Despite these reservations, of which Littré was perfectly well aware, he several times praised Bernard's methods and the principles that inspired them. His 1856 article on Magendie ends with an acknowledgment of Bernard's superiority over his teacher. The thirteenth edition of the *Dictionnaire de médecine* contains a number of articles, obviously written by Littré, which refer implicitly or explicitly to Bernard. While the article on "Observation" seems rather to summarize the views of Comte, those on "Experience" and "Experimentation" are condensations of the views of Bernard.

"Experimentation" ends with the same comparison that Bernard borrowed from Georges Cuvier: the observer listens, the experimenter questions. The article on "Medicine" mentions Bernard's name in the discussion of experimental medicine. In Bernard's teaching and conception of life, Littré no doubt saw arguments capable of supporting his own personal conviction that biological phenomena could not be reduced to physics and chemistry. ["Littré," *Actes du Colloque Emile Littré, 1801-1881*, pp. 279-80]

Chapter Twelve

Claude Bernard

A Philosophical Physiologist

[99] A philosophical physiologist: arranged in that order, the two words cry out for an immediate correction. Philosophical here does not mean inclined toward metaphysics. Claude Bernard never claimed — as a physiologist and in the name of physiology — to go beyond experience. He had no patience with the idea of metaphysiology, meaning the claim to know not just the laws, or invariants, of the organic functions but the very essence of that plastic force which we refer to as life. But neither did he ever intend to limit biological science to the mere reporting of experimental results. Rather, by "philosophical physiologist" I mean a physiologist who, at a given stage in the evolution of a well-established science, explicitly recognizes the fact that science is above all a method of study and research, and who sets himself the express task, the personal responsibility that can be assigned to no one else, of providing that method with a foundation. In this sense, the philosophical work of the physiologist Claude Bernard provided the foundation for his scientific work. Just as nineteenth-century mathematicians set themselves the task of exploring the foundations of mathematics, so too did a physiologist take it upon himself to establish the foundations of his discipline. In both

cases, scientists assumed responsibility for what had previously – in the time of Descartes as well as of Plato and Aristotle – been the task of philosophy. But the foundational work of the mathematicians was very different from that of Bernard. Work on the foundations of mathematics has continued ever since; it has become an integral part of mathematics itself. By contrast, the trail blazed by Claude Bernard has been neglected by later physiologists – so neglected, in fact, that when physiologists today feel the need to justify distinctive aspects of their work, they frequently, and sometimes anachronistically, rely on the work of Bernard himself. ["Claude Bernard," *Dialogue*, pp. 556-57]

[100] In the few lines that Claude Bernard devoted to Francis Bacon (laudatory by convention and critical by conviction, though less so than Bernard's contemporary von Liebig), he noted that "there were great experimentalists before there was a doctrine of experimentalism." There can be no doubt that he meant this maxim to apply to himself. An explanation can be found in his notebooks: "Everyone follows his own path. Some undergo lengthy preparation and follow the path laid out for them. I took a twisting route to science and, abandoning the beaten path, exempted myself from all the rules." What rules did this man who had learned the experimental method in the shadow of François Magendie think he had exempted himself from? The answer can be gleaned from the names of two physiologists whom he quotes on several occasions: Hermann von Helmholtz, toward whom he was always respectful, and Emile Du Bois-Reymond, for whom his admiration was less unalloyed. The rules Bernard had in mind were those of mathematical physics:

> It has been said that I found what I wasn't looking for, whereas Helmholtz found only what he was looking for. This is correct, but exclusionary prescriptions are harmful. What is physiology? Phys-

ics? Chemistry? Who knows? It is better to do anatomy. [Johannes] Müller, [Friedrich] Tiedemann and [Daniel Friedrich] Eschricht were disgusted and turned to anatomy.

In other words, what Bernard wanted was a way of doing research in physiology based on assumptions and principles stemming from physiology itself, from the living organism, rather than on principles, views and mental habits imported from sciences as prestigious, and as indispensable to the working physiologist, as even physics and chemistry.

There is a chronological fact whose importance cannot be overstated: Claude Bernard alluded to the distinctive character of physiological experimentation in public for the first time on December 30, 1854, in the third lecture of a course on experimental physiology applied to medicine, which he delivered at the Collège de France in his last appearance there as Magendie's substitute. In that lecture, he reviewed the experiments and the conclusions reported in the doctoral thesis he had defended the year before on a newly discovered function of the liver in humans and animals — the ability to synthesize glucose. "It is surprising," Bernard noted, "that an organic function of such importance and so readily observed was not discovered sooner." The reason for this failure, he showed, was that nearly all previous physiologists had attempted to study dynamic functions with methods borrowed from anatomy, physics and chemistry; such methods, though, were incapable of yielding new knowledge about physiological phenomena. The only way to explain an organic function is to observe it in action in the only place where it meaningfully exists, to wit, within the organism. From this, Bernard derived a principle of which the *Introduction*, published eleven years later, might fairly be called the elaboration:

> Neither anatomy nor chemistry can answer a question of physiology. What is crucial is experimentation on animals, which makes it possible to observe the mechanics of a function in a living creature, thus leading to the discovery of phenomena that could not have been predicted, which cannot be studied in any other way.

The lectures at the Collège de France followed Bernard's completion of work on his doctorate, so the assertion that "there were great experimentalists before there was a doctrine of experimentalism" and the insistence on having left "the beaten path" were more than literary flourishes; they were generalizations of the lessons Bernard had drawn from his own intellectual adventure. Nothing else is worthy of the name "method." As Gaston Bachelard has written in *The New Scientific Spirit*, "Concepts and methods alike depend on empirical results. A new experiment may lead to a fundamental change in scientific thinking. In science, any 'discourse on method' can only be provisional; it can never hope to describe the definitive complexion of the scientific spirit."[71] Notwithstanding Bachelard's dialectical insistence, it is by no means clear that Bernard himself did not succumb to the belief that he was describing the "definitive constitution of the scientific spirit" in physiology. Yet he clearly understood, and taught, that physiology would have to change because it had seen something new, something so new that it forced Bernard to agree with the judgment that some had uttered in criticism of his work: that he had found what he was not looking for. Indeed, one might even go so far as to say that he had found the opposite of what he was looking for. [*Etudes*, pp. 144–46]

The Implications of a Paradoxical Discovery
[101] The importance, then and now, of the *Leçons sur les phénomènes de la vie communs aux animaux et aux végétaux* stems first

of all from the fact that, behind this plain title, Bernard systematically pursued the consequences of a discovery that was a surprise to him and a paradox to his contemporaries. That discovery was set forth in the doctoral thesis he defended on March 17, 1853: "Recherches sur une nouvelle fonction du foie considéré comme organe producteur de matière sucrée chez l'homme et chez les animaux." This thesis dethroned the dogma according to which animals, being incapable of synthesizing the nutrients they need, must ingest vegetable matter in order to obtain them. Bernard, in his work on glycogenesis, showed that that the liver can synthesize glucose and, therefore, that animals need not obtain this substance from plants. [Preface, *Leçons*, p. 9]

[102] For our purposes, it is not important that Bernard obtained his result by dint of flaws in his chemical analytic techniques and rough approximations in his measurements. The fact that he detected no glucose in the portal vein but did detect it in the superhepatic vein led him to conclude – and then to verify – that the liver not only secretes bile but also produces the glucose that is essential to sustaining living tissue and enabling various parts of animal organisms, in particular the muscles, to do their work. Yet Bernard's faith in his verification procedure, the famous "clean liver" experiment, was also greater than the accuracy of his methods warranted. His genius, however, was to have grasped at once the significance, implications and consequences of his discovery.

First, he understood that he had taken the first step toward the solution of a problem that dated back to the eighteenth century: What was the function of the so-called ductless glands (or blood vessel glands) such as the thyroid? Bernard solved this problem through a series of experiments intended to demonstrate the new concept of "internal secretion" (1855), a phrase that only a few years earlier would have been taken as a contradiction in terms, an impossibility as unthinkable as a square circle.

Second, and more important, Bernard understood that he had hit upon an argument capable of exploding a theory firmly established in the minds of contemporary chemists. Whatever misgivings one may have about illustrative comparisons, a comparison here is irresistible. When Galileo observed spots on the sun, he delivered a decisive blow to the old Aristotelian distinction between the sublunary world, supposedly susceptible to generation and corruption, and the supralunary world, supposedly eternal and incorruptible. He taught mankind to see analogous things in analogous ways. Similarly, when Claude Bernard discovered the glycogenic function of the liver, he delivered a decisive blow to the old distinction between the plant and animal kingdoms, according to which plants can and animals cannot synthesize simple organic compounds, in particular hydrocarbons. He taught the human eye to see life in a new way, without distinction between plant and animal.

In the fortieth lesson of the *Cours de philosophie positive*, Auguste Comte had written in 1838 that while there were hundreds of ways to live, there was probably only one way to die a natural death. In 1853, Claude Bernard proved that there was no division of labor among living things: plants were not essential as suppliers of the glucose without which animals cannot live. The two kingdoms do not form a hierarchy, and there is no teleological subordination of one to the other. This discovery paved the way for a general physiology, a science of the life functions, and this discipline immediately gained a place in the academy alongside comparative physiology. From Bernard's doctoral thesis to the last courses he gave as professor of general physiology at the Muséum (published in 1878 as *Leçons sur les phénomènes de la vie communs aux animaux et aux végétaux*), his work was all aimed at proving the validity of a single guiding principle, which might be called philosophical, or, to use a term less suspect to the scientific mind, metaphysiological. That idea can be summed up in a sentence first written in 1878: "There is but one way of life, one physiology, for all living things." ["Claude Bernard," *Dialogue*, pp. 560–62]

[103] In the eighteenth century, Immanuel Kant argued that the conditions under which physical science was possible were the transcendental conditions of knowledge in general. Later, in Part Two of the *Critique of Practical Reason*, entitled "The Critique of Teleological Judgment," he modified this view, acknowledging that organisms were totalities whose analytic decomposition and causal explanation were subordinate to an idea of finality, the governing principle of all biological research. According to Kant, there could be no "Newton of a blade of grass." In other words, the scientific status of biology in the encyclopedia of knowledge could never compare with that of physics. Before Claude Bernard, biologists were forced to choose between identifying biology with physics, in the manner of the materialists and mechanists, or radically distinguishing between the two, in the manner of the French naturalists and German nature philosophers. The Newton of the living organism was Claude Bernard, in the sense that it was he who realized that living things provide the key to deciphering their own structures and functions. Rejecting both mechanism and vitalism, Bernard was able to develop techniques of biological experimentation suited to the specific nature of the object of study. It is impossible not to be struck by the contrast, probably unwitting, between the following two passages. In *Leçons sur les phénomènes physiques de la vie* (Lessons of December 28 and 30, 1836), François Magendie wrote, "I see the lung as a bellows, the trachea as an air tube, and the glottis as a vibrating reed.... We have an optical apparatus for our eyes, a musical instrument for our voices, a living retort for our stomachs." Bernard, on the other hand, in his *Cahier de notes*, wrote, "The larynx is a larynx,

and the lens of the eye is the lens of the eye: in other words, the mechanical and physical conditions necessary for their existence are satisfied only within the living organism." Thus, while Bernard took from Lavoisier and Laplace by way of Magendie what he himself called the idea of "determinism," he was the sole inventor of the biological concept of the "internal environment," the concept that finally enabled physiology to become a deterministic science on a par with physics but without succumbing to fascination with the physical model. [*Etudes*, 148–49]

The Theoretical Foundations of the Method

[104] The unusual, and at the time paradoxical, nature of what Bernard had "inadvertently" discovered was what enabled him to conceptualize his early results in such a way as to determine the course of all his future research. Without the concept of the inner environment, it is impossible to understand Bernard's stubborn advocacy of a technique that he did not invent but to which he lent new impetus: the technique of vivisection, which he was obliged to defend against both emotional outrage and the protests of Romantic philosophy. "Ancient science was able to conceive only of the external environment, but in order to place biological science on an experimental footing one must also imagine an *internal environment*. I believe that I was the first to express this idea clearly and to stress its importance in understanding the need for experimentation on living things." Note that the *concept* of the internal environment is given here as the theoretical underpinning of the *technique* of physiological experimentation. In 1857, Bernard wrote, "The blood is made for the organs. That much is true. But it cannot be repeated too often that it is also made by the organs." What allowed Bernard to propose this radical revision of hematology was the concept of internal secretions, which he had formulated two years earlier. After all, there is a consid-

erable difference between the blood's relation to the lungs and its relation to the liver. In the lungs, the organism interacts with the inorganic world through the blood, whereas in the liver the organism interacts with itself. The point is important enough that it bears repeating: without the idea of internal secretions, there could be no idea of an internal environment, and without the idea of an internal environment, there could be no autonomous science of physiology. [*Etudes*, pp. 147–48]

[105] The concept of the internal environment thus depended on the prior formulation of the concept of internal secretions; it also depended on cell theory, whose essential contribution Bernard accepted even as he grew increasingly skeptical of the theory of the formative blasteme. Cell theory's crucial contribution was its insistence on the autonomy of the anatomical components of complex organisms and their functional subordination to the morphological whole. Bernard squarely embraced cell theory: "This cell theory is more than just a word," he wrote in his *Leçons sur les phénomènes de la vie communs aux animaux et aux végétaux*. By so doing, he was able to portray physiology as an experimental science with its own distinctive methods. In fact, cell theory made it possible to understand the relation between the part and the whole, the composite and the simple, in a way that differed sharply from the mathematical or mechanical model: the cell revealed a type of morphological structure quite different from that of earlier "artifacts" and "machines." It became possible to imagine ways of analyzing, dissecting and altering living things using mechanical, physical or chemical techniques to intervene in the economy of an organic whole without interfering with its essential organic nature. The fifth of the *Leçons de physiologie opératoire* contains a number of crucial passages on this new conception of the relation between the parts and the whole. First, Bernard explains that "all organs and tissues are nothing but a combina-

tion of anatomical elements, and the life of the organ is the sum of the vital phenomena inherent in each type of element." Second, he points out that the converse of this proposition is false: "In attempting to analyze life by studying the partial lives of the various kinds of anatomical elements, we must avoid an error that is all too easy to make, which is to assume that the nature, form and needs of the total life of the individual are the same as those of the anatomical elements." In other words, Bernard's general physiology grew out of a combination of the concept of the internal environment with the theory of the cell, which enabled him to develop a distinctive experimental method, one that was not Cartesian in style yet conceded nothing to vitalism or Romanticism. In this respect, Bernard was radically different from both Georges Cuvier, the author of the letter to Mertrud that served as preface to Cuvier's *Leçons d'anatomie comparée*, and Auguste Comte, the author of the fortieth lesson of the *Cours de philosophie positive* and a faithful disciple of Blainville's introduction to the *Cours de physiologie générale et comparée*. For all three of these authors – Cuvier, Comte and Blainville – comparative anatomy was a substitute for experimentation, which they held to be impossible because the analytic search for the simple phenomenon inevitably, or so they believed, distorts the essence of the organism, which functions holistically. Nature, by exhibiting (in Cuvier's words) "nearly all possible combinations of organs in all the classes of animals," allowed the scientist to draw "very plausible conclusions concerning the nature and use of each organ." By contrast, Bernard saw comparative anatomy as a prerequisite for developing a general physiology on the basis of experiments in comparative physiology. Comparative anatomy taught physiologists that nature laid the groundwork for physiology by producing a variety of structures for analysis. Paradoxically, it was the increasing individuation of organisms in the animal series that made the analytical study of functions possible. In the *Principes de médecine expérimentale*, Bernard wrote, "For analyzing life phenomena, is it better to study higher or lower animals? The question has been examined frequently. Some say that the lower animals are simpler. I do not think so, and, in any case, one animal is as complete as the next. I think, rather, that the higher animals are simpler because they are more fully differentiated." Similarly, in *Notes détachées* he observed that "an animal higher up the scale exhibits more highly differentiated vital phenomena, which in some ways are simpler in nature, whereas an animal lower down the organic scale exhibits phenomena that are more confused, less fully expressed, and more difficult to distinguish." In other words, the more complex the organism, the more distinct the physiological phenomenon. In physiology, distinct means differentiated, and the functionally distinct must be studied in the morphologically complex. In the elementary organism, everything is confused because everything is confounded. If the laws of Cartesian mechanics are best studied in simple machines, the laws of Bernardian physiology are best studied in complex organisms. [*Etudes*, pp. 149–51]

Life, Death and Creation

[106] All of Bernard's work bears traces of the struggle that went on in his mind between his profound but not unconditional admiration for Xavier Bichat and his sincere gratitude for the lessons he had learned from François Magendie. Yet Bernard found a way to reconcile the two men's conflicting philosophies of biology without compromising either. He did this by persistently exploiting his own fundamental experiments and the new concepts he had been obliged to formulate in order to interpret his results. The upshot was a "fundamental conception of life" incorporating two lapidary propositions: "life is creation" (1865) and "life is death" (1875).

Life is death. By this Bernard meant that a working organism is an organism engaged in the process of destroying itself, and that its functions involve physical and chemical phenomena that can be understood in terms of the laws of (nonliving) matter.

Was this a mechanist position? Absolutely not. Now that chemistry was a positive science, the various forms of energy had been unified by a law of conservation, and the explanation of electrical phenomena had necessitated the formulation of the new concept of a "field" – it was no longer possible to be a strict mechanist. More than that, Bernard found in his concept of the internal environment yet another reason not to be a mechanist. Mechanism implied a geometric representation of things: the mechanist physicians of the eighteenth century had represented the organism as a machine composed of interlocking parts. But Bernard did not think of organisms as machines, although he continued to use the phrase *machine vivante* (without in any way being bound by the metaphor). The internal environment welds the parts together in a whole immediately accessible to each one. The organism is not rooted, as we represent it, in metric space. Indeed, the existence of the internal environment assures the "higher" organism – so-called because it possesses an internal environment – of an "obvious independence," "a protective mechanism," an "elasticity."[72] Thus, the relation of the organism to the environment is not one of passive dependence.

What is more, it was because Bernard was not a mechanist – and knew that he was not seen as one – that he always insisted that science in general, and his physiology in particular, were deterministic, and further, that he was the first (as indeed he was) to introduce the term "determinism" into the language of scientists and philosophers. The macroscopic organism's relative independence of the environment was ensured by the determinate dependence of its microscopic elements on the internal environment. Bernard thus rejected any attempt to portray his doctrine as a kind of vitalism or as somehow implying that life is exempt from the laws of physics and chemistry.

Life is creation. If Bernard was not a mechanist, was he not a materialist insofar as he attempted to base the laws of living things on those of inert matter? The answer is no, because he insisted that "life is creation." What did he mean by this?

The phrase "life is death" acknowledged the power of physical and chemical laws over what is *organic* in living organisms. The phrase "life is creation" acknowledged the distinctiveness of the organism's *organization*. Vital creation, organizing synthesis – these terms referred to that aspect of life that Bernard also called "evolution," though not in the Darwinian sense, since it referred only to ontogenetic development. It was the one phenomenon of life with no nonorganic analogue: "It is unique, peculiar to living things. This evolutionary synthesis is what is truly vital in living things."[73] Bernard applied the term "organic creation" to both chemical synthesis, or the constitution of protoplasm, and morphological synthesis, or the reconstitution of substances destroyed by the functioning of the organism. Creation or evolution was the living expression of the organism's need to structure matter. ["Claude Bernard," *Dialogue*, pp. 566-68]

[107] In Bernard's most carefully written texts – the *Introduction*, the *Rapport* and *La Science expérimentale* – he distinguished between *laws*, which are general and applicable to all things, and *forms* or *processes*, which are specific to organisms. This specificity is sometimes termed "morphological," sometimes "evolutionary." In fact, in Bernard's lexicon, evolution refers to the regular development of an individual from inception to maturity. The mature form is the secret imperative of the evolution. In the *Introduction* he states that "specific, evolutionary, physiological conditions are the *quid proprium* of biological science," and the *Rapport* confirms

this view: "It is obvious that living things, by nature evolutionary and regenerative, differ radically from inorganic substances, and the vitalists are correct to say so."[74] The difference between biology and the other sciences is that biology takes account of the guiding principle of vital evolution, of the "idea that expresses the nature of the living being and the very essence of life."[75]

The notion of an organic guiding principle may well have been the guiding principle of Claude Bernard's philosophy of biology. That may be why it remained somewhat vague, masked by the very terms it used to express the idea of organization – vital idea, vital design, phenomenal order, directed order, arrangement, ordering, vital preordering, plan, blueprint, and formation, among others. Is it too audacious to suggest that with these concepts, equivalent in Bernard's mind, he intuitively sensed what we might nowadays call the antirandom character of life – antirandom in the sense not of indeterminate but of negative entropy? A note in the *Rapport* seems to support this interpretation:

> If special material conditions are necessary to create specific phenomena of nutrition or evolution, that does not mean that the law of order and succession that gives meaning to, or creates, relations among phenomena comes from matter itself. To argue the contrary would be to fall into the crude error of the materialists.

In any case, there can be no doubt that Bernard, in the *Introduction*, identified physical nature with disorder, and that he regarded the properties of life as improbable relative to those of matter: "Here as always, everything comes from the idea that creates and guides all things. All natural phenomena express themselves by physicochemical means, but those means of expression are distributed haphazardly like characters of the alphabet in a box, from which a force extracts them in order to express the most diverse thought or mechanisms."[76] Recall, too, that heredity, which was still an obscure concept and beyond man's reach in 1876, nevertheless seemed to Bernard an essential element of the laws of morphology, of ontogenetic evolution.[77] Am I stretching words, then, or distorting Bernard's meaning, if I suggest that, in his own way and in defiance of the reigning supremacy of physical concepts in biology, he was formulating a concept similar to what today's biologists, educated by cybernetics, call the genetic code? The word "code," after all, has multiple meanings, and when Bernard wrote that the vital force has legislative powers, his metaphor may have been a harbinger of things to come. But he glimpsed only a part of the future, for he does not seem to have guessed that even information (or, to use his term, legislation) requires a certain quantity of energy. Although he called his doctrine "physical vitalism,"[78] it is legitimate to ask whether, given his notion of physical force and his failure to grant the "vital idea" the status of a force, he really went beyond the metaphysical vitalism that he condemned in Bichat. [*Etudes*, pp. 158-60]

The Idea of Experimental Medicine

[108] Just as certain philosophers believe in an eternal philosophy, many physicians even today believe in an eternal and primordial medicine, that of Hippocrates. To some, then, it may seem deliberately provocative that I date the beginning of modern medicine from the moment when experimental medicine declared war on the Hippocratic tradition. To do so is not to disparage Hippocrates. In fact, Claude Bernard made free use of Comte's law of three stages of human development. He acknowledged that "the stage of experimental medicine depended on a prior evolution."[79] Yet, while history shows Hippocrates to have been the founder of observational medicine, concern for the future is forcing medicine not to renounce Hippocrates but to divest itself of

his method.[80] The Hippocratic method was to rely on nature; observational medicine was passive, contemplative and descriptive. Experimental medicine is aggressive science. "With the aid of the *active experimental sciences*, man becomes an inventor of phenomena, a foreman in the factory of creation, and there is no limit to the power that he may obtain over nature."[81] By contrast, an observational science "predicts, watches, avoids, but actively changes nothing."[82] In particular, "observational medicine examines, observes and explains illnesses but does not touch disease.... When [Hippocrates] abandoned pure expectation to administer remedies, it was always to encourage nature's own tendencies, to hasten disease through its regular phases."[83] Bernard applied the designation "Hippocratic" to any modern doctor who failed to make curing his patient his top priority, and who was concerned above all to define and classify diseases — who chose diagnosis and prognosis over treatment. These were the nosologists: Thomas Sydenham, François Boissier de Sauvages de la Croix, Philippe Pinel, even René-Théophile Hyacinthe Laënnec, and all the others who held that diseases were essences that manifested themselves more often than not in impure form. In addition, Bernard branded as mere naturalists all the physicians, including Rudolph Virchow, who, since the time of Giovanni Battista Morgagni and Bichat, had looked for etiological relations between changes in anatomical structures and detectable symptoms in the hope of making pathological anatomy the basis of a new science of disease. For Bernard, who did not believe in the existence of distinct disease *entities*, the ultimate goal of experimental medicine was to demolish all nosologies and do away with pathological anatomy.[84] Instead of disease, there are only organisms in normal or abnormal conditions, and disease is just a disturbance in the organism's physiological functions. Experimental medicine is the experimental physiology of the morbid. "Physiological laws manifest themselves in pathological phenomena."[85] "Whatever exists pathologically must present and explain itself physiologically."[86] Thus, it follows that "the experimental physician shall bring his influence to bear on a disease once he knows its exact *determinism*, that is, its proximate cause."[87] It was indeed time to say farewell to expectant medicine. Pierre Jean George Cabanis had earlier distinguished between the Ancients' art of observation and the Moderns' art of experimentation. Bernard saw the history of scientific medicine in similar terms: "Antiquity does not seem to have conceived of the idea of experimental science or, at any rate, to have believed in its possibility."[88] But instead of linking medicine and observation to the Ancients, as Cabanis did, Bernard urged medicine to set out on the path of experimentation toward a future of domination and power. "To dominate living nature scientifically, to conquer it for the benefit of man: that is the fundamental idea of the experimental physician."[89] The idea of experimental medicine, the domination of living nature, was the opposite of the Hippocratic idea as expressed in the title of Toussaint Guindant's 1768 treatise, "La Nature opprimée par la médecine moderne."[90] [*Etudes*, p. 131]

[109] Bernard took from François Magendie not only the name of the new discipline he was about to create but also a certain idea of what its content should be: namely, that the subject matter and method of physiology should be the same as those of pathology. In one of his *Leçons sur les phénomènes physiques de la vie* (December 28, 1836), Magendie stated that "pathology is also physiology. For me, pathological phenomena are nothing but modified physiological phenomena." As a theoretical proposition, this was not a new idea: in the early part of the nineteenth century, even a modestly cultivated physician would have associated the idea that pathology is a subset of physiology with the still prestigious name of Albrecht von Haller. In the preface to his 1755

French translation of von Haller's *De partibus corporis humani sentientibus et irritabilibus* (1752), M. Tissot wrote, "If pathology's dependence on physiology were better known, there would be no need to belabor the influence that the new discovery ought to have on the art of healing. But unfortunately we lack a work entitled *The Application of Theory to Practice*, so I have ventured to express a few thoughts concerning the practical benefits of irritability." This statement is followed by a series of observations on the administration of opium, tonics, purgatives and so forth. To be sure, this was a mere "system," whereas Magendie claimed to be able to read, and to teach others to read, the natural identity of physiology and pathology in the facts themselves, independent of any interpretation. Yet it took a medical system, indeed the last of the medical systems according to Bernard,[91] to reveal the idea of experimental medicine, that is, the idea that the methods of the laboratory and the methods of the clinic are one and the same. Built on the ruins of the great nosologies, this idea turned medicine from a speculative system into a progressive science. The system Bernard had in mind, that which paved the way for a medicine without systems, was François-Joseph Victor Broussais's. [*Etudes*, p. 135]

[110] In recognizing that Broussais had demolished the idea of pathology as a science of disease distinct from the science of physiological phenomena, Bernard did not relinquish his own claims to originality, which lay in his having been the first to propose basing a scientific medicine on an experimental physiology. But what did he make of Magendie? In 1854, when he filled in for Magendie at the Collège de France, his first words were that "the scientific medicine I am supposed to teach does not exist." In 1865, he noted that "experimental or scientific medicine is now coming together on the basis of physiology...this development is now certain."[92] In the *Principes*, he summed up the twenty

years that had passed since his first course.⁹³ He was sure that progress had been made: "I am the founder of experimental medicine." Magendie had blazed a trail, according to Bernard, but he had neither set a destination nor developed a method. Nor could he have, because he lacked the means to build a bridge between the laboratory and the clinic, to prove that effective treatments could be deduced from the results of physiology. What sustained Bernard in his path-breaking enterprise was the awareness of just such a possibility, of just such a reality: "I think that there are now enough facts to prove clearly that physiology is the basis of medicine, in the sense that a certain number of pathological phenomena can now be traced back to physiological phenomena, and it can be shown, moreover, that the same laws govern both."⁹⁴ Stated more clearly, Bernard's claim to have *founded* a discipline, even though he credits others with having the idea first and obtaining the earliest results, rests on the physiopathology of diabetes, that is, ultimately, on the discovery of the glycogenic function of the liver. [...] For Bernard, the experimental explanation of the mechanism of diabetes demonstrated the validity of the principles set forth in the *Introduction* of 1865: the principle of the identity of the laws of health and disease; the principle of the determinism of biological phenomena; and the principle of the specificity of biological functions, that is, the distinction between the internal and external environments. To found experimental medicine was to demonstrate the consistency and compatibility of these principles. That done, Bernard went on to rescue the new discipline from its detractors, the old-fashioned systematists inextricably wedded either to ontology or to vitalism, by showing them that these same principles could explain the very phenomena on which they based their objections. Magendie's style was very different from Bernard's: Magendie had asserted truths, refuted errors, pronounced judgments – for him, life was a mechanical phenomenon and vitalism an aberration. The discovery of internal secretions, the formulation of the concept of the internal environment, the demonstration of certain regulatory mechanisms and stabilized parameters in the composition of that environment — these things enabled Bernard to be a determinist without being a mechanist, and to understand vitalism as an error rather than a folly. In other words, he found a way to change perspectives in the discussion of physiological theory. When Bernard proclaimed, with a self-confidence that could easily be mistaken for smugness, that there would be no more revolutions in medicine, it was because he lacked the means to describe philosophically what he was conscious of having achieved. He did not know what to call his idea of experimental medicine; he did not know how to say that he had brought about a Copernican revolution. Once it could be shown that the internal environment afforded the organism a certain autonomy with respect to changing conditions in the external environment, it also became possible not only to refute the misconceptions of vitalism but to explain how they had come about in the first place. And once it could be shown that the processes responsible for the symptoms of a disease such as diabetes exist in the normal as well as the pathological state, it became legitimate to claim that the proper approach to understanding disease was to understand health. At that moment, the culture's attitude toward disease changed. When people believed that diseases were essences with a nature all their own, their only thought was, as Bernard said, "to be wary of them," that is, to strike a compromise with them. But when experimental medicine claimed the ability to determine the conditions of health and defined disease as a deviation from those conditions, attitudes toward disease changed: mankind now rejected illness and sought to stamp it out. Thus, experimental medicine was but one form of the demiurgic dream that afflicted all the industrialized societies of the mid-nineteenth century, when science, through its applications, became a social force. That is why Bernard was immediately recognized by his contemporaries as one of those who symbolized the age: "He was not [merely] a great physiologist, he was Physiology," Jean-Baptiste Dumas told Victor Duruy on the day of Bernard's funeral, thereby transforming the man into an institution.

It may even be that Bernard, in all modesty, identified himself with physiology. When he staked his claim as the founder of experimental medicine, he simply demonstrated his awareness that it was his own research which had enabled him to refute the various objections raised against the new discipline.

Bernard knew that he had invented neither the term nor the project of experimental medicine but, by reinventing the content, he had made the idea his own: "Modern scientific medicine is therefore based on knowledge of the life of the elements in an *internal environment*. Thus, it relies on a different conception of the human body. These ideas are mine, and this viewpoint is essentially that of experimental medicine."[95] However, no doubt remembering that he had written in the *Introduction* that "art is *I*, science is *we*," he added: "These new ideas and this new point of view did not spring full-blown from my imagination. They came to me, as I hope to show, purely because of the evolution of science. My ideas are therefore far more solid than if they had been my own personal views and nothing more."[...]

At several points in the foregoing account, I have written that "Claude Bernard did not know how to say" this or that. Someone might object that I am substituting for what he actually said what I think he should have said. I am perfectly willing to concede that I do not share the admiration of some commentators for Bernard as a writer; perhaps my critics will concede that, in attempting to situate Bernard's *Introduction* historically and conceptualize it epistemologically, I have given him precisely the credit he deserves, since everything I say is borrowed from him. As Victor Cousin, a philosopher I do not customarily quote, once put it, "Fame is never wrong. The only problem is finding out what constitutes a claim upon it." [*Etudes*, pp. 138-41]

The Limits of Bernardian Theory

[111] There can be no doubt that the accumulation of knowledge in such basic disciplines as pathological anatomy, histology and histopathology, physiology and organic chemistry necessitated painful revision of many of the attitudes toward disease that the eighteenth century bequeathed to the nineteenth. Of all the disciplines, it was physiology that most directly challenged the naturalistic paradigm, which rightly or wrongly claimed the authority of a Hippocratic tradition revamped to suit contemporary tastes. While insisting on the fundamental identity of the normal and the pathological, physiology promised to deduce modes of treatment from knowledge of their practical effects. Being an experimental science, like physics and chemistry, whose results and techniques it used, physiology was not only not antagonistic to the idea of a scientifically based medicine but actually called for the rationalization of medical practice. The term "rationalism" was in fact widely used to characterize the medicine of the future; one of the first to use the term in this way was Charles Schützenberger in Strasbourg, who in 1844 advocated the application to medicine of what he called "experimental rationalism," which as late as 1879 he still preferred to Bernard's "experimental medicine."[96] In 1846, the German Jakob Henle published a *Handbuch der rationellen Pathologie*. At the time, Claude Bernard was still a young doctor, and it was not until the 1860s that he took up the term "rationalism," for example in his *Principes de médecine expérimentale* (first published in 1947) and in his notes

for a proposed work on problems raised by the practice of medicine (preserved at the Collège de France). "Scientific empiricism is the opposite of rationalism and radically different from science. Science is based on the rationalism of the facts.... Medical science is the science in which we rationally and experimentally explain diseases in order to predict or alter their progress."[97] Another formulation is even clearer: "Medicine is the art of healing, but it must become the science of healing. The art of healing is empiricism. The science of healing is rationalism."[98] In a work devoted to epistemology, the author will perhaps be allowed to express a preference for the term "rationality" over "rationalism," which is out of place beyond the history of philosophy. In any case, Emile Littré and Charles Robin's *Dictionnaire de médecine* contains an article on "rationalism" that is really a definition of "rational," where it is stated that a rational treatment of an illness is one based on principles of physiology and anatomy, and not on mere empiricism. This definition of a rational therapy is repeated verbatim in the 1878 *Dictionnaire de la langue française* under "rationality."[...]

There is no exemplary figure, no classical period, in the history of rationality. The nineteenth century taught the twentieth that every problem requires an appropriate method for its solution. In medicine as in other fields, rationality reveals itself after the fact; it is not given in advance but reflected in the mirror of success. Bernard sometimes found it difficult to accept that not every rational method had to resemble his, which he considered paradigmatic. His criticisms of Rudolph Virchow and cellular pathology were harsh. Although he approved of Louis Pasteur's refutation of the theory of spontaneous generation, he never imagined how fruitful germ theory would prove in treating disease. An obsession with the dogma that all diseases are nervous in origin proved to be an obstacle to rational understanding of infection and contagion. While it is correct, as Bernard claimed, that the nerves exert an influence on infectious disease, it would have been better if he had never written that "a nervous paralysis can produce a septic disease."[99] Here the physiopathological type of rationality leads to an explanation of symptoms, but it was Pasteur and Heinrich Hermann Robert Koch who developed a different type of rationality capable of answering questions of etiology. Extreme physiologism had its limits: for proof one need only consider the rear-guard action waged by Elie de Cyon against the triumphant Pasteurians in his study of Etienne-Jules Marey, the author of a little-known work entitled *Essai de théorie physiologique du choléra* (1865).[100] Marey was perfectly well aware that "the search for an absolutely effective medication or certain prophylaxis" would require the identification of what he still called a microscopic parasite.[101] The adverb "absolutely" and the adjective "certain" reflect the Bernardian conception of rationality: the veneration of determinism led to outright rejection and scorn for attempts to introduce concepts of probability and statistics into medicine. But at least Marey was fully aware that knowledge of the role of the vasomotor nervous system in circulation and calorification was not enough to suggest an anticholera therapy more "rational" than the many medications already tested empirically on the intestinal and pulmonary forms of the disease.

The publication of Marey's article may be taken as a recognition of the limits of Bernardian rationality. Meanwhile, the man who boasted of its universal validity could write, "I do not believe that medicine can change the laws of human mortality or even of the mortality of a nation,"[102] and elsewhere, "Medicine must act on individuals. It is not destined to act on collectivities or peoples."[103] [*Etudes*, p. 393-96]

Part Five

Problems

CHAPTER THIRTEEN

Knowledge and the Living

Science and Life

The Vitalist Imperative

[112] Vitalism as defined by the eighteenth-century Montpellier physician Paul-Joseph Barthez explicitly claimed to be a continuation of the Hippocratic tradition. This Hippocratic ancestry was probably more important than the doctrine's other forebear, Aristotelianism, for while vitalism borrowed much of its terminology from Aristotle, its spirit was always Hippocratic. Barthez put it this way in his *Nouveaux éléments de la science de l'homme* (1778):

> By man's "vital principle" I mean the cause of all the phenomena of life in the human body. The name given to that cause is of relatively little importance and may be chosen at will. I prefer "vital principle" because this suggests a less circumscribed notion than the term *impetum faciens* (το ενορμων) that Hippocrates used or than any of the other terms that have been used to denote the cause of the life functions.

Vitalism was in one respect a biology for physicians skeptical of the healing powers of medication. According to the Hippocratic theory of *natura medicatrix*, the defensive reaction of the organism is more important than diagnosing the cause of the disease. By the same token, prognosis, though dependent on diagnosis, is the dominant art. It is as important to anticipate the course of a disease as to determine its cause. Because nature is the first physician, therapy is as much a matter of prudence as of boldness. Vitalism and naturalism were thus inextricably associated. Medical vitalism reflected an almost instinctive wariness of the healing art's powers over life. There is an analogy to be drawn here: the contrast between nature and art is reminiscent of Aristotle's contrast between natural movement and violent movement. Vitalism was an expression of the confidence among the living in *life*, of the mind's capacity, as living consciousness of life, to identify with the living as like with like.

These remarks suggest the following observation: vitalism reflected an enduring life-imperative in the consciousness of living human beings. This was one reason for the vagueness and nebulousness that mechanist biologists and rationalist philosophers saw as defects of vitalist doctrine. If vitalism was above all an imperative, it was only natural that it should have some difficulty expressing itself in determinate formulations. [*Connaissance*, p. 86]

[113] Indeed, Emanuel Radl recognized that vitalism was an imperative rather than a method and more of an ethical system, perhaps, than a theory.[1] Man, he argued, can look at nature in two ways. He *feels* that he is a child of nature and has a sense of belonging to something larger than himself; he sees himself in nature and nature in himself. But he also *stands before* nature as before an undefinable alien object. A scientist who feels filial, sympathetic sentiments toward nature will not regard natural phenomena as strange and alien; rather, he will find in them life, soul and meaning. Such a man is basically a vitalist. Plato, Aristotle, Galen, all medieval and most Renaissance scholars were in this sense vitalists. They regarded the universe as an organism, that

is, a harmonious system obedient to certain laws and dedicated to certain ends. They conceived of themselves as an organized part of this universe, a sort of cell in the universal organism, all of whose cells were unified by an internal sympathy. It therefore seemed natural to them that the fate of the partial organ should be bound up with the movements of the heavens.

Such an interpretation may well be fodder for the psychoanalysis of knowledge. That it may have some merit is suggested by its convergence with Walther Riese's comments on Constantin von Monakow's biological theories: "In von Monakow's neurobiology, man is a child of nature who never leaves its mother's breast."[2] There can be no doubt that, for the vitalists, the fundamental biological phenomenon was generation, which conjured up certain images and posed certain problems that, to one degree or another, influenced the representation of other phenomena. A vitalist, I would venture to suggest, is a person who is more likely to ponder the problems of life by contemplating an egg than by turning a winch or operating the bellows of a forge.

Vitalists were confident of the spontaneity of life and reluctant — in some cases horrified — to think of it as springing from a nature conceived of as a series of mechanical processes and, thus, paradoxically reduced to a congeries of devices similar to those which human beings had created in their quest to overcome the obstacles that nature had placed in their way. Typical of these attitudes was a man like Jean Baptiste van Helmont. [...]

Van Helmont denied Descartes's contention that the forces of nature are unified. Every being, he argued, has both its own individual force and the force of its species. Nature is an endless hierarchy of forces and forms. This hierarchy comprises seeds, leavens, principles and ideas. The living body is organized as a hierarchy of *archēs*. The term *archē*, first principle, borrowed from Paracelsus, described an organizing, commanding power, something rather more akin to the general of an army than to a workman. It marks a return to the Aristotelian conception of the body as subordinate to the soul in the same sense as the soldier is subordinate to his captain or the slave to his master.[3] Vitalism attacked the technological version of mechanism at least as much as, and perhaps even more than, it did the theoretical version. [*Connaissance*, pp. 88–89]

[114] It may seem absurd to argue that vitalism was in fact a fertile doctrine, particularly given the fact that it always portrayed itself as a return to ancient beliefs – a tendency quite evident in the naive penchant of many vitalists to borrow Greek terms for the rather obscure entities they felt obliged to invoke. The vitalism of the Renaissance was in one sense a return to Plato intended to counter the overly logicized medieval version of Aristotle. But the vitalism of van Helmont, Georg Ernst Stahl and Paul-Joseph Barthez has been called a return beyond Descartes to the Aristotle of *De anima*. For Hans Driesch, the case is patent. But how is this return to the Ancients to be interpreted? Was it a revival of older and consequently timeworn concepts, or was it a case of nostalgia for ontologically prior intuitions, for a more direct relation between intention and object? Archaeology stems as much from a nostalgia for original sources as from a love of ancient things. We are more apt to grasp the biological and human significance of a sharpened flint or adze than of an electric timer or a camera. In the realm of theory, one must be sure of a theory's background and development to interpret reversion as retreat or rejection as reaction or betrayal. Wasn't Aristotle's vitalism already a reaction against the mechanism of Democritus, and wasn't Plato's finalism in the *Phaedo* a reaction against the mechanism of Anaxagoras? In any case, there can be no doubt that vitalists were after a certain pretechnological, prelogical naiveté of vision, a vision of life as it was before man created tools and language to extend his

reach. That is what Théophile de Bordeu, the first great theorist of the Montpellier School, meant when he called van Helmont "one of those enthusiasts that every century needs in order to astound the scholastics."[4] [*Connaissance*, pp. 91–92]

The Technological Model
[115] The word "mechanism" comes from the Greek μηχανή, or device, which combines the two senses of ruse (or stratagem) and machine. Perhaps the two meanings are actually one. Is not man's invention and use of machines, his technological activity in general, what Hegel calls the "ruse of reason" in Section 209 of his *Logic*? This ruse consists in accomplishing one's own ends by means of intermediate objects acting upon one another in conformity with their own natures. The essence of a machine is to be a mediation or, as mechanics say, a link. A mechanism creates nothing, and therein lies its inertia (*iners*), yet it is a ruse whose construction necessarily involves art. As a scientific method and philosophy, mechanism is therefore an implicit postulate in any use of machines. The success of this human ruse depends on the lack of any similar ruse in Nature. Nature can be conquered by art only if she herself is not art: only a man named Ulysses (No-Man) is capable of devising a scheme to get the wooden horse inside the gates of Troy, and he succeeds only because his enemies are forces of nature rather than clever engineers. The ruses by which animals avoid traps are often adduced as objections to the Cartesian theory of the animal-machine. In the foreword to the *New Essays*, Leibniz offers the ease with which animals are trapped as evidence for Descartes's contention that they are capable only of responding to immediate sensations (what we would today call "conditioned reflexes"). Conversely, Descartes's hypothetical description in the *Meditations* of a deceptive God or evil genius effectively transforms man into an animal surrounded by traps. If

God uses human ruses against humankind, man descends from the status of living creature to that of mere inert object. Is the theory of the living machine just such a human ruse, which, if taken literally, would prove that there is no such thing as life? But why then, if animals are mere machines, if nature is merely one vast machine, does the domination of animals and nature cost human beings so much effort? [*Connaissance*, p. 87]

[116] Mechanist philosophers and biologists took machines as a given, or, if they studied the problem of machine-building at all, solved it by invoking human calculation. They relied upon the engineer or, ultimately, as it seemed to them, the scientist. Misled by the ambiguity of the term "mechanical," they looked upon machines as nothing more than reified theorems, theorems made concrete by the relatively trivial operation of construction, which, they believed, involved nothing more than the application of knowledge in full consciousness of its limits and full certainty of its effects. In my view, however, the biological problem of the organism-machine cannot be treated separately from the technological problem whose solution it assumes, namely, the problem of the relation between technology and science. The usual solution is to say that knowledge is prior to its applications both logically and chronologically, but I shall try to show that the construction of machines involving authentically biological notions cannot be understood without revising this view of the relation between science and technology. [...]

To a scrupulous observer, living creatures other than vertebrates rarely exhibit structures likely to suggest the idea of a mechanism (in the technical sense). To be sure, Julien Pacotte notes that the arrangement of the parts of the eye and the movement of the eyeball correspond to what mathematicians would call a mechanism.[5] Perhaps a few definitions are in order. A machine is a man-made object that depends, for its essential function(s), on one or more mechanisms. A mechanism is a configuration of moving solids whose configuration is maintained throughout its movement; or, to put it another way, a mechanism is an assembly of parts whose relation to one another changes over time but is periodically restored to an initial configuration. The assembly consists in a system of linkages with fixed degrees of freedom: for example, a pendulum or a cam-driven valve has one degree of freedom; a worm gear shaft has two. The material embodiment of these degrees of freedom consists in guides, that is, structures limiting the movement of solids in contact. The movement of any machine is thus a function of its structure, and the mechanism is a function of its configuration. The fundamental principles of a general theory of mechanisms (as defined here) can be found in any standard work, for example Franz Reuleaux's *Kinematics*[6] (which was translated from German into French in 1877). [...]

The point of this brief review of the fundamentals of kinematics is that it allows me to point up the paradoxical significance of the following problem: Why did scientists use machines and mechanisms as models for understanding organic structures and functions? One problem with any mechanical model is its source of energy. A machine, as defined above, is not self-contained: it must take energy from somewhere and transform it. We always think of moving machines as connected with some source of energy.[7]

For a long time, the energy that set kinematic machines in motion came from the muscular effort of humans or animals. In that stage, it was obviously tautological to explain the movement of a living thing by comparing it to the movement of a machine dependent on muscular effort for its source of energy. Historically, therefore, as has frequently been shown, there could be no mechanical explanation of life functions until men had constructed automata: the very word suggests both the miraculous quality of the object and its appearance of being a self-contained mechanism whose energy does not come, immediately at any rate, from the muscular effort of a human or animal. [*Connaissance*, pp. 102–104]

[117] Aristotle, I think, took a customary way of looking at animal organisms, a sort of cultural *a priori*, and raised it to the level of a concept of life in general. The vocabulary of animal anatomy is full of terms for organs, parts and regions of organisms based on technological metaphors or analogies.[8] The development of the anatomical vocabulary in Greek, Hebrew, Latin and Arabic shows that the perception of organic forms was shaped in part by technological norms.[9] This explains why physiology was traditionally regarded as subordinate to anatomy. For followers of Galen, physiology was the science of the use of the parts, *de usu partium*. From William Harvey to Albrecht von Haller and beyond, moreover, the science of organic functions was called *anatomia animata*. Claude Bernard was a forceful critic of this way of looking at things, though often with more rhetorical energy than practical consequences. As long as technology served as the source of models for explaining organic functions, the parts of the organism were likened to tools or machine parts.[10] The parts were rationally conceived as means to the organism's end, and the organism itself was conceived of as a static structure, the sum of its parts.

The standard histories may well overemphasize the contrast between Aristotelianism and Cartesianism, at least as far as their theories of life are concerned. To be sure, there is an irreducible difference between explaining animal movement as a consequence of desire and giving a mechanist explanation of desire itself. The principle of inertia and the conservation of momentum led to an irreversible revolution in natural science: with the theory of stored energy and deferred utilization, Descartes was able to refute the Aristotelian conception of the relation between nature and art. All that notwithstanding, it remains true that the use of mechanical models to represent living organisms implied that those organisms were conceived as necessary and invariant structures of their component parts. The implicit idea of order was that of the workshop. In part five of the *Discourse on the Method* Descartes discusses a work that he never published, *Le Monde* ("The World," though it was actually about man): "I showed there what kind of workshop the nerves and muscles of the human body must constitute in order that the animal spirits have the strength to move the limbs." Later, in discussing the behavior of animals, he says, "It is nature that acts within them, according to the disposition of their organs." Workshop, disposition: these were technological concepts before they became anatomical ones. From Andreas Vesalius, Descartes borrowed a concept that was actually in fairly wide use in the sixteenth and seventeenth centuries, that of the *fabrica corporis humani*. In a letter to Marin Mersenne, a reference to Vesalius followed this statement of principle: "The number and the orderly arrangement of the nerves, veins, bones and other parts of an animal do not show that nature is insufficient to form them, provided you suppose that in everything nature acts exactly in accordance with the laws of mechanics, and that these laws have been imposed on it by God."[11] This invocation of God the mechanic, apparently intended only to rule out any vital teleology, fully merits Raymond Ruyer's acerbic remark that the more people thought of organisms as automata, the more they thought of God as an Italian engineer. [...]

In short, both Aristotle and Descartes based the distinction between the organism and its parts on technologically conditioned perceptions of macroscopic animal structures. The technological model reduced physiology to a matter of deduction from anatomy: an organ's function could be deduced from the way it was put together. Although the parts were seen, in dynamic terms, as subordinate to the whole, just as the parts of a machine were subordinate to the whole machine, that functional subordination led to a view of the static structure of the machine as merely the sum of its parts. [*Etudes*, pp. 323–25]

The Social Model

[118] The foregoing conception was not seriously challenged until the first half of the nineteenth century, when two things happened. First, two basic disciplines, embryology and physiology, which had been struggling to define their own distinctive methods and concepts, achieved the status of experimental sciences. Second, there was a change in the scale of the structures studied by morphologists; or, to put it another way, cell theory was introduced into general anatomy.

Leaving aside the regeneration and reproduction of Abraham Trembley's famous plant-animals and Charles Bonnet's observation of parthenogenesis in plant lice, no biological phenomenon was more difficult for eighteenth-century theorists to interpret in terms of technological models than that of morphological development, or the growth from seed to adult form. Historians of biology frequently associate the epigenetic view of development with mechanist biology; in so doing, they neglect the close and all but obligatory association of mechanism with preformationism. Since machines do not assemble themselves, and since there are no machines for creating (in the absolute sense) other machines, the living machine must in one way or another be associated with what eighteenth-century thinkers liked to call a *machiniste*, an inventor or builder of machines. If no such builder was perceptible in the present, then there must have been one at the inception: the theory of a seed within a seed and so on, ad infinitum, was thus a logical response to the problem that gave rise to the theory of preformation. Development then became a simple matter of increasing size, and biology became a kind of geometry, as Henri Gouhier once remarked about the concept of containment in Nicolas de Malebranche.

When Caspar Friedrich Wolff showed (in 1759 and 1768) that the development of an organism involved the emergence of a series of nonpreformed structures, however, it became necessary to restore responsibility for the organism's organization to the organism itself. That organism was not random or idiosyncratic, and anomalies were understood as failures to develop or to progress beyond a normally intermediate stage. Hence there must be some formative tendency, what Wolff called a *nisus formativus* and Johan Friedrich Blumenbach called a *Bildungstrieb*. In other words, it was necessary to assume an immanent plan of organogenesis.

These facts underlie Kant's theory of organic finality and totality as set forth in the *Critique of Judgment*. A machine, Kant says, is a whole whose parts exist for one another but not by one another. No part is made from any other; in fact, nothing is made of things of the same type as itself. No machine possesses its own formative energy.

A little more than a hundred years ago, Claude Bernard developed an identical theory in his *Introduction à l'étude de la médecine expérimentale*: "What characterizes the living machine is not the nature of its physicochemical properties, complex though they may be, but the creation of that machine, which develops before our eyes under conditions peculiar to itself and in accordance with a definite idea, which expresses the nature of the living thing and the essence of life itself."[12] Like Kant, Bernard gave the name "idea" to the morphological *a priori*, as it were, that determines the formation and shape of each part in relation to all the rest through a sort of reciprocal causation. And again like Kant, Bernard taught that natural organization cannot be thought of as being in any way akin to human agency. Stranger still, after ruling out, on explicit grounds, any possibility of a technological model of organic unity, Kant hastened to suggest organic unity as a possible model for social organization.[13] Bernard used the converse of the same analogy when he compared the unity of the multicellular organism to that of a human society. [*Etudes*, pp. 325-27]

[119] Claude Bernard *accepted* cell theory, as he had to in order to make experimentation in physiology possible. He *elaborated* the concept of the internal environment, and that, too, was a necessary condition for experimental physiology. The physiology of regulation (or homeostasis, as it has been called since Walter Bradford Cannon), together with cytologic morphology, enabled Bernard to treat the organism as a whole and to develop an analytic science of organic functions without brushing aside the fact that a living thing is, in the true sense of the word, a *synthesis*. Bernard's most important remarks on the subject that concerns us here can be found in his *Leçons sur les phénomènes de la vie communs aux animaux et aux végétaux*, based on lectures he gave at the Muséum in the final years of his life. The structure of the organism reflects the exigencies of life on a more basic level, that of the cell. The cell itself is an organism, either a distinct individual or a constituent of a larger "society" of cells forming an animal or plant. The term "society," which Rudolph Ludwig Karl Virchow and Ernst Heinrich Haeckel also seized upon at around the same time as Bernard, suggested a model for the organic functions very different from the technological model – namely, an economic and political one. Complex organisms were now thought of as totalities comprising virtually autonomous subordinate elements. "Like society, the organism is constructed in such a way that the conditions of elementary or individual life are respected."[14] Division of labor was the law for organisms as well as for societies. Conceived in terms of a technological model,

an organism was a set of strictly related basic mechanisms; conceived in terms of an economic and political model, though, an organism was a set of structures that grew increasingly complex and diverse as they assumed specialized responsibility for originally undifferentiated functions. Between the level of the elementary cell and that of man, Bernard explained, one finds every degree of complexity as organ combines with organ. The most highly developed animals possess multiple systems: circulatory, respiratory, nervous and so on.

Physiology was thus the key to organic totalization, the key that anatomy had failed to provide. The organs and systems of a highly differentiated organism exist not for themselves or for other organs and systems but for cells, the countless anatomical radicals, for which they create an internal environment whose composition is maintained in a steady state by a kind of feedback mechanism. By joining in association and instituting a kind of society, the basic elements obtain the collective means to live their separate lives: "If one could at every moment create an environment identical to that which the actions of nearby parts constantly create for a given elementary organism, that organism would live in freedom exactly as it lives in society."[15] The part depends on a whole that exists solely in order to maintain it. By referring all functions to the cell level, general physiology provided an explanation for the fact that the structure of the whole organism is subordinate to the functions of each part. Made *of* cells, the organism is also made *for* cells, for parts that are themselves less complicated wholes.

The use of an economic and political model enabled nineteenth-century biologists to understand what the use of a technological model had prevented their predecessors from grasping. The relation of the parts to the whole is one of *integration* (a concept that later met with success in neurophysiology), with the survival of the parts being the ultimate end: the parts were seen no longer as instruments or pieces but as *individuals*. At a time when what would later become cell theory was still at the stage of philosophical speculation and preliminary microscopic exploration, the term "monad" was often used for the atomic component of an organism; it was only later that "monad" lost out to "cell." Auguste Comte, in fact, rejected what he called the "theory of monads" and we now call cell theory.[16] The indirect but real influence of Leibniz on the early Romantic philosophers and biologists who dreamed up cell theory allows us to say of the cell what Leibniz said of the monad, namely, that it is a *pars totalis*. It is not an instrument or a tool but an individual, a *subject* in relation to its functions. Bernard frequently uses the term "harmony" to convey what he means by "organic totality." It is not too difficult to detect therein a faint echo of Leibniz's philosophy. And so, with the recognition of the cell as the basic morphological element of all organized substances, the meaning of the concept of organization changed: the whole was no longer a structure of interrelated organs but a totalization of individuals.[17] Simultaneously, the development of set theory changed the traditional mathematical meaning of the term "part," just as the development of cell theory changed its traditional anatomical meaning. [*Etudes*, pp. 329-31]

The Organism Is Its Own Model

[120] Did the technique of *in vitro* culture of explanted cells, which was perfected by Alexis Carrel in 1910 but invented by J.M.J. Jolly in 1903, offer experimental proof that the structure of the organism is an analogue of liberal society? Claude Bernard, who died thirty years earlier, had indeed suggested such an analogy, using the society of his own time as a model. The organism ensured that the conditions necessary to maintain the life of indi-

vidual cells were satisfied; Bernard had hypothesized that those conditions could also be satisfied when cells were taken out of their association with other cells, provided that an appropriate artificial environment was created. But what did it actually mean for the cell to live in freedom, that is, liberated from the inhibitions and stimulations stemming from its integration into the organism? In order for life in freedom to replicate life in society exactly, the cell would have to be provided with an environment that aged as it did. But then the life of the cell would proceed in parallel with changes in the artificial environment; it would not be independent. Furthermore, living in freedom rendered a cell unfit to return to society: the liberated part irrevocably lost its character of being part of a whole. Etienne Wolff remarks:

> No attempt to create an association of previously dissociated cells has been able to reconstitute structural unity. Analysis has never been succeeded by synthesis. By an illogical abuse of language, one often applies the term "tissue culture" to anarchic proliferations of cells that do not reflect either the structure or cohesion of the tissue from which they are taken.[18]

In other words, an organic element can be called an element only in its undissociated state. The situation recalls Hegel's observation in his *Logic* that it is the whole which creates the relation among its parts, so that without the whole there are no parts.

Experimental embryology and cytology thus corrected the concept of organic structure. Bernard had allowed himself to be unduly influenced by a social model, which all in all amounted to little more than a metaphor. In reaction against the use of mechanical models in physiology, Bernard wrote: "The larynx is the larynx, and the lens of the eye is the lens of the eye: in other words, the mechanical and physical conditions necessary for their existence are satisfied only within the living organism."[19] But what *can* be said about the use of mechanical models in biology can also be said about the use of social models. The concept of the organism as a regulative totality controlling developments and functions has remained a permanent feature of biological thought since the time when Bernard was among the first to demonstrate its experimental efficacy. Nevertheless, its fate is no longer bound up with that of the social model from which it originally drew support; an organism is not a society, although, like a society, it exhibits an organizational structure. In the most general sense, organization is the solution to the problem of converting competition into compatibility. For an organism, organization is a fact; for a society, organization is a goal. Just as Bernard said that "the larynx is the larynx," we can say that the model of an organism is the organism itself. [*Etudes*, pp. 331–33]

The Concept of Life

Aristotelian Logic

[121] It is surprising at first sight that there should be any question about the relation between concepts and life, for the theory of the concept and the theory of life have the same age and the same author. What is more, that author ascribed both to the same source. I am speaking, of course, of Aristotle, the logician of the concept and the systematic philosopher of living things. It was Aristotle the naturalist who based his system for classifying animals on structure and mode of reproduction, and it was the same Aristotle who used that system as a model for his logic. If reproduction played such a prominent role in Aristotle's classification, it was because perpetuation of the structural type, and therefore of behavior in the ethological sense, was the clearest indication of nature's purpose. For Aristotle, soul was not only the nature but also the form of the living thing. Soul was at once life's reality (*ousia*) and definition (*logos*). Thus, the concept of the living thing was, in the end, the living thing itself. The resemblance between the logical principle of noncontradiction and the biological law of specific reproduction may have been more than coincidental. Just as creatures give birth not to arbitrary offspring but to children of their own species, so in logic it is not possible

to assert an arbitrary predicate of a given object. The fixity of repetition of being constrains thought to identity of assertion. The natural hierarchy of cosmic forms requires the hierarchy of definitions in the realm of logic. The conclusion of a syllogism is necessary by virtue of the hierarchy according to which a species dominated by its genus becomes a dominating genus in relation to an inferior species. Knowledge is therefore the world made into thought in the soul, and not the soul thinking up the world. If the essence of a living thing is its natural form, it follows that, things being as they are, they are known as they are and for what they are. The intellect is identical with those things that are intelligible. The world is intelligible and, in particular, living things are intelligible, because the intelligible is in the world.

A first major difficulty in Aristotle's philosophy concerns the relation between knowledge and being, in particular between intelligence and life. If one treats intelligence as a function of contemplation and reproduction, if one gives it a place among the forms, however eminent, one thereby situates (that is, limits) the thought of order at a particular place in the universal order. But how can knowledge be at once mirror and object, reflector and reflection? If the definition of man as ζωον λογικον, or reasoning animal, is a naturalist's definition (in the same sense that Carolus Linnaeus defines the wolf as *canis lupus* or the maritime pine as *pinus maritima*), then science, and in particular the science of life, is an activity of life itself. One is then forced to ask what the organ of that activity is. And it follows that the Aristotelian theory of the active intellect, a pure form without organic basis, has the effect of separating intelligence from life; it lets something from outside (θυραθεν, in Aristotle's terms) enter the human embryo, as through a doorway, namely, the extranatural or transcendent power to make sense of the essential forms that individual beings embody. The theory thus makes the conception of concepts either something more than human or else something transcending life (*supravitale*).

A second difficulty, which is in fact an instance of the first, concerns the impossibility of accounting for mathematical knowledge in terms of a biological function. A celebrated passage of the *Metaphysics* states that mathematics has nothing to do with final causes,[20] which is equivalent to saying that there are intelligible things that are not forms in the proper sense of the word, and that knowledge of those things has nothing to do with knowledge of life. Hence, there is no mathematical model of the living. Although Aristotle describes nature as ingenious, creative and inventive, it should not then be conflated with the demiurge of the *Timaeus*. One of the most astonishing propositions of Aristotle's philosophy of biology is that it makes not the artisan but the art responsible for what is produced. What cures the patient is not the physician but health. It is because the form "health" is present in medical activity that medicine is, in fact, the cause of the cure. By art, Aristotle means the unreflective purpose of a natural logos. Meditating on the example of the physician who heals not because he is a physician but because he is inhabited and animated by the form "health," one might say that the presence of the concept in thought, in the form of an end represented as a model, is an epiphenomenon. Aristotle's anti-Platonism was reflected in his depreciation of mathematics: mathematics was denied access to the immanent activity of life, which is God's essential attribute, and it was only through knowledge (that is, imitation) of that immanent activity that man could hope to form an idea of God. [*Etudes*, pp. 336-38]

Nominalism

[122] A further difficulty of Aristotelianism concerns the ontological and gnoseological status of individuality in a science of life

based on concepts. If the individual is an ontological reality and not simply an imperfection in the realization of a concept, what is the significance of the order of beings represented in the classification by genus and species? If the concept of a living being ontologically presides over its conception, what mode of knowledge is the individual capable of? A system of living forms, if grounded in being, has the ineffable individual as its correlate; but an ontological plurality of individuals, if such a thing exists, has a concept, a fiction, as its correlate. There are two possibilities. Is it the universal that makes the individual *a* living thing as well as *this particular* living thing? If so, singularity is to life as the exception is to the rule. The exception confirms the rule, in the sense of revealing its existence and content, for the rule, the violation of the rule, is what makes the singularity apparent, indeed glaring. Or is it the individual that lends its color, weight and flesh to that ghostly abstraction, the universal? Without such a gift, "universal" would have no meaning in "life," and would be an empty word. The conflict between the individual and the universal as to their respective claims on "being" bears on life in all its forms: the vegetable as well as the animal, function as well as form, illness as well as "temperament." All approaches to life must be homogeneous. If living species exist, then the diseases of living things must also form species. If only individuals exist, then there are no species of disease, only sick individuals. If life has an immanent logic, then any science of life and its manifestations, whether normal or pathological, must set itself the task of discovering that logic. Nature then becomes an enduring set of latent relations that must be brought to light. Once uncovered, however, those relations offer a reassuring guarantee of validity to the naturalist's efforts to classify and to the physician's efforts to heal. In *The History of Madness* and *Birth of the Clinic*, Michel Foucault brilliantly demonstrated how the methods of botany served as a model for nineteenth-century physicians in developing their nosologies. "The rationality of what threatens life," he wrote, "is identical to the rationality of life itself." But there is rationality and rationality.

The matter of universals was, of course, an important issue in medieval philosophy, theology and politics. Here, however, I shall approach this question only indirectly, by way of a few brief remarks on nominalism in the philosophy of the seventeenth and eighteenth centuries. Nominalists over the ages have relied on a varied but unchanging arsenal of arguments. Because they were not always engaged in the same struggle, however, different nominalists chose different weapons from that arsenal. Yet all of them, from Ockham to Hume by way of Duns Scotus, Hobbes, Locke and Condillac, shared one common purpose – to show that universals are merely a way of using singular things and not in the nature of things themselves. Ockham called universals "suppositions" (that is, positions of substitution); Hobbes called them "arbitrary impositions"; Locke called them "representations instituted as signs." Yet all agreed that concepts were a human, which is to say, factitious and tendentious, processing of experience. We say "human" because we do not know if we have the right to say "intellectual." Holding that the mind is a tabula rasa does not give one the right to say that a tabula rasa is a mind. Nominalists look upon shared properties of individual things as an authentic equivalent to universals, but isn't doing so tantamount to donning a mask of false simplicity? A trap awaits those who take this path, the trap of similarity. A general idea, Locke says, is a general name signifying a similar quality perceived under a variety of circumstances; that quality is weighed by abstraction, that is, by "consideration of the common as distinct from the particular." It can then serve as a valid representation for all particular ideas belonging to the same type. Unlike Locke, Hume ascribes to the faculty of generalization not only a power to reproduce sensations in memory but also to transpose the order in which impressions are received; this power belongs to the imagination, which may be unfaithful to the lessons of experience. Nevertheless, he argues that similarity of ideas guides the imagination toward certain habits, or uniformities, in dealing with the environment. Habit telescopes together a whole host of individual experiences. If any one of these experiences is evoked by a name, the individual idea of that experience conjures up others, and we yield to the illusion of generality.

It is easy to see that there can be no comfortable nominalist position on the relation of concepts to life. For the nominalist, diverse things must exhibit some minimal degree of similarity before one can construct the concept of that similar property which is supposed to take the place of universal essences. Hence, what those eighteenth-century authors who were empiricists as to the content of their knowledge and sensualists as to the origin of its forms really give us is a mirror image of Aristotelianism, because they sought to find the knowing [*le connaître*] among the known, to learn about life within the order of life. Human beings, they say, are endowed with a power (which might equally well be taken for a measure of impotence) to invent classes and, thus, to arrange other living beings in an orderly fashion, but only on condition that those beings exhibit certain common characters or repeated traits. How can a nominalist speak of nature or natures? He can do what Hume did and invoke a human nature, which is to concede at least that there is uniformity among humans, even though Hume held that human nature was inventive and, more specifically, capable of adopting deliberate conventions. What does this accomplish? It introduces a cleavage in the system of living beings, because the nature of one of those beings is defined by an artifice, by the possibility of establishing a convention rather than expressing the order of nature. Hence, in Locke and Hume as in Aristotle, the problem of how concepts are conceived is solved in a way that disrupts the project of naturalizing knowledge of nature. [*Etudes*, pp. 339-42]

Transcendental Logic

[123] Philosophy is better than the history of science at revealing the significance of the disparities between the scientific techniques of naturalists and their implicit or explicit underlying philosophy. This can be seen in a masterful text by Kant on "the regulative use of the ideas of pure reason."[21] Here Kant introduced the image of a "logical horizon" to account for the regulative, but not constitutive, character of the rational principles of homogeneity (diversity grouped by genus) and variety (species within genera). A logical horizon, according to Kant, is a conceptual viewpoint that encompasses a certain region; within that horizon, there are multiple viewpoints, each determining further horizons of smaller ambit. A horizon can be decomposed only into other horizons, just as a concept can be analyzed only in terms of other concepts. To say that a horizon can be decomposed only into other horizons and not into discrete points is to say that species can be decomposed into subspecies but never into individuals. This is because to know something is to know it in terms of concepts, and the understanding knows nothing by intuition alone.

Kant's image of a logical horizon and his definition of a concept as a viewpoint encompassing a region do not mark a return to nominalism, nor do they constitute an attempt to justify concepts on the basis of their pragmatic value in achieving economy of thought. Reason itself prescribes such economy, according to Kant, and in so doing proscribes the idea of nature according to which there is no such thing as similarity, for in that case the logical law of species as well as the understanding itself would be simultaneously abolished. [...] Reason thus assumes the role of interpreting the requirements of the understanding in that realm where the science of life pursues the heuristic task of identifying and classifying species. Those requirements define a transcendental structure of knowledge. It might therefore appear that Kant's analysis finally manages to break out of the circle within which all previous naturalist theories of knowledge had remained confined. The conception of concepts cannot be merely one concept among others. The dichotomy that neither Aristotle nor the empiricist nominalists had been able to avoid was grounded, justified and exalted by Kant.

If, however, we have gained the legitimation of a possibility — that of knowledge through concepts — have we not perhaps lost the certainty that, among the objects of knowledge, there are some, at least, whose existence is a necessary manifestation of the reality of concretely active concepts? Put differently, have we not lost the certainty that living beings do in fact number among the objects of knowledge? In Aristotelian logic, the forms of reasoning mimic the hierarchy of living forms, hence there is a guaranteed correspondence between logic and life. Transcendental logic, which constitutes nature *a priori* as a system of physical laws, does not in fact succeed in constituting nature as the theater of living organisms. We gain a better understanding of the naturalist's research, but we do not arrive at an understanding of nature's ways; we gain a better understanding of the concept of causality, but we do not understand the causality of the concept. The *Critique of Judgment* attempts to give meaning to this limitation, which the understanding experiences as a fact. An organized being is one that is both its own cause and its own effect; it organizes itself and reproduces its organization; it forms itself and creates its own replica in accordance with a type. Its teleological structure, in which the interrelation of the parts is regulated by the whole, exemplifies a nonmechanical causality of the concept. We have no *a priori* knowledge of this type of causality. Forces that are also forms and forms that are also forces are indeed part of nature and in nature, but we do not know this through the understanding; we perceive it, rather, in experience. That is why the idea of a "natural end," which is essentially the idea of a self-constructing organism, is not a category in Kant but a regulative idea, which can be applied only in the form of maxims. To be sure, art provides an analogy whereby nature's mode of production can be judged. But we cannot hope to adopt the viewpoint of an archetypal intellect for which concept and intuition would be identical, an intellect capable of creating its own objects, for which concepts would be not only objects of knowledge but also, to use Leibniz's term, original roots of being. Kant holds that the fine arts are arts of genius, and he regards genius as nature dictating its law to art. Yet he refuses to permit himself to assume, in dogmatic fashion, a similar viewpoint, that of genius, in order to grasp the secret of nature's *operari*. Kant, in other words, refuses to identify the logical horizon of the naturalists with what one might call the poetic horizon of *natura naturans*. [*Etudes*, pp. 343-45]

The Bernardian Conception

[124] Claude Bernard described his reflections as a scientific theory of general physiology. They are interesting, however, precisely because Bernard did not divorce the study of functions from that of structures. In Bernard's day, moreover, the only structure known to be common to both animals and plants, hence the structure on which the study of life must henceforth focus, was the cell. Bernard also did not divorce the study of structures from the study of the origin of those structures. Thus, his general physiology is full of references to embryology, which ever since the work of Karl Ernst von Baer had served as a beacon for nineteenth-

century biologists, a source of concepts and methods for use by other disciplines. [...]

Bernard's general physiology was, first of all, a theory of the development of organs, and his basic conception of life would resolve, or at any rate recast in more meaningful terms, a problem that positivist biology had avoided and that mechanist biology had resolved through conflation of concepts: to wit, in what sense is an organism organized? The naturalists of the eighteenth century had been obsessed with the question. Indeed, it was not a question that lent itself to easy solution in terms of mechanical models. Preformationism, the theory that the growth of the adult organism from the original seed is simply a matter of enlargement of structures already contained in miniature in the seed – along with the logically derivative theory that seeds contain smaller seeds containing still smaller seeds and so on, ad infinitum – referred the whole issue of organization back to Creation. The rise of embryology as a basic science in the nineteenth century made it possible to reformulate the question. For Bernard, the question of organization and the obstacle it posed to explaining life in physical and chemical terms was what made general physiology a distinct science. [...]

Bernard was possessed by one idea: that the organized living thing is the temporary manifestation of an *idée directrice*, a guiding idea. The laws of physics and chemistry do not in themselves explain how they are brought to bear on the composition of a particular organism. This argument is developed at length in the *Leçons sur les phénomènes de la vie*:

> My experience has led me to a certain conception of things.... There are, I believe, of necessity, two orders of phenomena in living things: phenomena of vital creation or organic synthesis, and phenomena of death or organic destruction.... Only the first of these

two classes of phenomena is without direct analogue elsewhere; it is peculiar, specific, to living things. This evolutive synthesis is what is truly vital.

Hence, for Bernard, a functioning organism was an organism engaged in destroying itself. The functioning of an organ was a physicochemical phenomenon, that is, death. We can grasp such phenomena, we can understand and characterize them, and so we are inclined, misleadingly, to apply the name "life" to what is in fact a form of death. Conversely, organic creation and organization are plastic acts of synthetic reconstitution of the substances that the functioning organism requires. This organic creation, this constitution of protoplasm, is a form of chemical synthesis, and it is also a form of morphological synthesis, which brings the "immediate principles" of living matter together in a particular kind of mold. The existence of an "internal mold" (*le moule intérieur*) was in fact Buffon's way of explaining how an invariant form persists in the midst of that incessant turbulence which is life.

At first sight, one might think that Bernard is here separating two forms of synthesis that modern biochemistry has reunited, and that he has failed to recognize the fact that the cytoplasm itself is structured. Indeed, it is no longer possible to agree with Bernard that "at its simplest level, shorn of all the ancillary phenomena that mask it in most beings, life, contrary to what Aristotle believed, is independent of any specific form. It resides in a substance defined by its composition and not by its configuration: protoplasm."

On the contrary, modern biochemistry is based on the principle that configuration and structure are relevant even at the most basic level of chemical composition. Perhaps Bernard's error, though, was not as total as it may seem, for he says later that "protoplasm, however basic, is still not a purely chemical substance, a simple immediate principle of chemistry. It has an origin that eludes us. It is the continuation of the protoplasm of an ancestor." In other words, protoplasm has a structure, and that structure is hereditary. "Protoplasm itself is an atavistic substance. We do not see its birth, only its continuation." Now, recall that by evolution Bernard means the law that determines the fixed direction of constant change; this law governs the manifestations of life both in its inception and in its perpetuation. For Bernard, moreover, nutrition was identical with evolution in this sense. Thus, it can be argued that Bernard did not make absolute the distinction between matter and form, between chemical and morphological synthesis. He had at least an inkling that the chemical interchanges occurring within the protoplasm obey a structural imperative. He also saw the structure of the protoplasm as something whose reproduction required something beyond the known laws of physics and chemistry. It was a product of heredity which could not be duplicated in the laboratory. In his own words, this structure was "the manifestation here and now of a primitive impulse, a primitive action and *message*, which nature repeats according to a pattern determined in advance."

Clearly, Bernard seems to have sensed that biological heredity consists in the transmission of something that we now think of as coded information. "Message" is, semantically speaking, not far from "code." Nevertheless, it would be incorrect to conclude that this semantic analogy points to a genuine conceptual kinship. The reason has to do with a simultaneous discovery. In 1865, the same year that Bernard's *Introduction à l'étude de la médecine expérimentale* appeared, Gregor Mendel, an obscure monk who would never in his lifetime experience anything like the celebrity that was lavished on Claude Bernard, published his *Versuche über Pflanzenhybriden*. No concept analogous to those associated with today's theory of heredity can be imputed to Bernard, because the concept of heredity itself was totally new and unlike any ideas Bernard might have had about generation and evolution. We must be careful, therefore, not to see analogies in terms taken out of context. Nevertheless, one can still argue that the Bernardian "message" has a functional affinity with today's genetic code. That affinity is based on their common relation to the concept of information. Consider Bernard's repeated use of certain terms and phrases: message, guiding idea, vital design, vital preordainment, vital plan, directed process. If genetic information is a coded program for protein synthesis, then Bernard's repeated use of such converging metaphors would appear to reflect an attempt to pinpoint a biological reality for which no adequate concept had yet been formulated.

To put it in slightly different terms, Bernard used concepts associated with a psychological concept of information to account for phenomena that we now interpret in terms of a physical concept of information. [...] Construction, growth, restoration and the self-regeneration of the living machine – it is no accident that these terms occur in combination. Evolution in the Bernardian sense, the fundamental characteristic of life, is the inverse of evolution in the physicist's sense, namely, the series of states assumed by an isolated system governed by the second law of thermodynamics. Biochemists today say that organic individuality, or the constancy of a system in dynamic equilibrium, reflects life's general tendency to slow the increase of entropy, to resist evolution toward the more probable state of uniformity in disorder.

"The law of order and succession that bestows meaning and order on phenomena": the formulation is rather surprising for a biologist whom no one would accuse of indulgence toward the use of mathematical concepts and models in biology. The formula is actually quite close to Leibniz's definition of individual substance: *Lex seriei suarum operationum,* the law of the series in the mathematical sense of the term, a series of operations. This almost formal (logical) definition of (biological) heredity can now be interpreted in the light of the fundamental discovery of molecular biology, the structure of DNA, the key constituent of chromosomes, the carriers of heredity, whose very number is itself a specific hereditary characteristic. [*Etudes*, pp. 354-60]

Information Theory

[125] In 1954, James D. Watson and Francis Crick, who eight years later received the Nobel Prize for their work, showed that it was the ordering of a finite number of bases along a double helix joined by phosphates of sugar which constitutes the genetic information or program code determining how the cell synthesizes the building blocks of protein for new cells. It has since been shown that this synthesis takes place on demand, that is, as a function of information stemming from the environment – meaning, of course, the cellular environment. In 1965, another Nobel Prize was awarded for this further discovery. In changing the scale on which the characteristic phenomena of life – which is to say, the structuration of matter and the regulation of functions, including the structuration function – are studied, contemporary biology has also adopted a new language. It has dropped the vocabulary and concepts of classical mechanics, physics and chemistry, all more or less directly based on geometrical models, in favor of the vocabulary of linguistics and communications theory. Messages, information, programs, code, instructions, decoding: these are the new concepts of the life sciences. [...]

When we say that biological heredity is the communication of a certain kind of information, we hark back in a way to the Aristotelian philosophy with which we began. [...] To say that heredity is the communication of information is, in a sense, to acknowledge that there is a *logos* inscribed, preserved and transmitted in living things. Life has always done – without writing, long before writing even existed – what humans have sought to do with engraving, writing and printing, namely, to transmit messages. The science of life no longer resembles a portrait of life, as it could when it consisted in the description and classification of species; and it no longer resembles architecture or mechanics, as it could when it was simply anatomy and macroscopic physiology. But it does resemble grammar, semantics and the theory of syntax. If we are to understand life, its message must be decoded before it can be read.

This will no doubt have a number of revolutionary consequences, and it would take many chapters to explain not what they are but what they are in the process of becoming. To define life as a meaning inscribed in matter is to acknowledge the existence of an *a priori* objective that is inherently material and not merely formal. In this connection, it seems to me that the study of instinct in the manner of Nikolaas Tinbergen and Konrad Lorentz, that is, through the demonstration of the existence of innate patterns of behavior, is a way of demonstrating the reality of such *a prioris*. To define life as meaning is to force oneself to look for new discoveries. Here, the experimental invention consists only in the search for a key, but once that key is discovered, the meaning is found, not constructed. The models used in seeking organic meanings require a mathematics different from that known to the Greeks. In order to understand living things one needs a nonmetric theory of space, a science of order, a topology; one needs a nonnumerical calculus, a combinatorics, a statistical machinery. In this respect too there has been, in a sense, a return to Aristotle. He believed that mathematics was of no use in biology because it recognized no theory of space other than the geometry to which Euclid gave his name. A biological form, Aristotle argued, is not a pattern, not a geometrical form. He was correct. Within an organism there are no distances: the whole is immediately present to all the (pseudo-)parts. The essence of the living thing is that, insofar as it is living, it is immediately present to itself. Its "parts" (the very term is misleading) are immediately present to one another. Its regulatory mechanisms, its "internal environment," make the whole immediately present to each of its parts.

Hence, in a certain sense Aristotle was not wrong to say that a certain kind of mathematics, the only mathematics he knew about, was of no use in understanding biological forms, forms determined by a final cause or totality, nondecomposable forms in which beginning and end coincide and actuality outweighs potentiality. [...]

If life is the production, transmission and reception of information, then clearly the history of life involves both conservation and innovation. How is evolution to be explained in terms of genetics? The answer, of course, involves the mechanism of mutations. One objection that has often been raised against this theory is that many mutations are subpathological, and a fair number lethal, so the mutant is less viable than the original organism. To be sure, many mutations are "monstrous" – but from the standpoint of life as a whole, what does "monstrous" mean? Many of today's life forms are nothing other than "normalized monsters," to borrow an expression from the French biologist Louis Roule. Thus, if life has meaning, we must accept the possibility of a loss of that meaning, of distortion, of misconstruction. Life overcomes error through further trials (and by "error" I mean simply a dead end).

What, then, is knowledge? If life is concept, does recognizing that fact give the intelligence access to life? What, then, is knowledge? If life is meaning and concept, how do we conceive of the activity of knowing? Earlier, I alluded to the study of instinctive behavior, of behavior structured by innate patterns. An animal is formed by heredity so as to receive and transmit certain kinds of information. Information that an animal is not structurally equipped to receive might as well not exist as far as that animal is concerned. In what we take to be the universal environment, each species' structure determines its own particular environment, as Alex von Uexküll has shown. If man is also formed by heredity, how does one explain the history of knowledge, which is the history of error and of triumph over error? Must we conclude that man became what he is by mutation, by an error of heredity? In that case, life would by error have produced a living thing capable of making errors. In fact, human error is probably one with human errancy. Man makes mistakes because he does not know where to settle. He makes mistakes when he chooses the wrong spot for receiving the kind of information he is after. But he also gathers information by moving around, and by moving objects around, with the aid of various kinds of technology. Most scientific techniques, it can be argued, are in fact nothing more than methods for moving things around and changing the relations among objects. Knowledge, then, is an anxious quest for the greatest possible quantity and variety of information. If the *a priori* is in things, if the concept is in life, then to be a subject of knowledge is simply to be dissatisfied with the meaning one finds ready at hand. Subjectivity is therefore nothing other than dissatisfaction. Perhaps that is what life is. Interpreted in a certain way, contemporary biology is, somehow, a philosophy of life. [*Etudes*, pp. 360–64]

CHAPTER FOURTEEN

The Normal and the Pathological

Introduction to the Problem

[126] To act, it is necessary at least to localize. For example, how do we take action against an earthquake or hurricane? The impetus behind every ontological theory of disease undoubtedly derives from therapeutic need. When we see in every sick man someone whose being has been augmented or diminished, we are somewhat reassured, for what a man has lost can be restored to him, and what has entered him can also leave. We can hope to conquer disease even if doing so is the result of a spell, or magic, or possession; we have only to remember that disease happens to man in order not to lose all hope. Magic brings to drugs and incantation rites innumerable resources stemming from a profoundly intense desire for cure. Henry Ernst Sigerist has noted that Egyptian medicine probably universalized the Eastern experience of parasitic diseases by combining it with the idea of disease-possession: throwing up worms means being restored to health.[22] Disease enters and leaves man as through a door.

A vulgar hierarchy of diseases still exists today, based on the extent to which symptoms can — or cannot — be readily localized, hence Parkinson's disease is more of a disease than thoracic shingles, which is, in turn, more so than boils. Without wishing to detract from the grandeur of Louis Pasteur's tenets, we can say without hesitation that the germ theory of contagious disease has certainly owed much of its success to the fact that it embodies an ontological representation of sickness. After all, a germ can be seen, even if this requires the complicated mediation of a microscope, stains and cultures, while we would never be able to see a miasma or an influence. To see an entity is already to foresee an action. No one will object to the optimistic character of the theories of infection insofar as their therapeutic application is concerned. But the discovery of toxins and the recognition of the specific and individual pathogenic role of *terrains* have destroyed the beautiful simplicity of a doctrine whose scientific veneer for a long time hid the persistence of a reaction to disease as old as man himself.

If we feel the need to reassure ourselves, it is because one anguish constantly haunts our thoughts; if we delegate the task of restoring the diseased organism to the desired norm to technical means, either magical or matter of fact [*positive*], it is because we expect nothing good from nature itself.

By contrast, Greek medicine, in the Hippocratic writings and practices, offers a conception of disease which is no longer ontological, but dynamic, no longer localizationist, but totalizing. Nature (*physis*), within man as well as without, is harmony and equilibrium. The disturbance of this harmony, of this equilibrium, is called "disease." In this case, disease is not somewhere in man, it is everywhere in him; it is the whole man. External circumstances are the occasion but not the causes. Man's equilibrium consists of four humors, whose fluidity is perfectly suited to sustain variations and oscillations and whose qualities are paired by opposites (hot/cold, wet/dry); the disturbance of these humors causes disease. But disease is not simply disequilibrium or discordance; it is, perhaps most important, an effort on the part of nature to effect a new equilibrium in man. Disease is a generalized reaction designed to bring about a cure; the organism develops a disease in order to get well. Therapy must first tolerate and, if necessary, reinforce these hedonic and spontaneously therapeutic reactions. Medical technique imitates natural medicinal action (*vis medicatrix naturae*). To imitate is not merely to copy an appearance but, also, to mimic a tendency and to extend an intimate movement. Of course, such a conception is also optimistic, but here the optimism concerns the way of nature and not the effect of human technique.

Medical thought has never stopped alternating between these two representations of disease, between these two kinds of optimism, always finding some good reason for one or the other attitude in a newly explained pathogenesis. Deficiency diseases and all infectious or parasitic diseases favor the ontological theory, while endocrine disturbances and all diseases beginning with *dys-* support the dynamic or functional theory. However, these two conceptions do have one point in common: in disease, or better, in the experience of being sick, both envision a polemical situation – either a battle between the organism and a foreign substance, or an internal struggle between opposing forces. Disease differs from a state of health, the pathological from the normal, as one quality differs from another, either by the presence or absence of a definite principle, or by an alteration of the total organism. This heterogeneity of normal and pathological states persists today in the naturalist conception, which expects little from human efforts to restore the norm, and in which nature will find the ways toward cure. But it proved difficult to maintain the qualitative modification separating the normal from the pathological in a conception that allows, indeed expects, man to be able to compel nature and bend it to his normative desires. Wasn't it said repeatedly after Bacon's time that one governs nature only by obeying it? To govern disease means to become acquainted with its relations with the normal state, which the living man – loving life – wants to regain. Hence, the theoretical need, delayed by an absence of technology, to establish a scientific pathology by linking it to physiology. Thomas Sydenham (1624–1689) thought that in order to help a sick man, his sickness had to be delimited and determined. There are disease species just as there are animal or plant species. According to Sydenham, there is an order among diseases similar to the regularity Isidore Geoffroy Saint-Hilaire found among anomalies. Philippe Pinel justified all these attempts at classification of disease (nosology) by perfecting the genre in his *Nosographie philosophique* (1797), which Charles Victor Daremberg described as more the work of a naturalist than a clinician.

Meanwhile, Giovanni Battista Morgagni's (1682–1771) creation of a system of pathological anatomy made it possible to link the lesions of certain organs to groups of stable symptoms, such that nosographical classification found a substratum in anatomical analysis. But just as the followers of William Harvey and Albrecht von Haller "breathed life" into anatomy by turning it into physiology, so pathology became a natural extension of physiology. (Sigerist provides a masterful summary of this evolution of medical ideas.[23]) The end result of this evolutionary process is the formation of a theory of the relations between the normal and the pathological, according to which the pathological phenomena found in living organisms are nothing more than quantitative variations, greater or lesser according to corresponding physiological phenomena. Semantically, the pathological is designated as departing from the normal not so much by *a-* or *dys-* as by *hyper-* or *hypo-*. While retaining the ontological theory's soothing confidence in the possibility of technical conquest of disease, this approach is far from considering health and sickness as qualitatively opposed, or as forces joined in battle. The need to re-establish continuity in order to gain more knowledge for more effective action is such that the concept of disease would finally vanish. The conviction that one can scientifically restore the norm is such that, in the end, it annuls the pathological. Disease is no longer the object of anguish for the healthy man; it has become instead the object of study for the theorist of health. It is in pathology, writ large, that we can unravel the teachings of health, rather as Plato sought in the institutions of the State the larger and more easily readable equivalent of the virtues and vices of the individual soul. [*The Normal and the Pathological* (*NP*), pp. 11–13]

The Identity of the Two States

Auguste Comte and the "Broussais Principle"

[127] It was in 1828 that Auguste Comte took notice of François-Joseph Victor Broussais's treatise *De l'Irritation et de la folie* and adopted the principle for his own use. Comte credits Broussais, rather than Xavier Bichat, and before him, Philippe Pinel, with having declared that all diseases acknowledged as such are only symptoms and that disturbances of vital functions could not take place without lesions in organs, or rather, tissues. But above all, adds Comte, "never before had anyone conceived the fundamental relation between pathology and physiology in so direct and satisfying a manner." Broussais described all diseases as consisting essentially "in the excess or lack of excitation in the various tissues above or below the degree established as the norm." Thus, diseases are merely the effects of simple changes in intensity in the action of the stimulants which are indispensable for maintaining health. [*NP*, pp. 47–48]

[128] The fortieth lecture of the *Cours de philosophie positive* – philosophical reflections on the whole of biology – contains Comte's most complete text on the problem now before us. It is concerned with showing the difficulties inherent in the simple extension of experimental methods, which have proved their

usefulness in the physicochemical sphere, to the particular characteristics of the living:

> Any experiment whatever is always designed to uncover the laws by which each determining or modifying influence of a phenomenon affects its performance, and it generally consists in introducing a clear-cut change into each designated condition in order to measure directly the corresponding variation of the phenomenon itself.[24]

Now, in biology the variation imposed on one or several of a phenomenon's conditions of existence cannot be random but must be contained within certain limits compatible with the phenomenon's existence. Furthermore, the fact of functional *consensus* proper to the organism precludes monitoring the relation, which links a determined disturbance to its supposedly exclusive effects, with sufficient analytical precision. But, thinks Comte, if we readily admit that the essence of experimentation lies not in the researcher's artificial intervention in the system of a phenomenon which he intentionally tends to disturb, but rather in the comparison between a control phenomenon and one altered with respect to any one of its conditions of existence, it follows that diseases must be able to function for the scientists as spontaneous experiments which allow a comparison to be made between an organism's various abnormal states and its normal state.

> According to the eminently philosophical principle which will serve from now on as a direct, general basis for positive pathology and whose definitive establishment we owe to the bold and persevering genius of our famous fellow citizen, Broussais, the pathological state is not at all radically different from the physiological state, with regard to which – no matter how one looks at it – it can only constitute a simple extension going more or less beyond the higher or

lower limits of variation proper to each phenomenon of the normal organism, without ever being able to produce really new phenomena which would have to a certain degree any purely physiological analogues.[25]

Consequently every conception of pathology must be based on prior knowledge of the corresponding normal state, but conversely the scientific study of pathological cases becomes an indispensable phase in the overall search for the laws of the normal state. The observation of pathological cases offers numerous, genuine advantages for actual experimental investigation. The transition from the normal to the abnormal is slower and more natural in the case of illness, and the return to normal, when it takes place, spontaneously furnishes a verifying counterproof. In addition, as far as man is concerned, pathological investigation is more fruitful than the necessarily limited experimental exploration. The scientific study of morbid states is essentially valid for all organisms, even plant life, and is particularly suited to the most complex and therefore the most delicate and fragile phenomena which direct experimentation, being too brusque a disturbance, would tend to distort. Here Comte was thinking of vital phenomena related to the higher animals and man, of the nervous and psychic functions. Finally, the study of anomalies and monstrosities conceived as both older and less curable illnesses than the functional disturbances of various plant or neuromotor apparatuses completes the study of diseases: the "teratological approach" (the study of monsters) is added to the "pathological approach" in biological investigation.[26]

It is appropriate to note, first, the particularly abstract quality of this thesis and the absence throughout of any precise example of a medical nature to suitably illustrate his literal exposition. Since we cannot relate these general propositions to any example, we do not know from what vantage point Comte states that the pathological phenomenon always has its analogue in a physiological phenomenon, and that it is nothing radically new. How is a sclerotic artery analogous to a normal one, or an asystolic heart identical to that of an athlete at the height of his powers? Undoubtedly, we are meant to understand that the laws of vital phenomena are the same for both disease and health. But then why not say so and give examples? And even then, does this not imply that analogous effects are determined in health and disease by analogous mechanisms? We should think about this example given by Sigerist: "During digestion the number of white blood cells increases. The same is true at the onset of infection. Consequently this phenomenon is sometimes physiological, sometimes pathological, depending on what causes it."[27]

Second, it should be pointed out that despite the reciprocal nature of the clarification achieved through the comparison of the normal with the pathological and the assimiliation of the pathological and the normal, Comte insists repeatedly on the necessity of determining the normal and its true limits of variation first, before methodically investigating pathological cases. Strictly speaking, knowledge of normal phenomena, based solely on observation, is both possible and necessary without knowledge of disease, particularly based on experimentation. But we are presented with a serious gap in that Comte provides no criterion which would allow us to know what a normal phenomenon is. We are left to conclude that on this point he is referring to the usual corresponding concept, given the fact that he uses the notions of normal state, physiological state and natural state interchangeably.[28] Better still, when it comes to defining the limits of pathological or experimental disturbances compatible with the existence of organisms, Comte identifies these limits with those of a "harmony of distinct influences, those exterior as well as interior"[29] – with the result that the concept of the normal or physiological, finally clarified by this concept of *harmony*, amounts to a qualitative and polyvalent concept, still more aesthetic and moral than scientific.

As far as the assertion of identity of the normal phenomenon and the corresponding pathological phenomenon is concerned, it is equally clear that Comte's intention is to deny the qualitative difference between these two admitted by the vitalists. Logically to deny a qualitative difference must lead to asserting a homogeneity capable of expression in quantitative terms. Comte is undoubtedly heading toward this when he defines pathology as a "simple extension going more or less beyond the higher or lower limits of variation proper to each phenomenon of the normal organism." But in the end it must be recognized that the terms used here, although only vaguely and loosely quantitative, still have a qualitative ring to them. [*NP*, pp. 19-21]

Claude Bernard and Experimental Pathology

[129] In Bernard's work, the real identity – should one say in mechanisms or symptoms or both? – and continuity of pathological phenomena and the corresponding physiological phenomena are more a monotonous repetition than a theme. This assertion is to be found in the *Leçons de physiologie expérimentale appliquée à la médecine* (1855), especially in the second and twenty-second lectures of Volume Two, and in the *Leçons sur la chaleur animale* (1876). We prefer to choose the *Leçons sur le diabète et la glycogenèse animale* (1877) as the basic text, which, of all Bernard's works, can be considered the one especially devoted to illustrating the theory, the one where clinical and experimental facts are presented at least as much for the "moral" of a methodological and philosophical order which can be drawn from it as for their intrinsic physiological meaning.

Bernard considered medicine as the science of diseases, physiology as the science of life. In the sciences it is theory which illuminates and dominates practice. Rational therapeutics can be sustained only by a scientific pathology, and a scientific pathology must be based on physiological science. Diabetes is one disease which poses problems whose solution proves the preceding thesis. "Common sense shows that if we are thoroughly acquainted with a physiological phenomenon, we should be in a position to account for all the disturbances to which it is susceptible in the pathological state: physiology and pathology are intermingled and are essentially one and the same thing."[30] Diabetes is a disease that consists solely and entirely in the disorder of a normal function. "Every disease has a corresponding normal function of which it is only the disturbed, exaggerated, diminished or obliterated expression. If we are unable to explain all manifestations of disease today, it is because physiology is not yet sufficiently advanced and there are still many normal functions unknown to us."[31] In this, Bernard was opposed to many physiologists of his day, according to whom disease was an extraphysiological entity, superimposed on the organism. The study of diabetes no longer allowed such an opinion.

> In effect, diabetes is characterized by the following symptoms: polyuria, polydipsia, polyphagia, autophagia and glycosuria. Strictly speaking, none of these symptoms represents a new phenomenon, unknown to the normal state, nor is any a spontaneous production of nature. On the contrary, all of them preexist, save for their intensity, which varies in the normal state and in the diseased state.[32]

Briefly, we know that Bernard's genius lies in the fact that he showed that the sugar found in an animal organism is a product of this same organism and not just something introduced from the plant world through its feeding; that blood normally contains sugar, and that urinary sugar is a product generally eliminated by the kidneys when the rate of glycemia reaches a certain threshold. In other words, glycemia is a constant phenomenon independent of food intake to such an extent that it is the absence of blood sugar that is abnormal, and glycosuria is the consequence of glycemia which has risen above a certain quantity, serving as a threshold. In a diabetic, glycemia is not in itself a pathological phenomenon — it is so only in terms of its quantity; in itself, glycemia is a "normal and constant phenomenon in a healthy organism."[33]

> There is only one glycemia, it is constant, permanent, both during diabetes and outside that morbid state. Only it has degrees: glycemia below 3 to 4 percent does not lead to glycosuria; but above that level glycosuria results.... It is impossible to perceive the transition from the normal to the pathological state, and no problem shows better than diabetes the intimate fusion of physiology and pathology.[34]

[*NP*, pp. 30-32]
[130] Claude Bernard, unlike Broussais and Comte, supported his general principle of pathology with verifiable arguments, protocols of experiments and, above all, methods for quantifying physiological concepts. Glycogenesis, glycemia, glycosuria, combustion of food, heat from vasodilatation are not qualitative concepts but the summaries of results obtained in terms of measurement. From here on we know exactly what is meant when it is claimed that disease is the exaggerated or diminished expression of a normal function. Or at least we have the means to know it, for in spite of Bernard's undeniable progress in logical precision, his thought is not entirely free from ambiguity.

First of all, with Bernard as with Bichat, Broussais and Comte, there is a deceptive mingling of quantitative and qualitative concepts in the given definition of pathological phenomena. Sometimes the pathological state is "the disturbance of a normal mechanism consisting in a quantitative variation, an exaggeration or attenuation of normal phenomena,"[35] sometimes the diseased state is made up of "the exaggeration, disproportion, discordance of normal phenomena."[36] Who doesn't see that the term "exaggeration" has a distinctly quantitative sense in the first definition and a rather qualitative one in the second. Did Bernard believe that he was eradicating the qualitative value of the term "pathological" by substituting for it the terms disturbance, disproportion, discordance?

This ambiguity is certainly instructive in that it reveals that the problem itself persists at the heart of the solution presumably given to it. And the problem is the following: Is the concept of disease a concept of an objective reality accessible to quantitative scientific knowledge? Is the difference in value, which the living being establishes between his normal life and his pathological life, an illusory appearance that the scientist has the legitimate obligation to deny? If this annulling of a qualitative contrast is theoretically possible, it is clear that it is legitimate; if it is not possible, the question of its legitimacy is superfluous. [*NP*, pp. 35-36]

[131] By way of summary, in the medical domain, Claude Bernard, with the authority of every innovator who proves movement by marching, formulated the profound need of an era that believed in the omnipotence of a technology founded on science, and which felt comfortable in life in spite, or perhaps because of, romantic lamentations. An art of living – as medicine is in the full sense of the word – implies a science of life. Efficient therapeutics assumes experimental pathology, which in turn cannot be separated from physiology. "Physiology and pathology are identical, one and the same thing." But must it be deduced from this, with brutal simplicity, that life is the same in health and disease,

that it learns nothing in disease and through it? The science of opposites is one, said Aristotle. Must it be concluded from this that opposites are not opposites? That the science of life should take so-called normal and so-called pathological phenomena as objects of the same theoretical importance, susceptible of reciprocal clarification in order to make itself fit to meet the totality of the vicissitudes of life in all its aspects, is more urgent than legitimate. This does not mean that pathology is nothing other than physiology, and still less that disease, as it relates to the normal state, represents only an increase or a reduction. It is understood that medicine needs an objective pathology, but research which causes its object to vanish is not objective. One can deny that disease is a kind of violation of the organism and consider it as an event that the organism creates through some trick of its permanent functions, without denying that the trick is new. An organism's behavior can be in continuity with previous behaviors and still be another behavior. The progressiveness of an advent does not exclude the originality of an event. The fact that a pathological symptom, considered by itself, expresses the hyperactivity of a function whose product is exactly identical with the product of the same function in so-called normal conditions, does not mean that an organic disturbance, conceived as another aspect of the whole of functional totality and not as a summary of symptoms, is not a new mode of behavior for the organism relative to its environment.

In the final analysis, would it not be appropriate to say that the pathological can be distinguished as such, that is, as an alteration of the normal state, only at the level of organic totality, and when it concerns man, at the level of conscious individual totality, where disease becomes a kind of evil? To be sick means that a man really lives another life, even in the biological sense of the word. [*NP*, pp. 86-88]

Implications and Counterpositions

Life as a Normative Activity

[132] First of all there emerges from this theory the conviction of rationalist optimism that evil has no reality. What distinguishes nineteenth-century medicine (particularly before the era of Louis Pasteur) in relation to the medicine of earlier centuries is its resolutely monist character. Eighteenth-century medicine, despite the efforts of the iatromechanists and iatrochemists, and under the influence of the animists and vitalists, remained a dualist medicine, a medical Manichaeanism. Health and Disease fought over Man the way Good and Evil fought over the World. It is with a great deal of intellectual satisfaction that we take up the following passage in a history of medicine:

> Paracelsus was a visionary, [Jean Baptiste] van Helmont, a mystic, [Georg Ernst] Stahl, a pietist. All three were innovative geniuses but were influenced by their environment and by inherited traditions. What makes appreciation of the reform doctrines of these three great men very hard is the extreme difficulty one experiences in trying to separate their scientific from their religious beliefs.... It is not at all certain that Paracelsus did not believe that he had found the elixir of life; it is certain that van Helmont identified health with

salvation and sickness with sin; and in his account of *Theoria medica vera* Stahl himself, despite his intellectual vigor, availed himself more than he needed to of the belief in original sin and the fall of man.[37]

More than he needed to! says the author, quite the admirer of Broussais, sworn enemy at the dawn of the nineteenth century of all medical ontology. The denial of an ontological conception of disease, a negative corollary of the assertion of a quantitative identity between the normal and the pathological, is first, perhaps, the deeper refusal to confirm evil. It certainly cannot be denied that a scientific therapeutics is superior to a magical or mystical one. It is certain that knowledge is better than ignorance when action is required, and in this sense the value of the philosophy of the Enlightenment and of positivism, even scientistic, is indisputable. It would not be a question of exempting doctors from the study of physiology and pharmacology. It is very important not to identify disease with either sin or the devil. But it does not follow from the fact that evil is not a being that it is a concept devoid of meaning; it does not follow that there are no negative values, even among vital values; it does not follow that the pathological state is essentially nothing other than the normal state. [*NP*, pp. 103-104]

[133] It is true that in medicine the normal state of the human body is the state one wants to reestablish. But is it because therapeutics aims at this state as a good goal to obtain that it is called normal, or is it because the interested party, that is, the sick man, considers it normal that therapeutics aims at it? We hold the second statement to be true. We think that medicine exists as the art of life because the living human being himself calls certain dreaded states or behaviors pathological (hence requiring avoidance or correction) relative to the dynamic polarity of life, in the form of a negative value. We think that in doing this the living

human being, in a more or less lucid way, extends a spontaneous effort, peculiar to life, to struggle against that which obstructs its preservation and development taken as norms. The entry in the *Vocabulaire philosophique* seems to assume that value can be attributed to a biological fact only by "him who speaks," obviously a man. We, on the other hand, think that the fact that a living man reacts to a lesion, infection, functional anarchy by means of a disease expresses the fundamental fact that life is not indifferent to the conditions in which it is possible, that life is polarity and thereby even an unconscious position of value; in short, life is in fact a normative activity. *Normative*, in philosophy, means every judgment which evaluates or qualifies a fact in relation to a norm, but this mode of judgment is essentially subordinate to that which establishes norms. Normative, in the fullest sense of the word, is that which establishes norms. And it is in this sense that we plan to talk about biological normativity. We think that we are as careful as anyone as far as the tendency to fall into anthropomorphism is concerned. We do not ascribe a human content to vital norms but we do ask ourselves how normativity essential to human consciousness would be explained if it did not in some way exist in embryo in life. We ask ourselves how a human need for therapeutics would have engendered a medicine which is increasingly clairvoyant with regard to the conditions of disease if life's struggle against the innumerable dangers threatening it were not a permanent and essential vital need. From the sociological point of view, it can be shown that therapeutics was first a religious, magical activity, but this does not negate the fact that therapeutic need is a vital need, which, even in lower living organisms (with respect to vertebrate structure) arouses reactions of hedonic value or self-healing or self-restoring behaviors. [*NP*, pp. 126-27]

Pathology as the Basis of Physiology

[134] Conversely, the theory in question conveys the humanist conviction that man's action on his environment and on himself can and must become completely one with his knowledge of the environment and man; it must be normally only the application of a previously instituted science. Looking at the *Leçons sur le diabète* it is obvious that if one asserts the real homogeneity and continuity of the normal and the pathological it is in order to establish a physiological science that would govern therapeutic activity by means of the intermediary of pathology. Here the fact that human consciousness experiences occasions of new growth and theoretical progress in its domain of nontheoretical, pragmatic and technical activity is not appreciated. To deny technology a value all its own outside of the knowledge it succeeds in incorporating is to render unintelligible the irregular way of the progress of knowledge and to miss that overtaking of science by the power that the positivists have so often stated while they deplored it. If technology's rashness, unmindful of the obstacles to be encountered, did not constantly anticipate the prudence of codified knowledge, the number of scientific problems to resolve, which are surprises after having been setbacks, would be far fewer. Here is the truth that remains in empiricism, the philosophy of intellectual adventure, which an experimental method, rather too tempted (by reaction) to rationalize itself, failed to recognize. [...]

Here again, we owe to the chance of bibliographical research the intellectual pleasure of stating once more that the most apparently paradoxical theses also have their tradition which undoubtedly expresses their permanent logical necessity. Just when Broussais was lending his authority to the theory which established physiological medicine, this same theory was provoking the objections of an obscure physician, one Dr. Victor Prus, who was rewarded by the Société de Médecine du Gard in 1821 for a report entered in a competition whose object was the precise definition of the terms "phlegmasia" and "irritation" and their importance for practical medicine. After having challenged the idea that physiology by itself forms the natural foundation of medicine; that it alone can ever establish the knowledge of symptoms, their relationships and their value; that pathological anatomy can ever be deduced from the knowledge of normal phenomena; that the prognosis of diseases derives from the knowledge of physiological laws, the author adds:

> If we want to exhaust the question dealt with in this article we would have to show that *physiology, far from being the foundation of pathology, could only arise in opposition to it*. It is through the changes which the disease of an organ and sometimes the complete suspension of its activity transmit to its functions that we learn the organ's use and importance.... Hence an exostosis, by compressing and paralyzing the optic nerve, the brachial nerves, and the spinal cord, shows us their usual destination. Broussonnet lost his memory of substantive words; at his death an abcess was found in the anterior part of his brain and one was led to believe that that is the center for the memory of names.... Thus pathology, aided by pathological anatomy, has created physiology: every day pathology clears up physiology's former errors and aids its progress.[38]

[*NP*, pp. 104–107]
[135] There are some thinkers whose horror of finalism leads them to reject even the Darwinian idea of selection by the environment and struggle for existence because of both the term "selection," obviously of human and technological import, and the idea of advantage, which comes into the explanation of the mechanism of natural selection. They point out that most living

beings are killed by the environment long before the inequalities they can produce even have a chance to be of use to them because it kills above all sprouts, embryos or the young. But as Georges Teissier has observed, the fact that many organisms die before their inequalities serve them does not mean that the presentation of inequalities is biologically indifferent.[39] This is precisely the one fact we ask to be granted. There is no biological indifference, and consequently we can speak of biological normativity. There are healthy biological norms and there are pathological norms, and the second are not the same as the first.

We did not refer to the theory of natural selection unintentionally. We want to draw attention to the fact that what is true of the expression "natural selection" is also true of the old expression *vis medicatrix naturae*. Selection and medicine are biological techniques practiced deliberately and more or less rationally by man. When we speak of natural selection or natural medicinal activity we are victims of what Henri Bergson calls the "illusion of retroactivity" if we imagine that vital prehuman activity pursues goals and utilizes means comparable to those of men. But it is one thing to think that natural selection would utilize anything that resembles *pedigrees*, and *vis medicatrix*, cupping glasses and another to think that human technique extends vital impulses, at whose service it tries to place systematic knowledge which would deliver them from much of life's costly trial and error.

The expressions "natural selection" and "natural medicinal activity" have one drawback in that they seem to set vital techniques within the framework of human techniques when it is the opposite that seems true. All human technique, including that of life, is set within life, that is, within an activity of information and assimilation of material. It is not because human technique is normative that vital technique is judged such by comparison.

Because life is activity of information and assimilation, it is the root of all technical activity. In short, we speak of natural medicine in quite a retroactive and, in one sense, mistaken way, but even if we were to assume that we have no right to speak of it, we are still free to think that no living being would have ever developed medical technique if the life within him – as within every living thing – were indifferent to the conditions it met with, if life were not a form of reactivity polarized to the variations of the environment in which it develops. This was seen very well by Emile Guyénot:

> It is a fact that the organism has an aggregate of properties which belong to it alone, thanks to which it withstands multiple destructive forces. Without these defensive reactions, life would be rapidly extinguished.... The living being is able to find instantaneously the reaction which is useful vis-à-vis substances with which neither it nor its kind has ever had contact. The organism is an incomparable chemist. It is the first among physicians. The fluctuations of the environment are almost always a menace to its existence.[...] The living being could not survive if it did not possess certain essential properties. Every injury would be fatal if tissues were incapable of forming scars and blood incapable of clotting.[40]

By way of summary, we think it very instructive to consider the meaning that the word "normal" assumes in medicine, and the fact that the concept's ambiguity, pointed out by André Lalande, is greatly clarified by this, with a quite general significance for the problem of the normal. It is life itself and not medical judgment that makes the biological normal a concept of value and not a concept of statistical reality. For the physician, life is not an object but, rather, a polarized activity whose spontaneous effort of defense and struggle against all that is of negative value is extended by medicine by bringing to bear the relative but indispensable light of human science. [*NP*, pp. 129–31]

Nature Is the End Point of a Teleological Process

[136] In writing the *Introduction à l'étude de la médecine expérimentale*, Claude Bernard set out to assert not only that efficacious action is the same as science, but also, and analogously, that science is identical with the discovery of the laws of phenomena. On this point his agreement with Comte is total. What Comte in his philosophical biology calls the doctrine of the conditions of existence, Bernard calls "determinism." He flatters himself with having been the first to introduce that term into scientific French. "I believe I am the first to have introduced this word to science, but it has been used by philosophers in another sense. It will be useful to determine the meaning of this word in a book which I plan to write: *Du déterminisme dans les sciences*. This will amount to a second edition of my *Introduction à la médecine expérimentale*."[41] It is faith in the universal validity of the determinist postulate which is asserted by the principle "physiology and pathology are one and the same thing." At the very time that pathology was saddled with prescientific concepts, there was a physical chemical physiology which met the demands of scientific knowledge, that is, a physiology of quantitative laws verified by experimentation. Understandably, early-nineteenth-century physicians, justifiably eager for an effective, rational pathology, saw in physiology the prospective model which came closest to their ideal. "Science rejects the *indeterminate*, and in medicine, when opinions are based on medical palpation, inspiration, or a more or less vague intuition about things, we are outside of science and are given the example of this medicine of fantasy, capable of presenting the gravest perils as it delivers the health and lives of sick men to the whims of an inspired ignoramus."[42] But just because,

of the two – physiology and pathology – only the first involved laws and postulated the determinism of its object, it was not necessary to conclude that, given the legitimate desire for a rational pathology, the laws and determinism of pathological facts are the same laws and determinism of physiological facts. We know the antecedents of this point of doctrine from Bernard himself. In the lecture devoted to the life and works of François Magendie at the beginning of the *Leçons sur les substances toxiques et médicamenteuses* (1857), Bernard tells us that the teacher whose chair he occupies and whose teaching he continues "drew the feeling of real science" from the illustrious Pierre-Simon Laplace. We know that Laplace had been Antoine-Laurent Lavoisier's collaborator in the research on animal respiration and animal heat, the first brilliant success in research on the laws of biological phenomena following the experimental and measuring methods endorsed by physics and chemistry. As a result of this work, Laplace had retained a distinct taste for physiology and he supported Magendie. If Laplace never used the term "determinism," he is one of its spiritual fathers and, at least in France, an authoritative and authorized father of the doctrine designated by the term. For Laplace, determinism is not a methodological requirement, a normative research postulate sufficiently flexible to prejudice in any way the form of the results to which it leads: it is reality itself, complete, cast *ne varietur* in the framework of Newtonian and Laplacian mechanics. Determinism can be conceived as being *open* to incessant corrections of the formulae of laws and the concepts they link together, or as being *closed* on its own assumed definitive content. Laplace constructed the theory of closed determinism. Claude Bernard did not conceive of it in any other way, and this is undoubtedly why he did not believe that the collaboration of pathology and physiology could lead to a progressive rectification of physiological concepts. It is appropriate here to recall Alfred

North Whitehead's dictum: "Every special science has to assume results from other sciences. For example, biology presupposes physics. It will usually be the case that these loans really belong to the state of science thirty or forty years earlier. The presuppositions of the physics of my boyhood are today powerful influences in the mentality of physiologists.[43] [*NP*, pp. 107–109]

[137] The dynamic polarity of life and the normativity it expresses account for an epistemological fact of whose important significance Xavier Bichat was fully aware. Biological pathology exists but there is no physical or chemical or mechanical pathology:

> There are two things in the phenomena of life: (1) the state of health; (2) the state of disease, and from these two distinct sciences derive: physiology, which concerns itself with the phenomena of the first state, pathology, with those of the second. The history of phenomena in which vital forces have their natural form leads us, consequently, to the history of phenomena where these forces are changed. Now, in the physical sciences only the first history exists, never the second. Physiology is to the movement of living bodies what astronomy, dynamics, hydraulics, hydrostatics and so forth are to inert ones: these last have no science at all that corresponds to them as pathology corresponds to the first. For the same reason, the whole idea of medication is distasteful to the physical sciences. Any medication aims at restoring certain properties to their natural type: as physical properties never lose this type, they do not need to be restored to it. Nothing in the physical sciences corresponds to what is therapeutics in the physiological sciences.[44]

It is clear from this text that natural type must be taken in the sense of normal type. For Bichat, the natural is not the effect of a determinism, but the term of a finality. And we know well everything that can be found wrong in such a text from the viewpoint of a mechanist or materialist biology. One might say that long ago Aristotle believed in a pathological mechanics, since he admitted two kinds of movements: natural movements, through which a body regains its proper place where it thrives at rest, as a stone goes down to the ground, and fire, up to the sky; and violent movements, by which a body is pushed from its proper place, as when a stone is thrown in the air. It can be said that, with Galileo and Descartes, progress in knowledge of the physical world consisted in considering all movements as natural, that is, as conforming to the laws of nature, and that likewise progress in biological knowledge consisted in unifying the laws of natural life and pathological life. It is precisely this unification that Auguste Comte dreamed of and Claude Bernard flattered himself with having accomplished, as was seen above. To the reservations that I felt obliged to set forth at that time, let me add this. In establishing the science of movement on the principle of inertia, modern mechanics in effect made the distinction between natural and violent movements absurd, as inertia is precisely an indifference with respect to directions and variations in movement. Life is far removed from such an indifference to the conditions which are made for it; life is polarity. The simplest biological nutritive system of assimilation and excretion expresses a polarity. When the wastes of digestion are no longer excreted by the organism and congest or poison the internal environment, this is all indeed according to law (physical, chemical and so on), but none of this follows the norm, which is the activity of the organism itself. This is the simple fact that I want to point out when we speak of biological normativity. [*NP*, pp. 127–28]

The Normal and the Pathological as Qualitative Contrast

[138] Finally, as a result of the determinist postulate, it is the reduction of quality to quantity which is implied by the essential

identity of physiology and pathology. To reduce the difference between a healthy man and a diabetic to a quantitative difference of the amount of glucose within the body, to delegate the task of distinguishing one who is diabetic from one who is not to a renal threshold conceived simply as a quantitative difference of level, means obeying the spirit of the physical sciences which, in buttressing phenomena with laws, can explain them only in terms of their reduction to a common measure. In order to introduce terms into the relationships of composition and dependence, the homogeneity of these terms should be obtained first. As Emile Meyerson has shown, the human spirit attained knowledge by identifying reality and quantity. But it should be remembered that, though scientific knowledge invalidates qualities, which it makes appear illusory, for all that, it does not annul them. Quantity is quality denied, but not quality suppressed. The qualitative variety of simple lights, perceived as colors by the human eye, is reduced by science to the quantitative difference of wavelengths, but the qualitative variety still persists in the form of quantitative differences in the calculation of wavelengths. Hegel maintains that by its growth or diminution, quantity changes into quality. This would be perfectly inconceivable if a relation to quality did not still persist in the negated quality which is called quantity.[45]

From this point of view, it is completely illegitimate to maintain that the pathological state is really and simply a greater or lesser variation of the physiological state. Either this physiological state is conceived as having one quality and value for the living man, and so it is absurd to extend that value, identical to itself in its variations, to a state called pathological whose value and quantity are to be differentiated from and essentially contrasted with the first. Or what is understood as the physiological state is a simple summary of quantities, without biological value, a simple fact or system of physical and chemical facts, but as this state has no vital quality, it cannot be called healthy or normal or physiological. Normal and pathological have no meaning on a scale where the biological object is reduced to colloidal equilibria and ionized solutions. In studying a state that he describes as physiological, the physiologist qualifies it as such, even unconsciously; he considers this state as positively qualified by and for the living being. Now this qualified physiological state is not, as such, what is extended, identically to itself, to another state capable of assuming, inexplicably, the quality of morbidity.

Of course, this is not to say that an analysis of the conditions or products of pathological functions will not give the chemist or physiologist numerical results comparable to those obtained in a way consistent with the terms of the same analyses concerning the corresponding, so-called physiological functions. But it is arguable whether the terms "more" and "less," once they enter the definition of the pathological as a quantitative variation of the normal, have a purely quantitative meaning. Also arguable is the logical coherence of Bernard's principle: "The disturbance of a normal mechanism, consisting in a quantitative variation, an exaggeration, or an attenuation, constitutes the pathological state." As has been pointed out in connection with François-Joseph Victor Broussais's ideas, in the order of physiological functions and needs, one speaks of more and less in relation to a norm. For example, the hydration of tissues is a fact that can be expressed in terms of more and less; so is the percentage of calcium in blood. These quantitatively different results would have no quality, no value in a laboratory, if the laboratory had no relationship with a hospital or clinic where the results take on the value or not of uremia, the value or not of tetanus. Because physiology stands at the crossroads of the laboratory and the clinic, two points of view about biological phenomena are adopted there, but this does not mean that they can be interchanged. The substitution of quantitative progression for qualitative contrast in no way annuls this opposition. It always remains at the back of the mind of those who have chosen to adopt the theoretical and metric point of view. When we say that health and disease are linked by all the intermediaries, and when this continuity is converted into homogeneity, we forget that the difference continues to manifest itself at the extreme, without which the intermediaries could in no way play their mediating role; no doubt unconsciously, but wrongly, we confuse the abstract calculation of identities and the concrete appreciation of differences. [*NP*, pp. 110–12]

CHAPTER FIFTEEN

Normality and Normativity

The Value of Norms

[139] The state of any living thing in a given situation is, in general, always normal. Henri Bergson says there is no such thing as disorder; rather, there are two orders, one of which is substituted for the other without our knowledge and to our dismay. Similarly, we ought to say that there is no such thing as *abnormal*, if by the term we mean merely the absence of a previous positive condition or state. From the biological, social and psychological points of view, a pathological state is never a state without norms – such a thing is impossible. Wherever there is *life* there are norms. Life is a polarized activity, a dynamic polarity, and that in itself is enough to establish norms. The normal is therefore a universal category of life. Hence, it is by no means nonsensical to call the pathological "normal." But that is not grounds for denying the distinctiveness of the pathological, or for arguing that in biology the normal and the pathological are, but for minor quantitative differences, identical. The normal should not be opposed to the pathological, because under certain conditions and in its own way, the pathological is normal. There is a necessary contrast between health and disease. Health is more than normality; in simple terms, it is normativity. Behind all apparent normality,

one must look to see if it is capable of tolerating infractions of the norm, of overcoming contradictions, of dealing with conflicts. Any normality open to possible future correction is authentic normativity, or health. Any normality limited to maintaining itself, hostile to any variation in the themes that express it, and incapable of adapting to new situations is a normality devoid of normative intention. When confronted with any apparently normal situation, it is therefore important to ask whether the norms that it embodies are creative norms, norms with a forward thrust, or, on the contrary, conservative norms, norms whose thrust is toward the past. [MS *Normalité et normativité*, f. 1r]

Normality and Species

[140] In the biology of species, the problem of the normal and the pathological arises in connection with the problem of variations. Is an anomalous individual, that is, an individual in some respect at variance with a defined statistical type, a sick individual or a biological innovation? Is a fruit fly with no wings, or vestigial wings, sick? Biologists hostile to evolution or skeptical of mutationist explanations insist that mutations are recessive, often subpathological, and sometimes lethal. If, however, one holds that biological normality is determined by the interaction between structures and behaviors, on the one hand, and environmental conditions, on the other, there are ways of distinguishing (if not instantaneously at least retroactively) between the pathological normal and the normative normal. Phillipe L'Héritier and Georges Teissier's experiments on wingless drosophila, for example, proved the superiority of that variety in a drafty environment. [MS *Normalité et normativité*, f. 2r]

[141] Teissier reports another fact which shows that life, perhaps without looking for it, by using the variation of living forms, obtains a kind of insurance against excessive specialization without

reversibility, hence without flexibility, which is essentially a successful adaptation. In certain industrial districts in Germany and England the gradual disappearance of gray butterflies and the appearance of black ones of the same species has been observed. It was possible to establish that in these butterflies the black coloration was accompanied by an unusual vigor. In captivity the blacks eliminate the grays. Why isn't the same true in nature? Because their color stands out more against the bark of the trees and attracts the attention of birds. When the number of birds diminishes in industrial regions, butterflies can be black with impunity.[46]

In short, this butterfly species, in the form of varieties, offers two combinations of opposing characteristics, and they balance each other: more vigor is balanced by less security and vice versa. In each of the variations, an obstacle has been circumvented, to use a Bergsonian expression, a powerlessness has been overcome. To the extent that circumstances allow one such morphological solution to operate in preference to another, the number of representatives of each variety varies, and a variety tends more and more toward a species. [...]

Hence, finally, we see how an anomaly, particularly a mutation, that is, a directly hereditary anomaly, is not *pathological* because it is an anomaly, that is, a divergence from a specific type, which is defined as a group of the most frequent characteristics in their average dimension. Otherwise, it would have to be said that a mutant individual, as the point of departure for a new species, is both pathological, because it is a divergence, and normal, because it maintains itself and reproduces. In biology, the normal is not so much the old as the new form, if it finds conditions of existence in which it will appear normative, that is, displacing all withered, obsolete and perhaps soon to be extinct forms.

No fact termed normal, because expressed as such, can usurp the prestige of the norm of which it is the expression, starting from the moment when the conditions in which it has been referred to the norm are no longer given. There is no fact that is normal or pathological in itself. An anomaly or a mutation is not in itself pathological. These two express other possible norms of life. If these norms are inferior to specific earlier norms in terms of stability, fecundity, or variability of life, they will be called pathological. If these norms in the same environment should turn out to be equivalent, or in another environment, superior, they will be called normal. Their normality will come to them from their normativity. The pathological is not the absence of a biological norm; it is another norm, but one that is, comparatively speaking, pushed aside by life. [*NP*, pp. 81–82]

[142] No environment is *normal*. An environment is as it may be. No structure is normal in itself. It is the relation between the environment and the living thing that determines what is normal in both. A living thing is normal in the true sense when it reflects an effort on the part of life to maintain itself in forms and within norms that allow for a margin of variation, a latitude of deviation, such that as environmental conditions vary, one of those forms may prove to be more advantageous, hence more *viable*. An environment is normal when it allows a species to multiply and diversify in it in such a way as to tolerate, if necessary, changes in the environment.

If the relation between the environment and the living thing is such that neither can vary without compromising the viability of the living thing irreparably, the apparent normality of adaptation is in fact pathological. To be sick is to be unable to tolerate change. [MS *Normalité et normativité*, f. 2r]

Normality and Individuals

[143] From the standpoint of the biology of individuals, the problem of the normal and the pathological comes down to what Kurt

Goldstein calls "preferred behavior" and "catastrophic reaction." In responding to stimuli from the environment, an organism does not use every form of behavior it is capable of using but only certain preferred behaviors – preferred because they most fully express the nature of the organism and afford it the maximum possible order and stability. A sick individual is an individual locked in a struggle with its environment to establish a new order or stability. Recovery establishes a new norm, different from the old. During the course of the illness, the sick individual does everything possible to avoid catastrophic reactions. A catastrophic reaction is one that prevents rapid adaptation to changing environmental conditions. The concern with avoiding catastrophic reactions therefore reflects the organism's instinct of self-preservation. Self-preservation is not the most general characteristic of life; it is, rather, a characteristic of a reduced, diminished life. A healthy person is a person capable of confronting risks. Health is creative – call it normative – in that it is capable of surviving catastrophe and establishing a new order.

Goldstein's views overlap neatly with René Leriche's views of conception. Health becomes perceptible only in relation to disease, which reveals its essence by suggesting a possible transition to new norms. A person who cannot survive at high altitudes because of hypotension may be able to live normally at altitudes up to fifteen hundred feet. No one is obliged to live at altitudes above three thousand feet, but anyone may someday be forced to do so. In that case, anyone who cannot is "inferior." Man is a creature capable of changing or adapting to ambient conditions in order to survive. [MS *Normalité et normativité*, f. 2r, 3r]

[144] Health is a margin of tolerance for the inconstancies of the environment. But isn't it absurd to speak of the inconstancy of the environment? This is true enough of the human social environment, where institutions are fundamentally precarious, con-

ventions revocable and fashions as fleeting as lightning. But isn't the cosmic environment, the animal environment in general, a system of mechanical, physical and chemical constants, made of invariants? Certainly this environment, which science defines, is made of laws, but these laws are theoretical abstractions. The living creature does not live among laws but among creatures and events that vary these laws. What holds up the bird is the branch and not the laws of elasticity. If we reduce the branch to the laws of elasticity, we must no longer speak of a bird, but of colloidal solutions. At such a level of analytical abstraction, it is no longer a question of environment for a living being, nor of health nor of disease. Similarly, what the fox eats is the hen's egg and not the chemistry of albuminoids or the laws of embryology. Because the qualified living being lives in a world of qualified objects, he lives in a world of possible accidents. Nothing happens by chance, everything happens in the form of events. Here is how the environment is inconstant. Its inconstancy is simply its becoming, its history.

For the living being, life is not a monotonous deduction, a rectilinear movement; it ignores geometrical rigidity, it is discussion or explanation (what Goldstein calls *Auseinandersetzung*) with an environment where there are leaks, holes, escapes and unexpected resistances. Let us say it once more. We do not profess indeterminism, a position very well supported today. We maintain that the life of the living being, were it that of an amoeba, recognizes the categories of health and disease only on the level of experience, which is primarily a test in the affective sense of the word, and not on the level of science. Science explains experience but it does not for all that annul it.

Health is a set of securities and assurances (what the Germans call *Sicherungen*), securities in the present, assurances for the future. As there is a psychological assurance which is not presumption, there is a biological assurance which is not excess, and which is health. Health is a regulatory flywheel of the possibilities of reaction. Life often falls short of its possibilities, but when necessary can surpass expectations. [*NP*, pp. 197–98]

The Problem of Psychological Norms

The Child and the adult
[145] Childhood is a transitional state. It is normal for human beings to leave the state of childhood and abnormal to fall back into it. In childhood there is an intrinsic forward drive, a capacity for self-transcendence, that flourishes if the child is physically robust, intellectually perspicacious and allowed a certain freedom to pursue worthwhile goals. A child thinks constantly of imitating or rivaling what he sees adults doing: every day he thinks, "Tomorrow I will be a grown-up." Aristotle makes this magnificent observation: *anthropos anthropon genna*, man engenders man. This is true in terms of the material cause: it is man who supplies the seed from which the child is born. It is also true in terms of the formal cause: the embryo, the child and the adolescent develop toward adult human form. And it is true in terms of the final cause, an ideal of man and of the adult virtues that education instills in the child's mind. This last proposition should not be interpreted in too modern a sense, however. For the Ancients, and for Aristotle in particular, the essence of a thing was identical with its final form; the potential pointed toward the act, and movement ended in rest. The theory of forms telescoped the whole process of becoming into a typical privileged state. How a potential becomes an act, how a formal indeterminate becomes a form, would be unintelligible if form were not in every sense prior to potential and matter. Thus, humanity is transmitted from man to man, just as knowledge is transmitted from intelligence

telligence. Childhood, being a state of transition, is without human value. Greek pedagogy was therefore based on the identification of man with his typical finished form, his *acme*. In the child, the Greeks saw only the future soldier and future citizen. Plato shows no indulgence for the typical predilections and tendencies of childhood. Nothing was more alien to the ancient mind than the idea that childhood is, in each instance, a new beginning for mankind, a beginning whose innocence and enthusiasm are worthy of respect because of the implicit possibility of going further than man has ever gone before. Furthermore, the ancient family was based on strong paternal authority, and there was often violent conflict between fathers and sons owing to the father's domination of wife and children. Théodore de Saussure attached great importance to this fact in *Le Miracle grec*.[47] It can be argued, moreover, that the longer one remains ignorant of how children are made, the longer one remains a child; and one remains ignorant as long as one fails to contrast one's ideas with actual experience. At the root of the child's mentality is anxiety at not knowing why one is a child, that is, weak, powerless, dependent and attached to one's mother as a plant is attached to the nurturing soil. To remedy this anxiety the child dreams of vast magical powers, of a compensatory omnipotence. But contact with reality, which takes the form of conflict, cruelly demonstrates that such dreams are vain illusions. In other words, for political, philosophical and, in a more profound sense, religious reasons, the Ancients devalued childhood in a way that only accentuated those characteristics of childhood apt to provoke the contempt of adults. For the Ancients, the normal man was the normative man, and that meant quintessentially the adult. This is, moreover, a characteristic of all classical periods. The seventeenth-century French had basically the same idea. Descartes spoke of childish credulity and nursery tales in much the same manner as Plato. Jean de la Fontaine is famous for having said that he took pleasure in fairy tales, but his fables are hard on children. A certain value attached to the childish taste for the marvelous and for fiction, but it was a relative value; judged by logical norms, such things were considered absurd.

Paradoxically, it was the nineteenth century, which is often wrongly maligned for its alleged blind faith in science, that once again ascribed value not only to poetry but to childhood fantasy as well: witness Victor Hugo and Charles Baudelaire. (Every child is a genius in its way, and every genius is a child. [...] Genius is a deliberate reversion to childhood.) It was poets, long before psychologists, who proposed looking at the child's mentality as normal and valid, however distinct from the positive and utilitarian mentality of the bourgeois adult (as Baudelaire remarked, "To be *useful* has always seemed to me a most hideous thing"). Charles Dickens did in England what Hugo and Baudelaire did in France, especially in *Hard Times*. Artists, whose function is to dream for mankind beyond what is known, to scorn the real, to make the need for change imperative, found a treasure trove in the thought of children. When Eugène Delacroix said, "What is most real for me are the illusions I create," he was formulating the idea of a child. Then, with respect to the rehabilitation of childhood and many other things as well, contemporary psychology and philosophy came to the rescue: they provided poetic intuition with a discourse.

The study of the mentality of children began at roughly the same time as the study of primitive mentality. In French-speaking countries, the former discipline is epitomized by the name of Jean Piaget, the latter by Lucien Lévy-Bruhl. There can be no doubt that the methodological implications of Piaget's research were initially the same as those of Lévy-Bruhl: Piaget compared the thought of the child to that of a contemporary cultivated adult, an adult whose culture was of the sort that Piaget regarded as normative for his time, that is, for which scientific and rationalist values stood at the top of the hierarchy. Compared with the rational mentality, children's thinking could be characterized by adjectives beginning with the prefix *a-*, indicating some sort of lack or absence. Note, however, that Piaget's adult is what Max Weber and Karl Jaspers call an "ideal type." To be sure, it can be argued that this normal type is not only normative but average and characteristic of the majority. But the "mentality" of an age is a social fact, determined by education. If, in fact, in surveys, the ideal type turns out to reflect the average, it is because compulsory education has established certain norms. Here again, man engenders man, and if the norms imposed on many generations of children included a systematic devaluation of childhood, it should come as no surprise that, in comparing today's children to today's adults, it turns out that children lack many of the traits inculcated in adults. The problem of mentalities is inextricably interwined with that of education, and the problem of education is inextricably intertwined with that of generations. At any given point in time, those who happen to be adults are former children who were raised by other adults. It takes a generation to test the validity of educational ideas. And it takes fifty to sixty years (two generations) for philosophical values to become rooted as habits. Piaget's adults more or less unwittingly betray superficial tokens of respect for the positivist values of the period 1860–90, which gained favor with the educational reformers of the late nineteenth and early twentieth centuries. [MS *Le Normal et le problème des mentalités*, II, f. 1r, 2r, 3r]

[146] There is a characteristic gap between a child's desires and his means of realizing those desires. The child therefore creates a world of representations in which desires have the ability immediately to create objects presumed capable of satisfying

them. The child can experience pleasure only with permission or by delegation. He is strictly dependent on adults to meet its vital needs. Thus, to obey is to live. At first, there is no difference between social obligation and physical necessity. Adults, then, are both compensation for and inescapable reminders of the child's helplessness. Freudian psychology had the great merit of revealing the true essence of the child's thought. The child lives in illusion because he lives in desire, and because he feels desire long before gratification is physically possible. So long as it is impossible to act on the world in certain ways, desire and reality fail to coincide. And so long as desire sees no possibility of satisfaction, there is also no possibility of expression. The child cannot admit that he wants to grow up in order to subject his father to paternal law and the world to the law of the world, that is, to dominate men and domesticate things. He cannot admit this as long as he does not know, beyond what he is told and what he is not told (which comes to the same thing), how to act on things and men. The content of the child's thought is his ignorance of the biological reaiity of childhood. That ignorance lasts as long as the child remains unaware of copulation as his inception and fate, and so long as he is forbidden, whether by organic immaturity or social taboo, to engage in copulation himself.

In fairy tales and fantasies, the child seeks to satisfy a need for pleasure and to assert a power for which he still lacks the means. The wealth of imagination compensates for the poverty of realization.

What we Moderns call "adult" in man is his awareness of the gap between desire and reality. The adult does not rely on myth for the gratification of desire. In the adult, responsibility for the gratification of desires that present-day reality places out of reach can be delegated instead to *play* or *art*, that is, to illusions conscious of their practical value as well as their theoretical irreality.

The adult does not necessarily believe in the inevitability of progress, of knowledge and industry. Adults know that there are epistemological obstacles to progress and areas over which theory is powerless, yet they do not feel compelled on that account to seek compensation by harking back to a mode of thought that believes totally in the realization of desire in a normative reality.

It is normal to believe that there are possibilities other than those contained in science and technology at a particular point in time. It is normal for the child's generosity to persist into adulthood. But it is abnormal, because historically regressive, to suggest that the puerility of myth is superior to science and technology. The modern adult has limits that must be overcome, but they *cannot be overcome by returning to a mode of thought which ignores precisely that there are limits to desire in reality and obstacles to value in existence.*

To be sure, childhood deserves to be treated as a norm by adults – or, rather, not as a norm, precisely, but as a normative requirement, something to be transcended. This normative superhumanity of childhood is not to be confused with the responses that a child itself may adopt to his temporary powerlessness, responses that the child wishes with all his might to replace with true solutions, that is, solutions that are both verifiable and effective. [...]

In short, because the child is not a complete being, he exhibits a generosity that compensates for his avidity: this generosity can be proposed as normal because it is *normative*, that is, an affirmation of value.

Because the child is a helpless creature, however, he is credulous. *Credulity* is not normal in humans because it is not normative; it consists in taking for granted what has yet to be constructed.

In the end, the most perceptive rehabilitation of childhood is that of the poet. The poet is a visionary, a seer, but he sees what does not exist. We see what is. The poet does not so much describe what exists as point to values. The *poetic* consciousness is a correlative of the scientific consciousness, but also its inverse. Poetry is a poetic function, not a noetic one.

To hold out childhood as an ideal to adult humankind is to demonstrate that childhood is a *promise* and not a fact. Man must remain a child in the sense that he deserves to become the complete man of which children dream. [MS *Le Normal et le problème des mentalités*, II, f. 5r, 6r]

Primitive mentality
[147] Théodule Armand Ribot, following Auguste Comte, criticized introspective psychology as the psychology of the *civilized, adult, healthy white male*. Psychology's contempt for modes of thought different from that of the respectable, cultivated male reflected a hidden assumption that the respectable, cultivated male's mode of thought was somehow valid and normal. Montaigne wondered on what basis we judged the natives of the countries we colonized to be savages, but his skepticism was widely dismissed. Erasmus wrote *In Praise of Folly*, but it was regarded as no less fantastic than the plays of Shakespeare in which madmen were portrayed as wise. And Rousseau taught in *Emile* that the child is a complete human being, different from the adult not only in possessing less knowledge and experience but also in having an entirely different attitude toward life. But since Rousseau was accused of having abandoned his own children, his teaching was deemed utopian.

The seventeenth century identified man with his *acme*, or maturity, and Descartes held that "the prejudices of our childhood are the first and principal cause of our errors." Since we were "children before becoming adults," our reason was not as pure as if we had never made use of our senses. Before Philippe and Jean Etienne Dominique Esquirol, the insane were ;cted to punishment in lieu of treatment. Asylums were still ..... terrifying than prisons. To be sure, the eighteenth century witnessed the first glimmerings of relativism. When Montesquieu asked "How can anyone be a Persian?" he encouraged his contemporaries to recognize that such a thing was indeed perfectly possible. It became possible to submit Western society to the judgment of an Oriental and human psychology to the judgment of a mythical superman. But Montesquieu's *Persian Letters* and Voltaire's *Micromégas* were mere philosophical entertainments. Strange as it may seem, the prejudice that established the civilized white man as the standard of reference for all mankind grew out of a philosophy famous for condemning all prejudice. But Enlightenment philosophy found fault more with the *pre-* of prejudice than with the illusory certainty of its *judgment*: a prejudice was the judgment of a previous age. Yesterday's judgment was declared to be error because it survived only as a weapon of combat against the new. Diderot's purpose in rehabilitating the primitive, in the *Supplement to Bougainville's Voyage*, was essentially to discredit Christianity. The Christian religion was hoist on its own petard: whatever preceded the advent of truth was doomed to disappear. Historical precedence established logical perspective. Tolerance raises a similar problem: tolerance is the recognition of a plurality of values, the refusal to erect any value as a norm; intolerance is normative imperialism. But try as one will, a plurality of norms is comprehensible only as a hierarchy. Norms can coexist on a footing of equality only if drained of the normative intention that called them into existence as codified, normative decisions embodied in institutions, customs, dogmas, rites and laws. A norm cannot be normative without being militant, that is, intolerant. In intolerance, in aggressive normativity, there is of course hatred, but in tolerance there is contempt. Values tolerate what they deem to be valueless. The relativism and tolerance of the eighteenth century were inseparable from the essentially normative idea of progress. But progress was not conceived in terms of a relation of values; it was identified with the final value in a series, the one that transcended the others and in terms of which they were judged. That is why tolerance was the value in the name of which one became intolerant, and relativity the value in the name of which one became absolute. [MS *Le Normal et le problème des mentalités*, I, f. 1r]

[148] Positivism took the theories of Baron Turgot and Marquis de Condorcet on the progress of the human spirit and recast them in the form of a law, the law of three stages (theological, metaphysical and positive). In other words, it tried to force psychological speculation into the Procrustean bed of natural science. In formulating a *law* of progress, Comte was treating mind as if it were a natural object. At the same time he was declaring that sociology (or, as he saw it, the science of mind) was independent of biology in terms of object and method. The positive spirit was declared to be the ultimate form of the human spirit; theology and metaphysics were devalued, the first as a primitive form of spirit, the second as a transitional form. These forms impeded the development of spirit's full potential, so spirit rejected them. Dissatisfied with fictions, spirit created science. Hence, scientific thought was the normal (that is, the normative or ideal) state of thought. Positivism portrayed itself as the normal culmination of an ever closer and more faithful approximation to the intellectual norm. For Comte, theological thinking was like the thinking of children. With this simile, Comte ascribed positive value to maturity: that of the individual as well as that of the human race. And the maturation of the race, he implied, was just as inevitable and necessary as the maturation of the child.

Meanwhile, in Germany, Hegel's dialectic encouraged students of philosophy to see Hegelian philosophy as the culmination of the arduous advent of the Idea and the German bourgeois state as the normal form of all society. And in England, Spencer's evolutionism, taking up where Mill's positivism left off, further accentuated the philosophical belief that superiority and posteriority are one and the same. Anterior, less complex and inferior became synonymous.

Little by little a diffuse dogma took shape: namely, that the intellectually primitive and the intellectually puerile are two forms of a single infirmity. At around the same time, moreover, research in embryology showed that certain anatomical anomalies were the result of arrested development. A club foot, a harelip, a testicular ectopia – each of these conditions is the perpetuation after birth of a state through which every fetus or embryo passes while still in the uterus. What is abnormal is the halting of development at an intermediate stage. What is normal at one moment in time becomes abnormal later.

When Lucien Lévy-Bruhl published *Fonctions mentales dans les sociétés inférieures* in 1910, his initial use of the term "prelogical" to characterize the "primitive" mode of thought suggested an implicit depreciation. Philosophical opinion was divided. Some philosophers were delighted to discover that the theory of *mentalités* provided arguments to justify a normative conception of the history of thought. At last, there were criteria for choosing sides in philosophical combat, for distinguishing between fruitful new ideas and survivals of the past, for separating the backward-looking from the forward-looking. Léon Brunschvicg, for example, used both Lévy-Bruhl and Piaget to argue in favor of his own doctrine concerning the Ages of Intelligence and to disparage Aristotle's philosophy on the grounds that it remained confined within the mental framework of a primitive or a child of six.

Meanwhile, other philosophers, sensing that what Lévy-Bruhl

was really arguing was that primitive thought was not prelogical but heterogeneous, and sensing, too, that champions would soon come forward to defend the merits of forms of thought "different" from modern science, sought to restore continuity: the primitive, they argued, was not as alien to our logic as some claimed, nor was modern thought as fully logical as some believed. The transition from one form of mentality to another involved a certain loss of content (modern thought is not as rich as primitive thought) as well as the consolidation of a certain disposition (modern thought is more methodical). We can easily understand what the primitive is: it is what we become when we abandon the critical spirit, the precious prize of an always vulnerable conquest (thesis of Belot and Parodi, discussion at the Société Française de Philosophie after publication of Lévy-Bruhl's books).

Nevertheless, both groups of philosophers preserved the essential rationalist and positivist norms: reason is superior to mysticism; noncontradiction is superior to participation; science is superior to myth; industry is superior to magic; faith in progress is superior to the progress of faith. [MS *Le Normal et le problème des mentalités*, I, f. 2r, 3r]

[149] Rationalism and positivism thus depreciated mythical thinking. Despite the rationalist attitudes implicit in Christianity, moreover, the theologians recognized that this depreciation of myth was all-encompassing. Phenomenological theologians therefore decided that only one reaction was possible: all mythological and religious systems would have to be rescued *en bloc*.

Modern mythology portrays itself as restoring the value of myth in the face of rationalist depreciation. To grant recognition to other value systems is tantamount to restricting the value of rationalism. In the end, normative tolerance proves to be a depreciation of the positivist depreciation of myth. It is impossible to save the content of any religion without saving the content of all religions.... In order to save a religion that had, admittedly, abandoned the Inquisition and the stake, it was necessary to save other religions with their whirling dervishes and human sacrifices: for if it is true that primitive mentality is a totalizing structure, the rehabilitation of the mythic mentality is also the rehabilitation of savagery in all its forms. The friend of primitive mentality will object that the modern mentality is not hostile to the bombing of civilian populations. But no one is saying that the modern mentality or, for that matter, any constituted norm must be preferred over primitive mentality. *The modern mentality is not a structure but a tendency.* To prefer it is simply to prefer a tendency, a normative intention. [...]

The primitive and modern mentalities are not coexisting absolutes but successive relatives. Technology is clearly progress when it demonstrates the failure of magic; science is clearly progress when it grows out of the inadequacy of technology. The modern mentality has certain advantages over previous norms, advantages from which it derives relative but not absolute value.

Modernity is not *normal* in the sense of having achieved a definitive superior state. It is *normative*, however, because it strives constantly to outdo itself. Henri Bergson got at least one thing right: a true mechanics may not exist, but a true mysticism is a contradiction in terms. Despite Bergson's objective sympathy for the primitive mentality, his philosophy is in no sense a reactionary revaluation of irrationality. [...]

Modern man is experiencing a crisis in the sense that domination and mastery of the environment elude his grasp. But the resolution of that crisis does not lie in the past. It does not exist in ready-made form but remains to be invented.

The modern is modern only because it has found solutions to problems that the primitive seldom posed. Modernity poses different problems. Modern values are provisional. But the changes that have brought those values to consciousness are *normative*, and a normative direction is normally worth pursuing. [MS *Le Normal et le problème des mentalités*, I, f. 6r, 7r]

Normative invention
[150] In the evolution of the individual, the mentality of adulthood comes after the mentality of childhood; in the evolution of mankind, the modern mentality follows the primitive mentality. But when we refer to adulthood or modernity as normal, we do not mean simply that they succeed earlier stages of existence. Each of these states is normal in the sense that it effectively devalues another state hobbled by internal conflict: between desire and reality, or between power and science. To be sure, just because the modern recognizes these conflicts and to a limited degree resolves them, it does not thereby constitute the final stage of evolution. The expectation that today's understandings will be transcended is a normal feature of the modern mentality. Hence there is no remedy for modernity's ills in merely returning to old norms. The only true remedy lies in the invention of new norms. Generosity of spirit is to be imitated, but belief in the efficacy of immediate solutions must be rejected. Normativity is inherent in the kinds of change that brought modernity to consciousness. It is this normativity that must in the normal course of things be perpetuated.

To sum up, all normality must be judged with reference to the possibility of devaluation in a normative sense. Therein lies the only method for detecting *mystification*.

Pathology can sometimes mimic health. If sickness is often a refuge for an individual in conflict with himself, others or the environment, revolution is often a means of avoiding necessary innovation and reform. Time cannot settle the question of what a person's or a society's norms ought to be: neither yesterday nor tomorrow is an infallible oracle. Norms and values are tested by situations calling for normative invention. One can respond to a challenge either by seeking refuge or exercising creative ingenuity; often the two responses seem deceptively similar. Yet there is one sure criterion for identifying creativity: a willingness to put norms to the test, to ascertain their value fairly and without trying to make them seem artificially normal. The normal is that which is normative under given conditions, but not everything that is normal under given conditions is normative. It must always be permissible to test the normal by varying the ambient conditions. It is in this sense that *the history of the world is the judgment of the world*. [MS *Normalité et normativité*, f. 4r]

The Problem of Social Norms

[151] The Latin word *norma*, which, etymologically speaking, bears the weight of the initial meaning of the terms "norms" and "normal," is the equivalent of the Greek ὀρθός. Orthography [French, *orthographe*, but long ago *orthographie*], orthodoxy, orthopedics, are normative concepts prematurely. If the concept of orthology is less familiar, at least it is not altogether useless to know that Plato guaranteed it[48] and the word is found, without a reference citation, in Emile Littré's *Dictionnaire de la langue française*. Orthology is grammar in the sense given it by Latin and medieval writers, that is, the regulation of language usage.

If it is true that the experience of normalization is a specifically anthropological or cultural experience, it can seem normal that language has proposed one of its prime fields for this experience. Grammar furnishes prime material for reflection on norms. When Francis I in the edict of Villers-Cotterêt ordains that all judicial acts of the kingdom be drawn up in French, we are dealing with an imperative.[49] But a norm is not an imperative to do something under pain of juridical sanctions. When the grammarians of the same era undertook to fix the usage of the French language, it was a question of norms, of determining the reference, and of defining mistakes in terms of divergence, difference. The reference is borrowed from usage. In the middle of the seventeenth century this is Claude Favre de Vaugelas's thesis: "Usage is that to which we must subject ourselves entirely in our language."[50] Vaugelas's works turn up in the wake of works of the Académie française, which was founded precisely to embellish the language. In fact in the seventeenth century the grammatical norm was the usage of cultured, bourgeois Parisians, so that this norm reflects a political norm: administrative centralization for the benefit of royal power. In terms of normalization there is no difference between the birth of grammar in France in the seventeenth century and the establishment of the metric system at the end of the eighteenth. Cardinal Richelieu, the members of the National Convention and Napoleon Bonaparte are the successive instruments of the same collective demand. It began with grammatical norms and ended with morphological norms of men and horses for national defense,[51] passing through industrial and sanitary norms.

Defining industrial norms assumes a unity of plan, direction of work, stated purpose of material constructed. The article on "Gun-carriage" in the *Encyclopédie* of Diderot and d'Alembert, revised by the Royal Artillery Corps, admirably sets forth the motifs of the normalization of work in arsenals. In it we see how the confusion of efforts, the detail of proportions, the difficulty and slowness of replacements, useless expense, are remedied. The standardization of designs of pieces and dimension tables, the imposition of patterns and models have as their consequence the precision of separate products and the regularity of assembly. The "Gun-carriage" article contains almost all the concepts used in a modern treatise on normalization except the term "norm." Here we have the thing without the word.

The definition of sanitary norms assumes that, from the political point of view, attention is paid to populations' health considered statistically, to the healthiness of conditions of existence and to the uniform dissemination of preventive and curative treatments perfected by medicine. In Austria Maria Theresa and Joseph II conferred legal status on public health institutions by creating an Imperial Health Commission (*Sanitäts-Hofdeputation*, 1753) and by promulgating a *Haupt Medizinal Ordnung*, replaced in 1770 by the *Sanitäts-normativ*, an act with forty regulations related to medicine, veterinary art, pharmacy, the training of surgeons, demographical and medical statistics. With respect to norm and normalization here, we have the word with the thing.

In both of these examples, the norm is what determines the normal starting from a normative decision. As we are going to see, such a decision regarding this or that norm is understood only within the context of other norms. At a given moment, the experience of normalization cannot be broken down, at least not into projects. Pierre Guiraud clearly perceived this in the case of grammar when he wrote: "Richelieu's founding of the Académie française in 1635 fit into a general policy of centralization of which the Revolution, the Empire, and the Republic are the heirs.... It would not be absurd to think that the bourgeoisie annexed the language at the same time that it seized the instruments of production."[52] It could be said in another way by trying to substitute an equivalent for the Marxist concept of the ascending class. Between 1759, when the word "normal" appeared, and 1834, when the word "normalized" appeared, a normative class had won the power to identify – a beautiful example of ideological illusion – the function of social norms, whose content it determined, with the use that that class made of them.

That the normative intention in a given society in a given era cannot be broken down is apparent when we examine the relations between technological and juridical norms. In the rigorous and present meaning of the term, technological normalization consists in the choice and determination of material, the form and dimensions of an object whose characteristics from then on become necessary for consistent manufacture. The division of labor constrains businessmen to a homogeneity of norms at the heart of a technical-economic complex whose dimensions are constantly evolving on a national or international scale. But technology develops within a society's economy. A demand to simplify can appear urgent from the technological point of view, but it can seem premature from the industrial and economic point of view as far as the possibilities of the moment and the immediate future are concerned. The logic of technology and the interests of the economy must come to terms. Moreover, in another respect, technological normalization must beware of an excess of rigidity. What is manufactured must finally be consumed. Certainly, the logic of normalization can be pushed as far as the normalization of needs by means of the persuasion of advertising. For all that, should the question be settled as to whether need is an object of possible normalization or the subject obliged to invent norms? Assuming that the first of these two propositions is true, normalization must provide for needs, as it does for objects characterized by norms, margins for divergence, but here without quantification. The relation of technology to consumption introduces into the unification of methods, models, procedures and proofs of qualification, a relative flexibility, evoked furthermore by the term "normalization," which was preferred in France in 1930 to "standardization," to designate the administrative organism responsible for enterprise on a national scale.[53] The concept of normalization excludes that of immutability, includes the anticipation of a possible flexibility. So we see how a technological norm gradually reflects an idea of society and its hierarchy of values, how a decision to normalize assumes the representation of a possible whole of correlative, complementary or compensatory decisions. This whole must be finished in advance, finished if not closed. The representation of this totality of reciprocally relative norms is planning. Strictly speaking, the unity of a Plan would be the unity of a unique thought. A bureaucratic and technocratic myth, the Plan is the modern dress of the idea of Providence. As it is very clear that a meeting of delegates and a gathering of machines are hard put to achieve a unity of thought, it must be admitted that we would hesitate to say of the Plan what La Fontaine said of Providence, that it knows what we need better than we do.[54] Nevertheless – and without ignoring the fact that it has been possible to present normalization and planning as closely connected to a war economy or the economy of totalitarian regimes – we must see above all in planning endeavors the attempts to constitute organs through which a society could estimate, foresee and assume its needs instead of being reduced to recording and stating them in terms of accounts and balance sheets. So that what is denounced, under the name of rationalization – the bogey complacently waved by the champions of liberalism, the economic variety of the cult of nature – as a mechanization of social life perhaps expresses, on the contrary, the need, obscurely felt by society, to become the organic subject of needs recognized as such.

It is easy to understand how technological activity and its normalization, in terms of their relation to the economy, are related to the juridical order. A law of industrial property, juridical protection of patents or registered patterns, exists. To normalize a registered pattern is to proceed to industrial expropriation. The requirement of national defense is the reason invoked by many states to introduce such provisions into legislation. The universe of technological norms opens onto the universe of juridical norms. An expropriation is carried out according to the norms of law. The magistrates who decide, the bailiffs responsible for carrying out the sentence, are persons identified with their function by virtue of norms, installed in their function with the delegation of competence. Here, the normal descends from a higher norm through hierarchized delegation. In his *Reinen Rechtslehre*,[55] Hans Kelsen maintains that the validity of a juridical norm depends on its insertion in a coherent system, an order of hierarchized norms, drawing their binding power from their direct or indirect reference to a fundamental norm. But there are different juridical orders because there are several fundamental, irreducible norms. If it has been possible to contrast this philosophy of law with its powerlessness to absorb political fact into juridical fact, as it claims to do, at least its merit in having brought to light the relativity of juridical norms hierarchized in a coherent order has been generally recognized. So that one of Kelsen's most resolute critics can write: "The law is the system of conventions and norms destined to orient all behavior inside a group in a well-defined manner."[56] Even while recognizing that the law, private as well as public, has no source other than a political one, we can admit that the opportunity to legislate is given to the legislative power by a multiplicity of customs which must be institutionalized by that power into a virtual juridical whole. Even in the absence of the concept of juridical order, dear to Kelsen, the relativity of juridical norms can be justified. This relativity can be more or less strict. There exists a tolerance for nonrelativity which does not mean a gap in relativity. In fact the norm of norms remains convergence. How could it be otherwise if law "is only the regulation of social activity"?[57][...]

The correlativity of social norms – technological, economic, juridical – tends to make their virtual unity an organization. It is not easy to say what the concept of organization is in relation to that of organism, whether we are dealing with a more general structure than the organism, both more formal and richer; or whether we are dealing with a model which, relative to the organism held as a basic type of structure, has been singularized by so many restrictive conditions that it could have no more consistency than a metaphor.

Let us state first that in a social organization, the rules for adjusting the parts into a collective which is more or less clear as to its own final purpose — be the parts individuals, groups or enterprises with a limited objective — are external to the adjusted multiple. Rules must be represented, learned, remembered, applied, while in a living organism the rules for adjusting the parts among themselves are immanent, presented without being represented, acting with neither deliberation nor calculation. Here there is no divergence, no distance, no delay between rule and regulation. The social order is a set of rules with which the servants or beneficiaries, in any case, the leaders, must be concerned. The order of life is made of a set of rules lived without problems.[58] [*NP*, pp. 248–50]

[152] We shall say otherwise — certainly not better, probably less well — namely that a society is both machine and organism. It would be only a machine if the collective's ends could not only be strictly planned but also executed in conformity with a program. In this respect, certain contemporary societies with a socialist form of economy tend perhaps toward an automatic mode of functioning. But it must be acknowledged that this tendency still encounters obstacles in facts, and not just in the ill-will of skeptical performers, which oblige the organizers to summon up their resources for improvisation. It can even be asked whether any society whatsoever is capable of both clearsightedness in determining its purposes and efficiency in utilizing its means. In any case, the fact that one of the tasks of the entire social organization consists in its informing itself as to its possible purposes —

with the exception of archaic and so-called primitive societies where purpose is furnished in rite and tradition just as the behavior of the animal organism is provided by an innate model – seems to show clearly that, strictly speaking, it has no intrinsic finality. In the case of society, regulation is a need in search of its organ and its norms of exercise.

On the other hand, in the case of the organism the fact of need expresses the existence of a regulatory apparatus. The need for food, energy, movement and rest requires, as a condition of its appearance in the form of anxiety and the act of searching, the reference of the organism, in a state of given fact, to an optimum state of functioning, determined in the form of a constant. An organic regulation or a homeostasis assures first of all the return to the constant when, because of variations in its relation to the environment, the organism diverges from it. Just as need has as its center the organism taken in its entirety, even though it manifests itself and is satisfied by means of one apparatus, so its regulation expresses the integration of parts within the whole though it operates by means of one nervous and endocrine system. This is the reason why, strictly speaking, there is no distance between organs within the organism, no externality of parts. The knowledge the anatomist gains from an organism is a kind of display in extensiveness. But the organism itself does not live in the spatial mode by which it is perceived. The life of a living being is, for each of its elements, the immediacy of the copresence of all. [*NP*, pp. 252-53]

[153] Social regulation tends toward organic regulation and mimics it without ceasing for all that to be composed mechanically. In order to identify the social composition with the social organism in the strict sense of the term, we should be able to speak of a society's needs and norms as one speaks of an organism's vital needs and norms, that is, unambiguously. The vital

needs and norms of a lizard or a stickleback in their natural habitat are expressed in the very fact that these animals are very natural living beings in this habitat. But it is enough that one individual in any society question the needs and norms of this society and challenge them — a sign that these needs and norms are not those of the whole society — in order for us to understand to what extent social need is not immanent, to what extent the social norm is not internal, and, finally, to what extent the society, seat of restrained dissent or latent antagonisms, is far from setting itself up as a whole. If the individual poses a question about the finality of the society, is this not the sign that the society is a poorly unified set of means, precisely lacking an end with which the collective activity permitted by the structure would identify? To support this we could invoke the analyses of ethnographers who are sensitive to the diversity of systems of cultural norms. Claude Lévi-Strauss says: "We then discover that no society is fundamentally good, but that none is absolutely bad; they all offer their members certain advantages, with the proviso that there is invariably a residue of evil, the amount of which seems to remain more or less constant and perhaps corresponds to a specific inertia in social life resistant to all attempts at organization."[59] [*NP*, pp. 255-56]

On the Normative Character of Philosophical Thought

[154] Philosophy is the love of Wisdom. One sees immediately that wisdom is for philosophy an Ideal, since love is desire for something that it is possible to possess. Thus, at the origin of the philosophical quest is the confession of a lack, the recognition of a gap between an existence and a need.

Wisdom is more than science in the strict and contemporary sense of the word, for science is a contemplative possession of reality through exclusion of all illusion, error and ignorance, whereas Wisdom is the use of principles of appreciation provided by science for the purpose of bringing human life into a state of practical and affective perfection, or happiness.

Wisdom is therefore the realization of a state of human fulfillment and excellence, a realization immediately derived from knowledge of an order of perfection. Wisdom is thus clearly a practical form of consciousness.

Now let us compare the etymological definition and ancient conception of philosophy with our commonsense image. In common parlance, philosophy is a certain disposition to accept events deemed necessary and inevitable, to subject prejudices and phantoms of the imagination to cold scrutiny and criticism, and to regulate one's conduct in accordance with firm personal principles of judgment and evaluation. It seems probable, moreover, that insofar as those principles are remote from everyday life, people are inclined to think of philosophy as utopian and idle speculation of no immediate use and therefore of no value. Common sense, then, seems to lead to two contradictory judgments concerning philosophy. On the one hand, it sees philosophy as a rare and therefore prestigious discipline and, if it lives up to its promises, as an important spiritual exercise. On the other hand, it deduces from the variety of competing philosophical doctrines that philosophy is inconsistent and fickle, hence a mere intellectual game. Yet this judgment, which tends to discredit philosophical speculation, is contradicted by the fact that philosophers throughout history have been the object of hostility and even persecution, sometimes by political leaders and sometimes by the masses themselves. If the teachings and examples of the philosophers are so widely feared, then the activity must not be entirely futile.

Now let us bring these scattered observations together. To deny that philosophy has any "utility" is to recognize that it reflects a concern with the ultimate meaning of life rather than with immediate expedients, with life's ends rather than its means. Just as we cannot focus simultaneously on objects close to us and objects far away, we also cannot interest ourselves simultaneously in ends and means. Now, it is usual – not to say normal – for people to interest themselves primarily in means, or what they take to be means, without noticing that means exist only in relation to ends and that, in accepting certain means, they unconsciously accept the ends that make them so. *In other words, they accept whatever philosophy happens to be embodied in the values and institutions of a particular civilization.* To accept, for example, that saving is a means to a better life is implicitly to accept a bourgeois system of values, a value system totally different from that of feudal times. This perversion of our attention is what caused Blaise Pascal to say, "It is a deplorable thing to see men deliberating always on means and never on ends," and further, "Man's sensitivity to small things and insensitivity to large ones [are] signs of a peculiar inversion of values." Philosophy is a corrective to this inversion, and if the commonsense criticism that philosophy is not useful, which is strictly accurate, is intended to suggest that it is therefore absolutely valueless, it errs only in its identification of value with utility. It is true that philosophy is justified only if it has value or is a value, but it is not true that utility is the only value: utility is valuable only in something that is a means to an end.

Insofar as philosophy is the search for a meaning of life (a justification of life that is neither pure living nor even the will to live but *savoir-vivre*, knowledge of what it is to live), it enters into competition and occasionally into conflict with political and religious institutions, which are collective systems for organizing human interests. Every social institution embodies a human interest; an institution is the codification of a value, the embodiment of value as a set of rules. The military, for example, is a social institution that fulfills a collective need for security or aggression.

Philosophy is an individual quest, however. In the *History of Philosophy* Hegel says, "Philosophy begins only where the individual knows itself as individual, for itself, as universal, as essential, as having infinite value qua individual." The individual can participate directly in the Idea (or, as we would say, in value) without the mediation of any institution. Philosophy is an asocial activity. There are no philosophical institutions. Schools are associations, not societies.

Philosophical judgment therefore cannot avoid casting itself as a competitor of both political judgment and religious judgment, which in any case are closely related. It is not unusual, moreover, for competition to turn into rivalry. Either philosophy reinforces communal beliefs, in which case it is pointless, or else it is at odds with those beliefs, in which case it is dangerous. "Philosophy," Aristotle said, "must not take orders, it must give them."

The upshot of this discussion is that the essence of philosophical speculation is to apply a normative corrective to human experience — but that is not all. Any technique is basically normative, because it sets forth or applies rules in the form of formulas, procedures, models and so on. But this normative character of technique is secondary and abstract: secondary because it has to do with means, and abstract because it is limited to searching for one kind of satisfaction. The multiplicity of techniques assumes a plurality of distinct needs. If philosophy is a normative discipline, moreover, it is primordially and concretely so. The best-known definitions of philosophy tend to stress one of these aspects over the other: either normative or concrete. Nevertheless, both adjectives figure in all the definitions. The Stoics emphasize the normative: in defining philosophy as *spiritual medicine*, they assume that passion and disease are one and the same.

Novalis says something slightly different when he calls philosophy a "higher pathology."[...]

Although it is true that ancient philosophy postulates the unity of value, it does so, I think, in an ontological sense, for the Ancients also held that the value of action is inferior to that of knowledge. Ancient philosophy was intellectualist. Knowledge of the universal order is enough to establish it. Virgil's line "*Felix qui potuit rerum cognoscere causas*" (Happy is the man who knows the causes of things) might serve as an epigraph to all ancient philosophies. No anti-intellectualist has been as clear on this point as Nietzsche: "A metamorphosis of being by knowledge: therein lies the common error of rationalists, Socrates foremost among them."[60] In *The Birth of Tragedy*, he calls Socrates the "father of theoretical optimism" and holds him responsible for the illusory belief that "thought, following the Ariadne's thread of causality, can penetrate the deepest abysses of being, that it has the power not only to know but to reform existence."[61] (Note, in passing, that Pascal and Schopenhauer showed Nietzsche the way to the path of theoretical pessimism.)

Given that modern philosophy cannot use ancient wisdom as a *model*, can it perhaps better serve the *intention* that animated the ancient lovers of wisdom? The connection between ancient and modern philosophy is deeper than a shared ideal; it is a shared need. The need that gave rise to ancient philosophy was for a mental organizing structure, a structure at once normative and concrete and thus capable of defining what the "normal" form of consciousness was. This need manifested itself in the troubling, unstable, painful and therefore abnormal character of ordinary experience.[...]

The ancient mind nevertheless lacked the notion of a *spiritual subject*, that is, an infinitely generous and creative power. Ancient philosophy treated the soul as subordinate to the idea and creation as subordinate to contemplation. It comprised a physics, a logic, an ethics, but no aesthetics. Ancient thought was spontaneously naturalistic. It had no notion of values that might not exist or that ought not to exist. It sought value in being, virtue in strength, soul in breath. Modern philosophy is conscious of the powers of mind. Even the knowledge of impotence has, since Kant, often been interpreted as a power of mind. Hence, there is no obstacle to modern philosophy's being a search for a concrete unity of values. Summarizing the foregoing analysis, then, I offer this definition: modern philosophy is primordial, concrete, normative judgment.

What is true of norms in general is therefore true of philosophy. The abnormal, being the a-normal, logically follows the definition of the normal. It is a logical negation. But it is the priority of the abnormal that attracts the attention of the normative, that calls forth a normative decision and provides an opportunity to establish normality through the application of a norm. A norm that has nothing to regulate is nothing because it regulates nothing. The essence of a norm is its role. Thus practically and functionally the normal is the operational negation of a state which thereby becomes the logical negation of that state; the abnormal, though logically posterior to the normal, is functionally first. Hence philosophy is inevitably a second stage or moment. It does not create values because it is called into being by differences among values. Historically, philosophy can be seen as an effort of mind to give value to human experience through critical examination and systematic appreciation of the values spontaneously embodied in civilizations and cultures. The sciences little by little create truth for humankind. Political and religious institutions little by little turn human actions into good works. The arts, by representing man's dreams, little by little reveal the extent of his ambitions. In the primitive mind these functions are intertwined, so that myth imperiously defines what is real, what powers men have, and how they relate to one another, and that is why philosophy takes myth as its first object of reflection. In the past, philosophy grew out of conflict among myths; today it grows out of the conflict among the various functions of mind.

Philosophy can succeed in its intention – to recover the unity of effort behind disparate acts of spontaneous creation – only by relating the various elements of culture and civilization: science, ethics, religion, technology, fine arts. To establish such relations is to choose among values. Criticism and hierarchy are therefore essential. Philosophy cannot adopt anything but a critical attitude toward the various human functions that it proposes to judge. Its goal is to discover the meaning of those functions by determining how they fit together, by restoring the unity of consciousness. The business of philosophy is therefore not so much to solve problems as to create them. In Léon Brunschvicg's words, philosophy is the "science of solved problems," that is, the questioning of received solutions. Now we can understand why philosophy has attracted hostile reactions through the ages: philosophy is a questioning of life and therefore a threat to the idea that everything necessary to life is already in our possession. The goal of philosophy is to search for reasons to live by seeking the end for which life is supposed to be the means. But to pursue such a goal is also to discover reasons not to live. Nothing is more at odds with life than the idea that an end to life may be a value and not simply an accident. Therein lies one source of philosophy's unpopularity. [MS *Du Caractère normatif de la pensée philosophique*, f. 1r, 2r, 3r, 4r, 5r, 6r]

Critical Bibliography

Camille Limoges

This bibliography is divided into two parts. Part One includes the titles of Georges Canguilhem's published works. Part Two is a selection of the most significant published reviews of and commentaries on these works. This bibliography is intended primarily as a working tool. It includes a substantial number of titles, published mainly before 1943, that are not found in the only other available bibliography (see below, Part Two, the penultimate entry under 1985).

Succinct biographical and contextual information, whenever relevant and available, is given under an entry. Each entry appears under the year of its publication, in many cases with the circumstances surrounding the origin of the text – for example, a public lecture or paper presented at a scholarly conference. Those books consisting of a collection of lectures and/or previously published papers are identified as such. When applicable, various editions are noted at the first mention of a title. Only new editions involving a different publisher or translation, and/or revisions or additions to the texts, are cited under the year of the new publication.

No doubt, had Georges Canguilhem been asked to provide his own bibliography, he would not have included a good number of the titles given here — not because of a wish to conceal any of them, but because Canguilhem has always maintained a strict distinction between the works of the author (*l'oeuvre*) and the "traces" of the intellectual and professorial career.

As a bibliographer who is also a researcher interested in French contemporary intellectual history, it has been my contention that an account as complete as possible of the printed "traces" of Canguilhem's remarkable intellectual trajectory was well worth pursuing. I am confident that many readers will share my opinion.

Part One

Works by Georges Canguilhem

Georges Canguilhem received his early education at the elementary school and then the high school of his native town, Castelnaudary, in southwestern France. In 1921, at the age of seventeen, he entered the *khâgne* — special classes that prepared students for the highly competitive entrance examinations to the Ecole Normale Supérieure — of the Lycée Henri IV in Paris. Canguilhem attended the Lycée from 1921 to 1924, taught by the philosopher Emile Chartier (better known under the pen name "Alain"). Alain taught the philosophy course from 1903 to 1933, interrupted only by World War I, when he voluntarily enlisted (he was too old to be drafted) and served in the artillery. In his readings of the great philosophical and literary texts, Alain led his students to analyze critically and to respect these writings, while emphasizing a neo-Kantian perspective, as well as his own staunch pacifism — an ethics based on a fundamental distrust of power ("*le citoyen contre les pouvoirs*") and of republican generosity. Alain deeply influenced Canguilhem's intellectual life during these years.

In 1924, Canguilhem entered the Ecole Normale Supérieure, where, as an unapologetic antimilitarist and pacifist, he remained faithful to Alain's teachings. With the rise of national socialism, Canguilhem in 1934–35 came to reject his politics of pacifism, and later became an active resistance member. Despite the change in his politics, Canguilhem's close attachment to Alain never wavered; he was at Alain's bedside upon his death, after suffering through a long illness, on June 2, 1951 (see Jean-François Sirinelli, *Génération intellectuelle: Khâgneux et normaliens dans l'entre-deux-guerres* [Paris: Fayard, 1988], pp. 330ff. and 464ff.).

Canguilhem's class of 1924 at the Ecole Normale Supérieure was particularly distinguished: it included Raymond Aron, Jean-Paul Sartre, and Paul Nizan, among other luminaries. The philosopher and mathematician Jean Cavaillès had entered the Ecole the previous year, and he and Canguilhem commenced a strong friendship that would continue for many years.

1926

"La Théorie de l'ordre et du progrès chez Auguste Comte," Diplôme d'études supérieures, Sorbonne.

Written under the supervision of Célestin Bouglé. At that time, the Diplôme d'études supérieures was pursued upon completion of undergraduate studies (*licence*), and before the "*agrégation*" examination, which students prepared at the Ecole Normale Supérieure. Some fifty years later Canguilhem wrote a short commemoration of Bouglé (1978).

1927

Canguilhem played a major role that year in the iconoclastic revue that Ecole Normale students organized and staged at the end of each academic year. He was one of the writers of the play "Le Désastre de Langson," a pun involving the name of the director of the Ecole Normale, Gustave Lanson, and Lang Son in Indochina, where a battle between the French and the Chinese had led to the dismissal of the Jules Ferry government in 1885. Two antimilitarist songs were considered particularly outrageous – "Sur l'Utilisation des intellectuels en temps de guerre" and "Complainte du capitaine Cambusat" (Cambusat was an officer responsible for the military instruction of the Ecole Normale students). Canguilhem was author of the first and coauthor of the second, with a group of fellow students including Sartre. Sirinelli has reprinted the text of both songs (see *Génération intellectuelle*, pp. 326-28) and provides substantial material about the context of these events. Lanson held Canguilhem and others responsible for these actions, and the inscription "PR" (for "revolutionary propaganda") was recorded in the military dossiers of the culprits – who were supposed to become officers at the end of their "military preparation" at the Ecole Normale (Sirinelli, p. 339). Canguilhem purposely failed the examination concluding this preparation in Spring 1927 by allowing the base of the machine gun he was supposed to dismount to fall on the foot of the examining officer (ibid., p. 465).

At this time, he was actively circulating a petition against the Loi Paul-Boncour, which had just been passed by the Assemblée nationale, on the mobilization of the country for wartime (see below, first entry under 1927).

None of this precluded intellectual work, though: Canguilhem ranked second that year in the highly competitive examination for the *agrégation de philosophie*. Paul Vignaux, who would become an eminent scholar in medieval philosophy, ranked first, and Canguilhem's friend Jean Cavaillès ranked fourth.

Canguilhem then did his military service for eighteen months, between November 1927 and April 1929 – not as an officer but first as a private and, later, in preparation for noncommissioned officers (*brigadier*).

C.G. Bernard [pseud.], "La Philosophie d'Hermann Keyserling," *Libres propos* (March 20, 1927), pp. 18-21.

Review of Maurice Boucher, *La Philosophie d'Hermann Keyserling* (Paris: Rieder, 1939). Between 1927 and 1929, Canguilhem sometimes used the pen name "C.G. Bernard" to sign articles in *Libres propos*. It is now quite difficult to find issues of *Libres propos, Journal d'Alain*; in fact, a complete series can only be found in a few French libraries. The first issue appeared on April 9, 1921, printed by the "Imprimerie coopérative 'La laborieuse'," in Nîmes. Michel Alexandre (1888-1952), then a lycée professor in that city, with his wife Jeanne, assumed most of the editorial burden of what was then a weekly publication. When Alexandre first met Alain he was twenty years old; he remained a devoted disciple throughout his life. *Libres propos* quickly attracted enough attention among French intellectuals that Gallimard decided to publish it under its prestigious "NRF" imprint in 1922-23 and 1924, when the journal ceased publication. A second series of *Libres propos* was published as a monthly from March 1927 to September 1935; see Jeanne Alexandre, ed., *En Souvenir de Michel Alexandre: Leçons, textes, lettres* (Paris: Mercure de France, 1956), pp. 499-514. In 1931-32, Canguilhem assumed the main editorial functions of *Libres propos* (see below, entry under 1931).

C.G. Bernard [pseud.], "La Mobilisation des intellectuels – Protestation d'étudiants," *Libres propos* (April 20, 1927), pp. 51-52. Followed on pp. 53-54 by a text signed "G. Canguilhem."

Printed on pages 46-48, under the title "La Déclaration d'*Europe*, no. du 15 avril," is the text of the protest, first published in *Europe*, signed by 160 intellectuals and academics, including Alain, and followed by the signatures of fifty-four students from the Ecole Normale, including Canguilhem, Raymond Aron, Jean Cavaillès, Charles Ehresmann, Jean Hyppolite, Henri-I. Marrou and Jean-Paul Sartre. The Loi Paul-Boncour, about the "general mobilization of the nation in wartime," had been voted by the Assemblée nationale on March 7, 1927. The law was denounced for stifling intellectual independence and freedom of opinion in wartime.

According to Sirinelli (pp. 341-42), Canguilhem initiated the petition at the Ecole Normale.

"Anniversaires. 1er juillet naissance de Leibniz," *Libres propos* (July 20, 1927), p. 185.

Extracts from Leibniz's works, followed by the mention "communicated by G.C."

"De la Vulgarisation philosophique. Une Edition du Discours de la méthode," *Libres propos* (July 19, 1927), pp. 200-201.

Review of Descartes's "mutilated text" edited by Paul Lemaire (Paris: Hatier, 1927).

C.G. Bernard [pseud.], "La Logique des jugements de valeur," *Libres propos* (Aug. 20, 1927), pp. 248-51.

Review of E. Goblot, *Traité de logique* (Paris: Collin, 1927).

"Essais – A la Manière de...," *Libres propos* (Oct. 20, 1927), pp. 343-45.

A pastiche of the work of Voltaire, whose name is facetiously used to sign the text. An appended note reveals the real authors to be Canguilhem and Sylvain Broussaudier, a fellow student at the Ecole Normale. Most of the text is published in Sirinelli, pp. 324-25. The pastiche mocks the Ecole's director, Gustave Lanson, and his reactions to the antimilitarist content of the revue.

"Montagnes et frontières," *Libres propos* (Nov. 20, 1927), pp. 401-402.

Emile Boutroux, *Des Vérités éternelles chez Descartes*, Thèse latine traduite par M. Georges Canguilhem, élève de l'Ecole Normale Supérieure. Préface de M. Léon Brunschvicg, de l'Institut (Paris: Librairie Félix Alcan, 1927).

A French translation of Emile Boutroux's 1874 Latin doctoral dissertation. A new edition was published in 1985. The 1927 edition includes a study by Léon Brunschvicg on Boutroux's philosophy, "La Philosophie d'Emile Boutroux," which is not included in the 1985 edition; it can, however, be found in Léon Brunschvicg, *Ecrits philosophiques* (Paris: Presses Universitaires de France, 1954), vol. 2, pp. 211-31.

1928

Canguilhem spent this entire year in the army as part of his eighteen-month military service. He is not known to have published anything during this period.

1929

Canguilhem completed his military service in April 1929; resuming his use of the pseudonym "C.G. Bernard," he started to publish again before being released.

C.G. Bernard [pseud.], "Commentaires et documents – Adresse à la Ligue des droits de l'homme," *Libres propos* (Feb. 20, 1929), pp. 78-79.

C.G. Bernard [pseud.], "Essais. Esquisse d'une politique de Paix. Préambule," *Libres propos* (March 20, 1929), pp. 135-38.

"Le Sourire de Platon," *Europe* 20 (1929), pp. 129-38.

>Review of Alain, *Onze chapitres sur Platon* (1928). The title of the review is taken from Alain's *Souvenirs sur Jules Lagneau*, where he had written, "we forget the smile of Plato." Jules Lagneau, who remains a symbol of the self-abnegation, devotion to philosophy and high moral standards maintained by some professors of the early Third Republic, had been Alain's philosophy teacher at the lycée. That same year, Canguilhem reviewed the posthumous publication of some of Lagneau's lectures (see below, two entries down).

"Maxime Leroy, *Descartes le philosophe au masque*," *Europe* 21 (1929), pp. 152-56.

>Review.

"Célèbres leçons de Jules Lagneau. Nîmes, La Laborieuse, 1928," *Libres propos* (April 20, 1929), pp. 190-91.

>Review.

"La Fin d'une parade philosophique. Le Bergsonisme, sous le pseudonyme François Arouet. Paris, Ed. 'Les Revues'," *Libres propos* (April 20, 1929), pp. 191-95.

>Review. The real name of the author of this attack on Henri Bergson was Georges Politzer, a communist philosopher who became a soldier in the resistance and was executed by the Nazis in 1942.

"Préjugés et jugement," *Libres propos* (June 20, 1929), p. 291.

"Circulaire adressée aux membres de l'Association [de Secours aux Anciens Elèves de l'Ecole normale supérieure]," *Libres propos* (July 20, 1929), pp. 326-30.

>Canguilhem is one of the twelve signatories of this circular (including Alain, Romain Rolland, Georges Bénézé, Raymond Aron and Félicien Challaye, who seems to have been the writer) protesting against the president of the Association, the mathematician Emile Picard. At the annual meeting of the Association, in January, he had condemned the eighty-

three Ecole Normale students who in November 1928 had signed a protest against the mandatory officer preparatory training ("préparation militaire supérieure"). The letter was followed by a "Déclaration" signed by 270 alumni of the Ecole Normale, including, beyond the twelve signatories of the circular, Jean Cavaillès and Paul Nizan. Picard resigned at the end of the year (see Sirinelli, *Génération intellectuelle*, pp. 492-93 and 518-19).

"Civilité puérile et honnête," *Libres propos* (Aug. 20, 1929), pp. 392-93.

"A la Gloire d'Hippocrate, père du tempérament," *Libres propos* (Aug. 20, 1929), pp. 297-98.

Review of Allenby, *Orientation des idées médicales* (Paris: Au Sans Pareil, 1929).

"Éloge de Philipp Snowden, par un Français," *Libres propos* (Sept. 20, 1929), pp. 434-35.

C.G. Bernard [pseud.], "Versailles-1919, par Friederich Nowak, Rieder, 1928," *Libres propos* (Oct. 20, 1929), pp. 496-97.

Review.

"Réflexions sur une crise ministérielle," *Libres propos* (Nov. 20, 1929), pp. 530-32.

"28 Décembre. Mort de Bayle, à Rotterdam (1706)," *Libres propos* (Dec. 20, 1929), p. 573.

Extract from Pierre Bayle's *Pensées diverses à l'occasion de la comète*, followed by the mention "communicated by G.C."

"Théâtre. Le Grand Voyage, pièce de Sheriff (traduite de l'anglais) au Théâtre Edouard VII," *Libres propos* (Dec. 20, 1929), p. 660.

1930

With Georges Bénézé, "Divertissement philosophique. Discussion sur le temps selon Kant," *Libres propos* (Jan. 20, 1930), pp. 41-43.

An exchange in two parts, the first signed by Canguilhem (pp. 41-42), the second by Bénézé (pp. 42-43).

"La Fin de l'éternel, par Julien Benda, NRF, 1929," *Libres propos* (Jan. 20, 1930), pp. 44-45.

In this favorable review of Benda's book, Canguilhem, when discussing the question of responsibility during World War I, wrote that the author had not sufficiently emphasized that those who have sinned less against "eternal" (reason) are more common among the people than among the learned ("*clercs*"). This statement proved to be quite controversial; see below, six entries down.

"Examen des examens − Le Baccalauréat," *Libres propos* (Feb. 20, 1930), pp. 88−90.

"Témoins − Deux témoignages d'officiers," *Libres propos* (March 20, 1930), pp. 137−41.

Review of André Bridoux, *Souvenirs de la maison des morts* (Lyon: "Le Van," 1929), and Maurice Constantin-Weyer, *P.C. de compagnie* (Paris: Rieder, 1930). Bridoux had been a student of Alain.

"24 avril. Naissance de Kant (1724)," *Libres propos* (April 20, 1930), pp. 169−70.

Extracts from Immanuel Kant's "Anthropology," followed by the mention "Fragments recueillis par G.C."

"Examen des examens. La Dissertation philosophique au baccalauréat (juillet et octobre 1929)," *Libres propos* (April 20, 1930), pp. 179−81.

"Une Conception récente du sentiment religieux et du sacré," *Europe* 24 (1930), pp. 288−92.

Review of R. Otto, *Le Sacré, l'élément non-rationnel dans l'idée du divin et sa relation avec le rationnel* (Paris: Payot, 1929).

With Michel Alexandre, "La Trahison des clercs −− Autour de Romain Rolland," *Libres propos* (April 20, 1930), pp. 188−92.

A response to Julien Benda's interpretation of a statement made by Canguilhem in his review of *La Fin de l'éternel* (see above, six entries up), in a recent issue of the periodical *La Nouvelle revue française*. This article includes a letter to Julien Benda, signed by Canguilhem and Michel Alexandre (for whom, see the first entry under 1927) to vindicate Romain Rolland's antinationalist attitude during World War I. Rolland's pacifist views were close to those of Alain and his disciples at the time.

"Autour de Romain-Rolland − Lettre adressée à Julien Benda par G. Canguilhem et M. Alexandre," *Europe* 23 (1930), pp. 302-304.

Reprint from the preceding title published in *Libres propos*. *Europe*, a periodical with clear commitments to the Left, was founded by Romain Rolland in 1923.

"Examen des examens. Le Baccalauréat," *Libres propos* (May 1930), pp. 238-41.
Discours prononcé par G. Canguilhem, agrégé de l'Université, professeur de philosophie, à la distribution des prix du lycée de Charleville, le 12 juillet 1930 (Charleville: typographie et lithographie P. Anciaux, 1930).

A very rare eight-page publication, by the lycée, with an "official" title page bearing the inscription "Ville de Charleville, Lycée Chanzy, Académie de Lille." It is the text of the address given by Canguilhem, the newcomer on the teaching staff of the lycée, at the closing ceremonies for the academic year, attended by graduating students, their parents, the prefect and other local notabilities. The address has been characterized by Sirinelli (*Génération intellectuelle*, p. 595 n.10) as "a model of discreet impertinence" (*impertinence feutrée*). Canguilhem had been appointed professor at the lycée of Charleville after his military service; he stayed there until Fall 1930, when he was appointed to the lycée of Albi.

"Deux histoires d'hérésie – Pensées liminaires," *Libres propos* (June 1930), pp. 272-73.

A defense of two lycée professors, Armand Cuvillier and Félicien Challaye, who were under attack for expressing anticolonialist and antimilitarist opinions; followed by the presentation of documents by Michel Alexandre and René Maublanc (pp. 273-81).

"De l'Introspection," *Libres propos* (Nov. 1930), pp. 522-23.

A review of Luigi Pirandello, *Une Personne et cent mille* (Paris: Gallimard, 1930).

"Foch et Clémenceau ou la mort en phrases," *Libres propos* (Nov. 1930), pp. 534-35.

Review of Raymond Recouly, *Mémorial de Foch* (Paris: Les Editions de France, 1929), and Georges Clémenceau, *Grandeurs et misères d'une victoire* (Paris: Plon, 1930).

"Prolétariat, matérialisme et culture," *Libres propos* (Dec. 1930), pp. 584–87.

    Review of Emmanuel Berl, *Mort de la morale bourgeoise* (Paris: Gallimard, 1930), and "Révocations," published in *Europe*, October 15, 1930. The end of the article is marked "à suivre"; the sequel was published in January 1931 (see below, first entry under 1931).

1931

Canguilhem had arranged a leave from teaching during the academic year 1931–32. As Michel Alexandre was overwhelmed by his teaching duties, Canguilhem assumed responsibility for editing *Libres propos* (see André Sernin, *Alain: Un Sage dans la cité* [Paris: Robert Laffont, 1985], p. 298).

"Humanités et marxisme. Prolétariat, matérialisme et culture," *Libres propos* (Jan. 1931), pp. 40–43.

    Part two of the article published in *Libres propos*, December 1930.

"La Guerre et la paix. Le Discours de M. Paul Valéry au Maréchal Pétain," *Libres propos* (Feb. 1931), pp. 93–94.

"Deux explications philosophiques de la guerre: Alain et Quinton," *Libres propos* (Feb. 1931), pp. 95–98.

    Signed "G.C." A comparison of Alain, *Mars ou la guerre jugée* (Paris: Gallimard, 1921) and René Quinton, *Maximes sur la guerre* (Paris: Grasset, 1930).

"Désarmement – Un Discours de travailliste," *Libres propos* (March 1931), pp. 128–29.

    Followed (on pages 129–36) by extracts from articles and interventions in the Assemblée nationale by Léon Blum, with a final paragraph of commentary signed "G.C."

With Michel Alexandre, "Révision des traités. Peuple sans foi ni loi," *Libres propos* (April 1931), pp. 182–83.

"Deux livres de Jean-Richard Bloch," *Libres propos* (April 1931), pp. 194–96.

    Review of Jean-Richard Bloch's *Destin du théâtre* (Paris: Gallimard, 1930) and *Destin des siècles* (Paris: Rieder, 1931).

"L'Affaire Dreyfus et la troisième république par C. Charensol," *Libres propos* (April 1931), p. 197.

A brief note on the book, signed "G.C.," published under this title (Paris: Kra, 1931).

"Documents et jugements. Contre la caporalisation des intellectuels – Une Protestation de Normaliens," *Libres propos* (July 1931), pp. 324–25.

New regulations had been enacted at the Ecole Normale Supérieure specifically forbidding collective action by students without prior authorization by its director, and threatening disciplinary action against students who refused to comply fully with the mandatory military training. Twenty-two alumni of the Ecole (including Canguilhem, Nizan, Romain Rolland and Sartre) had signed this protest, as had four students still at the Ecole, among them Simone Weil, also a disciple of Alain. This text was also published in *L'Université syndicaliste* in June 1931.

"Août 1914. Août 1931. Rêveries très positives du citoyen mobilisable," *Libres propos* (Aug. 1931), pp. 357–58.

"*L'Internationale sanglante des armements*, par O. Lehmann Rüssbuld (L'Eglantine, Bruxelles, 1930)," *Libres propos* (Sept. 1931), pp. 415–16.

Review, followed by an extract from the book, p. 417.

With Michel Alexandre, " 'Désarmement,' série de textes sur le 'problème naval franco-allemand'," *Libres propos* (Oct. 1931), p. 462.

A paragraph of introduction by Canguilhem and Michel Alexandre to a collection of documents on French-German naval rivalry, pp. 462–67.

"Le Coin des ruades," *Libres propos* (Oct. 1931), p. 483.

A brief statement by Canguilhem, in response to Georges Demartial's critique of Canguilhem's review of O. Lehmann Rüssbuld's book (*Libres propos*, Sept. 1931).

"Elections anglaises," *Libres propos* (Nov. 1931), pp. 510–11.

Signed "G.C."

"Incertitudes allemandes, par Pierre Viénot. Librairie Valois, 1931," *Libres propos* (Nov. 1931), pp. 514–16.

Review.

"France-Amérique – Sur le Voyage de Laval," *Libres propos* (Nov. 1931), pp. 519-20. Signed "G.C."

"Sociologie – Les Causes du suicide," *Libres propos* (Nov. 1931), pp. 525-30.

Review of Maurice Halbwachs, *Les Causes du suicide* (Paris: Alcan, 1930).

"Défense du citoyen – La Presse, le désarmement et le conflit sino-japonais," *Libres propos* (Dec. 1931), pp. 567-72.

Signed "G.C."

"Critique et philosophie: Sur le Problème de la création," *Libres propos* (Dec. 1931), pp. 583-88.

Review of Pierre Abraham, *Créatures chez Balzac* (Paris: Gallimard, 1931).

1932

"Lectures. Décadence de la nation française – Le Cancer américain, par Aron et Dandieu (Rieder, 1931)," *Libres propos* (Jan. 1932), pp. 42-44.

Review.

"La Paix sans réserve? Oui," *Libres propos* (Feb. 1932), pp. 99-104.

In November 1931, the pacifist Félicien Challaye had published an article entitled "La Paix sans réserve" in the journal *La Paix par le droit*, which *Libres propos* summarized in January 1932 (pp. 36-37). Théodore Ruyssen, though himself a pacifist, had published a critique of Challaye under the title "La Paix sans réserve? Non." Ruyssen's text is summarized in this issue on pages 93-94. In his article, Canguilhem sides with Challaye, as does a following article signed Jean Le Mataf (pp. 104-109). (For further events in this controversy, see below, two entries down.)

"Documents. France – Les Intellectuels et le désarmement," *Libres propos* (April 1932), pp. 201-203.

Discussion of a letter of Jean Guéhenno published under the same title in *Europe*, March 15, 1932. Signed "G.C."

"Sans plus de réserve qu'auparavant," *Libres propos* (April 1932), pp. 210-13.

Canguilhem's answer to Ruyssen, following the reply of the latter (pp. 207-10 of this issue, entitled "La Paix, oui. Mais par le droit") to Canguilhem's article "La Paix sans réserve? Oui" (see above, two entries up).

With Michel Alexandre, "Mentalité primitive," *Libres propos* (May 1932), pp. 256-58.

"Elections 1932," *Libres propos* (May 1932), pp. 259-61.

Signed "G.C."

"L'Agrégation de philosophie," *Méthode. Revue de l'enseignement philosophique* 1 (May 1932), pp. 17-21.

Méthode had recently been founded by Georges Bénézé (1888-1978), an older disciple of Alain. The journal disappeared in June 1933, after its sixth issue. Canguilhem's friends Jean Hyppolite and Raymond Aron also signed articles there. Canguilhem's article, a critique of the *agrégation* program and of the omission of Descartes, Kant, Hegel, Comte and Nietzsche from the required authors, exemplifies the concerns of the journal's collaborators. Canguilhem published three times in *Méthode*.

"Alain, *Propos sur l'éducation* (Paris: Rieder, 1932)," *Europe* 31 (1932), pp. 300-301.

Review.

"Autour de Lucien Herr," *Libres propos* (Sept. 1932), pp. 476-79.

Review of Lucien Herr, *Choix d'écrits*, 2 vols. (Paris: Rieder, 1932), and Charles Andler, *Vie de Lucien Herr* (Paris: Rieder, 1932). Lucien Herr (1864-1927) had for decades been the librarian of the Ecole Normale Supérieure and an influential intellectual adviser to its students, as well as an unsuccessful proponent of Hegel in France. Canguilhem published a brief personal account of Lucien Herr in 1977.

"Un livre scolaire...en Allemagne," *Libres propos* (Oct. 1932), pp. 538-39.

Review of a reader of Alain's works translated into German: *Eine Auswahl ans seinen Werken zur Einführung in sein Denken*, ed. Julius Schmidt (Berlin: Westermann, 1932).

La Paix sans aucune réserve, Thèse de Félicien Challaye, suivie d'une discussion entre Théodore Ruyssen, Félicien Challaye, Georges Canguilhem et Jean Le-Mataf, et des textes de Bertrand Russell et d'Alain sur "La vraie et la folle Résistance," Documents des 'Libres Propos' Cahier no. 1 (Nîmes: Imprimerie La Laborieuse, 1932).

According to Sirinelli (*Génération intellectuelle*, p. 596 n.13), this text includes Canguilhem's contribution to the controversy published in *Libres propos* in February 1932, under the new title "Seconde riposte, ou fondements du refus de toute guerre nationale" (see above, second entry under 1932), and a conclusion, "Finale en sept points."

Raymond Aron published a comment critical of this booklet in *Libres propos* (Feb. 1933, pp. 96-99), dissenting from Canguilhem's viewpoint (on this, see Aron's *Mémoires* [Paris: Julliard, 1983], pp. 56-58). For Canguilhem's reactions to Aron's critique, see below, fourth entry under 1933.

Aron, who had been introduced to Alain by Canguilhem when they were fellow students at the Ecole Normale, published several articles in *Libres propos*. Canguilhem and Aron were to be colleagues in Toulouse and, later, at the Sorbonne. Canguilhem was present at Aron's obsequies and gave an address sketching his career (see *Le Monde*, Oct. 21, 1983).

1933

Canguilhem was appointed to the lycée of Douai for the academic year 1932-33, following the period of leave he had taken to manage *Libres propos*. In the fall of 1933, he was sent to Valenciennes, where he remained for the next two academic years.

[Comments on] André Joussain, "L'Enseignement de la sociologie," following this article, *Méthode. Revue de l'enseignement philosophique* (Jan. 1933), pp. 10-11.
"R. Le Senne, *Le Devoir* (Alcan, 1930), *Méthode. Revue de l'enseignement philosophique* (Feb. 1933), pp. 25-27.
Review.
"Sur une Interprétation de l'histoire," *Libres propos* (March 1933), pp. 155-56.
A critique of an article by Jacques Ganuchaud published in *Libres propos* of November 1932. Ganuchaud's reply appeared in the next issue of *Libres propos*, pp. 219-20.

"Essais. Pacifisme et révolution," *Libres propos* (March 1933), pp. 157-59.

    A reply to Raymond Aron's critique, published in the previous issue of *Libres propos* (see above, final entry under 1932).

"Modeste Herriot vu par lui-même (et commenté)," *Libres propos* (April 1933), pp. 217-19.

    Signed "G.C." A critique of an article published by Edouard Herriot in the newspaper *Le Démocrate* of Lyon on April 15, 1933.

"De l'Objection de conscience à la conscience de l'objection," *Libres propos* (May 1933), pp. 272-75.

    A critique of the administrative circular signed by Minister Camille Chautemps against the emergence of the conscientious objector movement.

"Sur la Philosophie contemporaine – H. Serouya, *Initiation à la philosophie contemporaine* (La Renaissance du Livre), J. Benrubi, *Les Sources et les courants de la philosophie contemporaine en France* (Alcan)," *Europe* 33 (1933), pp. 451-53. Review.

1934

"Deux nouveaux livres français sur les origines de la guerre," *Libres propos* (Jan. 1934), pp. 40-44.

    Review of Camille Bloch, *Les Causes de la guerre mondiale* (Paris: Hartmann, 1933), and Jules Isaac, *1914 – Le Problème des origines de la guerre* (Paris: Rieder, 1933).

"Jean-Richard Bloch, *Offrande à la politique* (Coll. Europe, Rieder, 1933)," *Libres propos* (Jan. 1934), pp. 52-53.

    Review, signed "G.C."

"Héroïsme universitaire," *Libres propos* (March 1934), pp. 144-45.

    This is the last article Canguilhem wrote for *Libres propos*. Though he would remain personally close to Alain until the latter's death, Canguilhem began at this time to distance himself from Alain's pacifism, as he realized that "one could not negotiate with Hitler" (see Sirinelli, *Génération intellectuelle*, pp. 597-98).

1935

During the academic years 1933-35, Canguilhem taught at the lycée of Valenciennes. He was appointed to Béziers for the academic year 1935-36.

"Alain, *Les Dieux* (Nrf, 1934)," *Europe* 37 (1935), pp. 445-48.
Review. Extracts from this review were reprinted in the *Bulletin de l'Association des amis d'Alain* 20 (Dec. 1964), pp. 31-32.
Comité de Vigilance des Intellectuels Anti-fascistes, *Le Fascisme et les paysans* (Paris, 1935).

Canguilhem was the anonymous author of this sixty-two-page document, printed in Cahors. The Comité de Vigilance des Intellectuels Anti-fascistes was created in response to the February 1934 riots in Paris and the threat of fascism, and it remained in existence up to the war. Its leaders were the ethnologist Paul Rivet, who chaired the committee, the physicist Paul Langevin and Alain. During these years, Alain was often ill and unable to attend some meetings; Canguilhem's friend Michel Alexandre would substitute for him on these occasions (see Jeanne Alexandre, ed., *En Souvenir de Michel Alexandre: Leçons, textes, lettres* [Paris: Mercure de France, 1956], p. 520). Thus, Canguilhem himself was quite close to the action of the committee. The booklet has three parts: "Proposals for an Agricultural Policy," a two-part appendix consisting of the results of a survey on the "agricultural crisis," and "Notes on Agriculture in Fascist Italy and Germany," which dealt with the consequences of fascist totalitarianism in rural areas.

1936

Canguilhem was appointed to Toulouse as professor of the *classe de khâgne*, beginning in October 1936. He kept this teaching position until the beginning of the Vichy regime, and began his medical studies while teaching.

"P.-M. Schuhl, *Essai sur la formation de la pensée grecque* (Alcan, 1934)," *Europe* 40 (1936), pp. 426-28.
Review.

"Raymond Aron, *La Sociologie allemande contemporaine* (Alcan, 1935)," *Europe* 40 (1936), pp. 573-74.
Review.

1937

"Descartes et la technique," in *Travaux du IXe Congrès international de philosophie (Congrès Descartes)*, tome II (Paris: Hermann, 1937), pp. 77-85.
Canguilhem's first conference paper, reprinted in *Cahiers STS* 7 (1985), pp. 87-93; included in this reader.

1938

"Activité technique et création," in *Communications et discussions*, Société toulousaine de philosophie (Years 1937 and 1938), 2nd series, pp. 81-86.
A paper given at the meeting of the Société on February 26, 1938. A footnote by Canguilhem (p. 86) indicates that discussion of the "increasing importance of biology and '*sociologie technologique*' for philosophy" has been omitted from this printed version. Followed by a discussion on pages 86-89.

1939

With Camille Planet, *Traité de logique et de morale* (Marseille: Imprimerie F. Robert et fils, 1939).
This textbook has become extremely difficult to find; the Bibliothèque nationale in Paris has one copy. Planet was teaching at the lycée of Marseille while Canguilhem was at the lycée of Toulouse. Two other textbooks by the same authors, on psychology and aesthetics – the other subjects that

were then part of the philosophy program of the lycée – which were "forthcoming," were never published.

1940–1942

In the fall of 1940, Canguilhem took leave from his teaching at the lycée of Toulouse, refusing to teach in the reactionary context imposed by the Vichy regime. He wrote to the rector of the Académie de Toulouse: "I have not become an *agrégé de philosophie* to teach 'Labor, Family, Fatherland' " (the motto of the Vichy government). He then dedicated himself to his medical studies. Raymond Aron, who also was in Toulouse at the time, wrote of Canguilhem then: "Some, like my friend Canguilhem, were getting ready to take a modest part – which was glorious – in the resistance" (*Mémoires* [Paris: Julliard, 1983], p. 164).

In February 1941, Jean Cavaillès, who was teaching philosophy at the University of Strasbourg (then at Clermont-Ferrand in Auvergne), was called to the Sorbonne in Paris; he convinced Canguilhem to replace him in Clermont-Ferrand. Canguilhem was appointed in April 1941. With Cavaillès and Emmanuel d'Astier de la Vigerie, Canguilhem was a writer of the first tract of the resistance movement, *Libération*, in 1941 (see Sirinelli, *Génération intellectuelle*, p. 599; Gilles Lévy and François Cordet, *A nous, Auvergne! La Vérité sur la résistance en Auvergne 1940–1944* [Paris: Presses de la cité, 1990], p. 27).

1942

"Certificat de philosophie générale et de logique. Indications bibliographiques," *Bulletin de la Faculté des Lettres de Strasbourg* 20.3 (1942), pp. 110–12.

A bibliography (for students preparing the *Certificat* as part of their *Licence de philosophie*) that complements the bibliography published by Cavaillès in the *Bulletin* the previous year.

"Commentaire au troisième chapitre de *L'Evolution créatrice*," *Bulletin de la Faculté des Lettres de Strasbourg* 21 (1942), pp. 126–43 and 199–214.

Canguilhem explains in a footnote that this article on Bergson's book was published to help students prepare the program for the *agrégation de philosophie* for 1943, and that it was based on his lectures at the University of Strasbourg, in Clermont-Ferrand, in 1942.

1943

Essai sur quelques problèmes concernant le normal et le pathologique, Publications de la Faculté des Lettres de l'Université de Strasbourg, Fascicule 100. Clermont-Ferrand, Imprimerie "La Montagne," 1943.

Canguilhem's doctoral dissertation in medicine includes a one-page preface omitted from later editions. This book was published again, under the same title, in 1950 (Paris: Belles Lettres), with a "Préface de la deuxième édition," and many times under the title *Le Normal et le pathologique* from 1966 on (Paris: Presses Universitaires de France).

Extracts from the second revised edition of this book (1972) are included in this reader.

1943–1944

On the morning of November 25, 1943, the Gestapo invaded the building of the Faculté des Lettres of the University of Strasbourg in Clermont-Ferrand; two professors were killed, and many students and professors were arrested and deported to Germany (see Gabriel Maugain, "La Vie de la Faculté des Lettres de Strasbourg de 1939 à 1945," *Mémorial des années 1939-1945*, Publications de la Faculté des Lettres de l'Université de Strasbourg, Fascicule 103 [Paris: Société d'édition Les Belles Lettres, 1947], pp. 3-50, esp. pp. 32-40; "La Journée du souvenir: 25 novembre 1945," *Bulletin de la Faculté des Lettres de Strasbourg* 22-23 [Jan. 1945], pp. 25-31). Canguilhem escaped the arrest and continued his action with the underground. Under the name "Lafont" he became an assistant to Henry Ingrand, the leader of the resistance in Auvergne. In early 1944, Canguilhem assumed important underground political functions in the directorate of the Unified Resistance Movements (see Lévy and Cordet, *A nous, Auvergne!*, pp. 140-97).

In June 1944, Canguilhem participated in one of the major battles between the resistance and the German forces, at Mont Mouchet, in the mountains of Auvergne, south of Clermont-Ferrand; he operated a field hospital and organized its evacuation under fire (see Henry Ingrand, *La Libération de l'Auvergne* [Paris: Hachette, 1974], pp. 97-102). One of his colleagues in these heroic acts was Dr. Paul Reiss of the Faculté de Médecine at Strasbourg, who was killed by the Nazis. During the summer, Canguilhem was sent to Vichy as the permanent representative of Henry Ingrand, who had then become Commissaire de la République, which was responsible for the administration and security of the entire region. According to Ingrand, Canguilhem fulfilled "delicate and still dangerous functions" (ibid., p. 149).

Canguilhem received the Military Cross and the Médaille de la Résistance in 1944.

At the end of the war, he was offered the position of inspecteur général de philosophie, responsible for overseeing the quality of teaching in the lycées and for grading professors. He rejected the offer and returned to his position at the Faculté des Lettres of the University of Strasbourg.

1945

"Jean Cavaillès, Résistant," *Bulletin de la Faculté des Lettres de Strasbourg* 22-23 (Dec. 1945), pp. 29-34.

Cavaillès, who had been a fellow student at the Ecole Normale, and always a close friend, had been assassinated by the Nazis in January 1944 (see Gabrielle Ferrières, *Jean Cavaillès, philosophe et combattant (1903-1944)* [Paris: Presses Universitaires de France, 1950]).

1946

"La Théorie cellulaire en biologie. Du Sens et de la valeur des théories scientifiques," *Mélanges 1945, IV. Etudes philosophiques*, Publications de la Faculté des Lettres de Strasbourg, Fascicule 107 (Paris: Les Belles Lettres, 1946), pp. 143-75.

Included, with revisions, in *La Connaissance de la vie* (1952); extracts from this article are included in this reader.

"Georges Friedmann, *Leibniz et Spinoza* (Paris: Gallimard, 1946)," *Bulletin de la Faculté des Lettres de Strasbourg* 25 (1946), pp. 43-47.

Review.

1947

"Milieu et normes de l'homme au travail," *Cahiers internationaux de sociologie* 23 (1947), pp. 120-36.

An essay on Georges Friedmann's *Problèmes humains du machinisme industriel* (Paris: Gallimard, 1946).

"Jean Cavaillès (1903-1944)," in *Mémorial des années 1939-1945*, Publications de la Faculté des Lettres de Strasbourg, Fascicule 103 (Paris: Les Belles Lettres, 1947), pp. 141-58.

With C. Ehresmann, "Avertissement des éditeurs," in Jean Cavaillès, *Sur la Logique et la théorie de la science* (Paris: Presses Universitaires de France, 1947), pp. ix-xiii.

Second edition, Presses Universitaires de France, 1960; third edition, Paris, Vrin, 1976; fourth edition, Vrin, 1987.

"Maurice Halbwachs, l'homme et l'oeuvre," in *Mémorial des années 1939-1945*, Publications de la Faculté des Lettres de Strasbourg, Fascicule 103 (Paris: Les Belles Lettres, 1947), pp. 229-41.

"Note sur la situation faite en France à la philosophie biologique," *Revue de métaphysique et de morale* 52 (1947), pp. 322-32.

1948

Canguilhem had returned to teaching in Strasbourg in 1944; from 1948 until 1955 he was inspecteur général de philosophie.

1949

"Préface," in Immanuel Kant, *Essai pour introduire en philosophie le concept de grandeur négative* (Paris: Vrin, 1949).
Translation, introduction and notes by Roger Kempf.

"Présentation," in "Mathématiques et formalisme (Inédit présenté par G. Canguilhem)," by Jean Cavaillès, *Revue internationale de philosophie* 3.8 (1949), p. 158.
Posthumous publication of an article Canguilhem found among the papers left by Cavaillès (pp. 159-64).

"Hegel en France," *Revue d'histoire et de philosophie religieuse* 28-29 (1948-49), pp. 282-97.
Extracts from this article were republished in *Magazine littéraire* 293 (Nov. 1991), pp. 26-29.

Kurt Goldstein, "Remarques sur le problème épistémologique de la biologie," in *Congrès international de philosophie des sciences. Paris 1949*, vol. 1 (Paris: Hermann, 1951), pp. 141-43.
Translated from the English by Georges Canguilhem and Simone Canguilhem.

1950

"Essais sur quelques problèmes concernant le normal et le pathologique," Publications de la Faculté des Lettres de Strasbourg, Fascicule 100 (2nd ed., Paris: Les Belles Lettres, 1950).
With a new "Préface de la deuxième édition."

1951

"Le normal et le pathologique," in René Leriche, ed., *Somme de médecine contemporaine* (Paris: Les Editions médicales de la Diane française, 1951), vol. 1, pp. 27-32.

Included, with revisions, in *La Connaissance de la vie* (1952).

1952

Besoins et tendances, Textes choisis et présentés par Georges Canguilhem (Paris: Hachette, 1952).

A reader, edited by Canguilhem, of extracts taken from the works of biologists and philosophers, which was published in the collection "Textes et documents philosophiques" under his direction. This collection includes other titles edited by Gilles Deleuze, Jean Brun, Francis Courtès, Robert Pagès and Jacques Guillerme, as well as a two-volume *Introduction à l'histoire des sciences*, published in 1970–71, edited by Canguilhem with students attending his seminars at the Institut d'histoire des sciences at the time (see below, 1970 and 1971). Most volumes include a five-page "Présentation de la collection" signed by Canguilhem.

La Connaissance de la vie (Paris: Hachette, 1952).
Includes:

"Avertissement," mentioning that some of the essays included have been revised since their first publication or oral presentation (pp. 5–6); an "Introduction: La Pensée et le vivant," published here for the first time (pp. 7–12); "L'Expérimentation en biologie animale," a lecture given at the Centre international pédagogique de Sèvres in 1951 (pp. 15–45);

"La Théorie cellulaire" (pp. 49–98), first published in 1946 in the *Mélanges 1945* of the Faculté des Lettres de Strasbourg;

"Aspects du vitalisme" (pp. 101–23), "Machine et organisme" (pp. 124–59) and "Le Vivant et son milieu" (pp. 160–93), three lectures given at the Collège philosophique in Paris in 1946–47, following an invitation from its organizer, the philosopher Jean Wahl;

"Le normal et le pathologique" (pp. 194–212), previously published, in 1951, in the first volume of the *Somme de médecine contemporaine*, edited by the surgeon René Leriche, then professor at the Collège de France; and three appendices: "Note sur le passage de la théorie fibrillaire à la théorie cellulaire" (pp. 213-15), "Note sur les rapports de la théorie cellulaire et de la philosophie de Leibniz" (pp. 215-17) and "Extraits du *Discours sur l'anatomie du cerveau* tenu par Sténon en 1665 à messieurs de l'Assemblée de chez monsieur Thévenot, à Paris" (pp. 217-18).

The second edition, "révisée et augmentée," was published by Vrin in 1965, and has since been reprinted many times; extracts from the fifth edition (1989) are included in this reader. The book was translated into Italian and Spanish in 1976.

"La Création artistique selon Alain," *Revue de métaphysique et de morale* 57 (1952), pp. 171-86.

1953

"La Signification de l'enseignement de la philosophie," in *L'Enseignement de la Philosophie. Une enquête internationale de l'UNESCO* (Paris: UNESCO, 1953), pp. 17-26.

Preceded, pp. 13-15, by a "Déclaration commune des experts," signed by Guido Calogero, Georges Canguilhem, Eugen Fink, Donald Mackinnon, Ibrahim Madkour, Gustave Monod, Merritt Moore, N.A. Nikam and Humberto Pinera Llera.

Canguilhem's text is the general presentation of the work done by the experts.

1955

Canguilhem succeeded Gaston Bachelard in the fall of 1955 as professor of philosophy at the Sorbonne, in Paris, as well as director of the Institut d'histoire des sciences et des techniques of the University of Paris. He remained there until his retirement in 1971.

La Formation du concept de réflexe aux XVIIe et XVIIIe siècles (Paris: Presses Universitaires de France, 1955).

Canguilhem's dissertation for the Doctorat ès Lettres, prepared under the direction of Gaston Bachelard. A second edition was published by Vrin in 1977. The book was translated into Spanish in 1975 and Japanese in 1988. Extracts from the second edition of this book are included in this reader.

"Le Problème des régulations dans l'organisme et dans la société," *Cahiers de l'Alliance Israélite universelle* 92 (Sept.–Oct. 1955), pp. 64–81.

The lecture, pp. 64–73, is followed by a discussion, pp. 73–81.

"Organismes et modèles mécaniques: Réflexions sur la biologie cartésienne," *Revue philosophique* 145 (1955), pp. 281–99.

Not a review but an analysis of Descartes's "Sixth Meditation," with a discussion of Martial Guéroult's interpretation of it in the second volume of his *Descartes selon l'ordre des raisons* (Paris: Aubier, 1953).

1956

"La Pensée de René Leriche," *Revue philosophique* 146 (1956), pp. 313–17.

A summary review of Leriche's intellectual contributions, following the famous surgeon's death. Canguilhem had discussed Leriche in *Le Normal et le pathologique* and had published an article in a book edited by Leriche in 1951.

1957

"Sur une Epistémologie concordataire," in G. Bouligand et al., *Hommage à Gaston Bachelard: Etudes de philosophie et d'histoire des sciences* (Paris: Presses Universitaires de France, 1957), pp. 3–12.

"Fontenelle, philosophe et historien des sciences," *Annales de l'Université de Paris* 27 (1957), pp. 384–90.

Reprinted in *Etudes d'histoire et de philosophie des sciences* (1968).

1958

Canguilhem was elected a corresponding member of the International Academy of the History of Science in 1958, and he became a full member in 1960. He served as vice president of the academy from 1971 to 1977.

"La Physiologie animale au XVIIIe siècle," in René Taton, ed., *Histoire générale des sciences*, vol. 2 (Paris: Presses Universitaires de France, 1958), pp. 593-619.
Unchanged in the various reprints of the work; included in this reader.

"La Philosophie biologique d'Auguste Comte et son influence en France au XIXe siècle," *Bulletin de la Société française de philosophie* 52 (1958), pp. 13-26.
Reprinted in *Etudes d'histoire et de philosophie des sciences* (1968); extracts from this article are included in this reader.

"Qu'est-ce que la psychologie?" *Revue de métaphysique et de morale* 63.1 (1958), pp. 12-25.
Lecture given at the Collège philosophique on December 18, 1956. Followed by "Remarques sur 'Qu'est-ce que la psychologie?' " by R. Pagès (pp. 128-34), and a concluding "Note" by Canguilhem. Published again in the *Cahiers pour l'analyse* in 1966 (reprinted in 1967) and in *Etudes d'histoire et de philosophie des sciences* (1968).
Translated into English in 1980. Included in this reader.

1959

"Pathologie et physiologie de la thyroïde au XIXe siècle," *Thalès* 9 for 1952-58 (1959), pp. 77-92.
Based on a lecture given at the Faculté de Médecine, University of Strasbourg, on January 10, 1958. Reprinted in *Etudes d'histoire et de philosophie des sciences* (1968).
Thalès had as its subtitle "Recueil des travaux de l'Institut d'histoire des sciences et des techniques de l'Université de Paris." The first volume (1934) appeared in 1935, published by the "Librairie Félix Alcan," the major French publisher for philosophy in Paris at the time, under the editorship of Abel Rey, the founder of the Institut d'histoire des sciences et des techniques.

Other volumes were published for the years 1935, 1936 and 1937–38, then publication was interrupted by the war. It reappeared in 1949 (volume 5, dated 1948), published by the Presses Universitaires de France. Three other volumes were published, related to the years when Gaston Bachelard was director of the Institut: in 1951 (dated 1949–50), 1953 (1951) and 1955 (1952). The last volumes appeared under the editorship of Canguilhem: volume 9 (1952–58) in 1959, volume 10 (1959) in 1960, volume 11 (1960) in 1962, volume 12 (1966) in 1968; volumes 13 (1969) and 14 (1970–71), the last to appear, did so as special issues of the *Revue d'histoire des sciences*, another journal published by the Presses Universitaires de France, in 1970 and 1972.

Reprinted in *Etudes d'histoire et de philosophie des sciences* (1968); extracts from this article are included in this reader.

"Avertissement," *Thalès* 9 for 1952–58 (1959), p. 1.

Unsigned. Announcement of the journal's reappearance after a hiatus of several years.

"Thérapeutique, expérimentation, responsabilité," *Revue de l'enseignement supérieur* 2 (1959), pp. 130–35.

Reprinted in *Etudes d'histoire et de philosophie des sciences* (1968).

"Les Concepts de 'lutte pour l'existence' et de 'sélection naturelle' " en 1858: Charles Darwin et Alfred Russel Wallace," *Conférences du Palais de la Découverte* (Paris: 1959), série D, no. 61.

Public lecture given at the Palais de la Découverte, in Paris, on January 10, 1959. Reprinted in *Etudes d'histoire et de philosophie des sciences* (1968).

Review of Maurice Daumas, ed., *Histoire de la science* (Paris: Gallimard, 1957), *Archives internationales d'histoire des sciences* 12 (1959), pp. 76–82.

1960

Canguilhem became a member of the Commission de philosophie, d'épistémologie et d'histoire des sciences of the Comité national of the Centre national de la recherche scientifique (CNRS) that year. He remained a member of the commission until his retirement in 1971, chairing it from 1967 to 1971.

"L'Homme et l'animal au point de vue psychologique selon Darwin," *Revue d'histoire des sciences* 13.1 (1960), pp. 81-94.

Reprinted in *Etudes d'histoire et de philosophie des sciences* (1968).

Review of *The Autobiography of Charles Darwin* (New York: Dover, 1958), *Archives internationales d'histoire des sciences* 13 (1960), p. 157.

Review of Bentley Glass, Owsei Temkin, William L. Straus Jr., eds., *Forerunners of Darwin 1745-1859* (Baltimore: Johns Hopkins University Press, 1959), *Archives internationales d'histoire des sciences* 13 (1960), pp. 157-59.

Review of Alvar Ellegard, *Darwin and the General Reader* (Göteborg: Almqvist & Wicksells, 1958), *Archives internationales d'histoire des sciences* 13 (1960), p. 159.

Review of Conway Zirkle, *Evolution, Marxian Biology and the Social Scene* (Philadelphia: University of Pennsylvania Press, 1959), *Archives internationales d'histoire des sciences* 13 (1960), pp. 159-60.

1961

"L'Ecole de Montpellier jugée par Auguste Comte," *Le Scalpel* 114.3 (1961), pp. 68-71.

Paper presented at the "XVIe Congrès international d'histoire de la médecine" (Montpellier, September 22-28, 1958). Reprinted in *Etudes d'histoire et de philosophie des sciences* (1968); included in this reader.

"La Physiologie en Allemagne," "Jeunes écoles de la seconde période," "Techniques et problèmes de la physiologie au XIXe siècle," in René Taton, ed., *Histoire générale des sciences*, tome III: *La Science contemporaine*, vol. 1, *Le XIXe siècle* (Paris: Presses Universitaires de France, 1961), pp. 475-78, 478-80, 480-84.

Unchanged in the various editions of the book; included in this reader.

"Nécessité de la 'diffusion scientifique'," *Revue de l'enseignement supérieur* 3 (1961), pp. 5-15.

[Comments following the lecture of] Olivier Costa de Beauregard, "Le Dilemme objectivité-subjectivité de la mécanique statistique et l'équivalence cybernétique entre information et entropie," *Bulletin de la Société française de philosophie* 53 (1961), pp. 208–10 and 216.

[Claude Bonnefoy, "Rien ne laissait voir que Sartre deviendrait 'Sartre,' " *Arts, Lettres, Spectacles, Musique*, Jan. 11–17, 804 (1961), pp. 13–14.]

Includes segments of an interview with Canguilhem concerning Sartre at the time they were both students at the Ecole Normale.

1962

With G. Lapassade, J. Piquemal, J. Ulmann, "Du Développement à l'évolution au XIXe siècle," *Thalès* 11 for 1960 (1962), pp. 1–65.

Canguilhem conducted a weekly seminar at the Institut d'histoire des sciences et des techniques during the academic years 1958–59 and 1959–60, to mark the centenary of the publication of Darwin's *Origin of Species* (as explained by Canguilhem in the "Avant-propos," p. 1). The article, jointly signed by the four authors, was reprinted as a small book, *Du Développement à l'évolution au XIXe siècle* (Paris: Presses Universitaires de France, 1985).

"La Monstruosité et le monstrueux," *Diogène* 40 (1962), pp. 29–43.

Based on a lecture given at the Institut des hautes études de Belgique, in Brussels, on February 9, 1962. Reprinted in the second edition of *La Connaissance de la vie* (1965).

[Comments in] *Agrégation, Philosophie, 1962: Rapport de M. Etienne Souriau, président du jury* (Paris: Ministère de l'éducation nationale, Institut Pédagogique National, 1962), pp. 3–4.

Mimeographed.

1963

"The Role of Analogies and Models in Biological Discoveries," in Alistair Cameron Crombie, ed., *Scientific Change* (London: Heinemann, 1963), pp. 507–20.

Paper presented at the Symposium on the History of Science, at the University of Oxford, held on July 9-15, 1961, under the auspices of the Division of History of Science of the International Union of the History and Philosophy of Science. Reprinted in *Etudes d'histoire et de philosophie des sciences* (1968), under the title "Modèles et analogies dans la découverte en biologie."

"Introduction. La Constitution de la physiologie comme science," in Charles Kayser, ed., *Physiologie* (Paris: Editions médicales Flammarion, 1963), vol. 1, pp. 11-48.

Reprinted in *Etudes d'histoire et de philosophie des sciences* (1968), and in the second edition of Kayser's *Physiologie* (Paris: Flammarion, 1970), pp. 11-50; extracts from this article are included in this reader.

"L'histoire des sciences dans l'oeuvre épistémologique de Gaston Bachelard," *Annales de l'Université de Paris* 1 (1963), pp. 24-39.

Reprinted in *Etudes d'histoire et de philosophie des sciences* (1968). Translated into Italian in 1969, and German in 1979.

"Dialectique et philosophie du non chez Gaston Bachelard," *Revue internationale de philosophie* 66 (1963), pp. 441-52.

Reprinted in *Etudes d'histoire et de philosophie des sciences*. Translated into Italian in 1969.

"Gaston Bachelard et les philosophes," *Sciences* 24 (March-April 1963), pp. 7-10.

Reprinted in *Etudes d'histoire et de philosophie des sciences*. Translated into Italian in 1969.

1964

"Histoire des religions et histoire des sciences dans la théorie du fétichisme chez Auguste Comte," in *Mélanges Alexandre Koyré*, vol. 2: *L'Aventure de l'esprit* (Paris: Hermann, 1964), pp. 64-87.

Contribution to the Festschrift in honor of the historian of science Alexandre Koyré (1892-1964). Reprinted in *Etudes d'histoire et de philosophie des sciences* (1968).

"Le Concept de réflexe au XIXe siècle," in K.E. Rothschuh, ed., *Von Boerhaave bis Berger: Die Entwicklung der kontinentalen Physiologie im 18. und 19. Jahrhundert* (Stuttgart: Fischer, 1964), pp. 157-67.

In French. Paper presented at a symposium held in Münster on September 18-20, 1962. Reprinted in *Etudes d'histoire et de philosophie des sciences* (1968).

"Galilée: La Signification de l'oeuvre et la leçon de l'homme," *Archives internationales d'histoire des sciences* 17 (1964), pp. 209-22.

Lecture given at the Institut Italien, in Paris, on June 3, 1964, on the occasion of the four hundredth anniversary of Galileo's birth. Reprinted in *Etudes d'histoire et de philosophie des sciences* (1968).

[Comments in] "Point de vue philosophique sur l'inadaptation dans le monde contemporain," *Recherches et débats* (March 1964), pp. 109-58 and 134-39.

Canguilhem's comments are part of the discussion on a paper presented by Pierre Colin bearing the above-mentioned title.

1965

La Connaissance de la vie (2nd ed., Paris: Vrin, 1965).

Reprint of the first edition, published by Hachette in 1952, with a new "Avertissement," some additional references and the addition of the study "La Monstruosité et le monstrueux," first published in 1962. This edition has been reprinted many times.

Extracts from the second edition of this book (1989) are published in this reader.

"L'Homme de Vésale dans le monde de Copernic: 1543," in *Commémoration solennelle du quatrième centenaire de la mort d'André Vésale, 19-24 octobre 1964* (Brussels: Palais des Académies, 1965), pp. 145-54.

Reprinted in *Etudes d'histoire et de philosophie des sciences* (1968).

"L'Idée de médecine expérimentale selon Claude Bernard," *Conférences du Palais de la Découverte* (Paris: Université de Paris, 1965), série D, no. 101.

Public lecture given at the Palais de la Découverte, in Paris, on February 6, 1965. Reprinted in *Etudes d'histoire et de philosophie des sciences*

(1968); extracts from this article are included in this reader.

"Gottfried Koller, *Das Leben des Biologen Johannes Müller 1801-1858*," Isis 56 (1965), p. 110.

Review.

"Théophile Cahn, *La Vie et l'oeuvre d'Etienne Geoffroy Saint-Hilaire*," Isis 56 (1965), pp. 244-46.

Review.

Agrégation de philosophie, 1965: Rapport de M. Georges Canguilhem, président du jury (Paris: Ministère de l'éducation nationale, Institut Pédagogique National, 1965).

Mimeographed.

"Philosophie et Science," *Revue de l'enseignement philosophique* 15.2 (Dec. 1964-Jan. 1965), pp. 10-17.

An exchange with Alain Badiou, broadcast on French educational television, January 23, 1965.

"Philosophie et Vérité," *Revue de l'enseignement philosophique* 15.4 (April 1965-May 1965), pp. 11-21.

An exchange, in the wake of the discussion with Alain Badiou in January 1965 (see above entry), on French educational television, with A. Badiou, D. Dreyfus, M. Foucault, J. Hyppolite, P. Ricoeur, broadcast on March 27, 1965.

1966

Le Normal et le pathologique (Paris: Presses Universitaires de France, 1966).

Reprint of the second edition, with its preface, published by Les Belles Lettres in 1950, and including a new second part: "Nouvelles réflexions concernant le normal et le pathologique (1963-66)," pp. 169-222, and a brief "Avertissement" (p. i). The "Nouvelles réflexions" correspond in part to a course given by Canguilhem at the Sorbonne the preceding year (M. Fichant, "Georges Canguilhem et l'idée de la philosophie" [1993], p. 38).

This edition appeared in the "Collection Galien," edited by Canguilhem, and which included studies in the history and philosophy of biology and medicine. Among the titles appearing in this series were works by several of his students, Yvette Conry, François Dagognet, Michel Foucault and Camille Limoges. *Le Normal et le pathologique* was reprinted in that collection (the fourth and fifth editions were identical to this one) until 1984, when the "Collection Galien" ceased to exist. The text then appeared, unrevised, in the new collection "Quadrige" (Presses Universitaires de France).

The book was translated into Spanish in 1971, German in 1974, Italian in 1975, English and Portuguese in 1978 and Japanese in 1987.

"Le Tout et la partie dans la pensée biologique," *Les Etudes philosophiques*, n.s., 21.1 (1966), pp. 3-16.

Reprinted in *Etudes d'histoire et de philosophie des sciences* (1968); extracts from this article are included in this reader.

"Préface," in Claude Bernard, *Leçons sur les phénomènes de la vie communs aux animaux et aux végétaux* (Paris: Vrin, 1966), pp. 7-14.

Included in this reader.

"Le Concept et la vie," *Revue philosophique de Louvain* 64 (May 1966), pp. 193-223.

Based on two public lectures given at the Ecole des sciences philosophiques et religieuses of the Faculté universitaire Saint-Louis in Brussels, on February 22 and 24, 1966. Reprinted in *Etudes d'histoire et de philosophie des sciences* (1968). Extracts from this article are included in this reader.

Agrégation de philosophie, 1966: Rapport de M. Georges Canguilhem, président du jury (Paris: Ministère de l'éducation nationale, Institut Pédagogique National, 1966).

Mimeographed.

"Qu'est-ce que la psychologie?" *Cahiers pour l'analyse* 2 (March 1966), pp. 112-26.

Mimeographed reprint of the article, followed by "Remarques sur 'Qu'est-ce que la psychologie?'" (pp. 128-34) by R. Pagès, and the concluding "Note" by Canguilhem, all already published in the *Revue de métaphysique et de morale* in 1958. Reprinted in the 1967 edition of the *Cahiers pour l'analyse*, then published by the Editions du Seuil, and again in *Etudes d'his-*

toire et de philosophie des sciences (1968). The *Cahiers pour l'analyse*, which first appeared in mimeograph form, were published by the Cercle d'épistémologie de l'Ecole Normale Supérieure, a group of students close to Louis Althusser.

Published in English in 1980; included in this reader.

Review of "M.D. Grmek, ed., Claude Bernard, *Cahier de notes (1850-1860)* (Paris: Gallimard, 1965)," *Revue d'histoire des sciences* 19 (1966), pp. 405-406.

["Du Singulier et de la singularité en épistémologie biologique," *Revue internationale de philosophie* (1966), p. 325.]

The summary of a lecture Canguilhem gave to the Société belge de philosophie on February 10, 1962.

1967

"Théorie et technique de l'expérimentation chez Claude Bernard," in Etienne Wolf, ed., *Philosophie et méthodologie scientifiques de Claude Bernard* (Paris: Masson, Fondation Singer-Polignac, 1967), pp. 23-32.

Paper presented at an international colloquium organized for the celebration of the centenary of the publication of Claude Bernard's *Introduction à l'étude de la médecine expérimentale*, in 1965. Reprinted in *Etudes d'histoire et de philosophie des sciences* (1968). Translated into German in 1979. Extracts from this article are included in this reader.

"Un Physiologiste philosophe: Claude Bernard," *Dialogue* 5.4 (1967), pp. 555-72.

Lecture given at the Département de philosophie, Université de Montréal, in the fall of 1966; included in this reader.

"Mort de l'homme ou épuisement du Cogito?" *Critique* 242 (July 1967), pp. 599-618.

Essay/review of Michel Foucault, *Les Mots et les choses* (Paris: Gallimard, 1966). Also published in Italian (see below, two entries down).

"Du Concept scientifique à la réflexion philosophique," in *Cahiers de philosophie*, published by the Groupe d'études de philosophie de l'Université de Paris. UNEF-FGEL. no. 1 (Jan. 1967), pp. 39-69.

A lecture by Canguilhem (pp. 39-52), followed by a discussion.
"Morte dell'uomo o estinzione del cogito?" in Michel Foucault, *Le Parole e le cose* (Milan: Rizzoli, 1967), pp. 432-33.
Italian translation of the text first published in French.

1968

"Claude Bernard et Xavier Bichat," *Actes du XIe Congrès international d'histoire des sciences (1965)* 5 (1968), pp. 287-92.
Published in *Etudes d'histoire et de philosophie des sciences* (1968).
Etudes d'histoire et de philosophie des sciences (Paris: Vrin, 1968).
Includes:

"Avant-propos" (p. 7);

"L'Objet de l'histoire des sciences" (pp. 9-23), previously unpublished, based on a lecture given at the invitation of the Canadian Society for the History and Philosophy of Science, in Montreal, on October 28, 1966; republished in Italian and German in 1979; Canguilhem had given a series of lectures on "La fonction et l'objet de l'histoire des sciences" at the Ecole Normale Supérieure in 1984;

"L'Homme de Vésale dans le monde de Copernic" (pp. 27-35), published in 1964, reprinted as a pamphlet in 1991;

"Galilée: la signification de l'oeuvre et la leçon de l'homme" (pp. 37-50), published in 1964;

"Fontenelle, philosophe et historien des sciences" (pp. 51-58), published in 1957;

"La Philosophie biologique d'Auguste Comte et son influence en France au XIXe siècle" (pp. 61-74), published in 1958;

"L'Ecole de Montpellier jugée par Auguste Comte" (pp. 75-80), published in 1961;

"Histoire des religions et histoire des sciences dans la théorie du fétichisme chez Auguste Comte" (pp. 81-98), published in 1964;

"Les Concepts de 'lutte pour l'existence' et de 'sélection naturelle' en 1858: Charles Darwin et Alfred Russel Wallace" (pp. 98-111), published in 1959;

"L'Homme et l'animal du point de vue psychologique selon Charles Darwin" (pp. 112-25), published in 1960;

"L'Idée de médecine expérimentale selon Claude Bernard" (pp. 127-42), published in 1965;

"Théorie et technique de l'expérimentation chez Claude Bernard" (pp. 143-55), previously unpublished;

"Claude Bernard et Bichat" (pp. 156-62), based on a paper published in the proceedings of the XIth International Congress for the History of Science, in Warsaw and Cracow, on August 28, 1965;

"L'Evolution du concept de méthode de Claude Bernard à Gaston Bachelard" (pp. 163-71), previously unpublished, based on a lecture given at the invitation of the Société de philosophie de Dijon, on January 24, 1966;

"L'Histoire des sciences dans l'oeuvre épistémologique de Gaston Bachelard" (pp. 173-86), published in 1963;

"Gaston Bachelard et les philosophes" (pp. 187-95), published in 1963;

"Dialectique et philosophie du non chez Gaston Bachelard" (pp. 196-207), published in 1963;

"Du Singulier et de la singularité en épistémologie biologique" (pp. 211-25), previously unpublished, based on a paper presented to the Société belge de philosophie, in Brussels, on February 10, 1962, translated into German in 1979;

"La Constitution de la physiologie comme science" (pp. 226-73), published in 1963;

"Pathologie et physiologie de la thyroïde au XIXe siècle" (pp. 274-304), published in 1959;

"Modèles et analogies dans la découverte en biologie" (pp. 305-18), published in English in 1963;

"Le Tout et la partie dans la pensée biologique" (pp. 319-33), published in 1966;

"Le Concept et la vie" (pp. 335-64), published in 1966 [this article is sometimes erroneously cited as "La Nouvelle connaissance de la vie," which is actually the title of the subsection of the book to which this article belongs];

"Qu'est-ce que la psychologie?" (pp. 365-81), first published in 1956; reprinted here without the comment by R. Pagès and the following "Note" by Canguilhem, both of which can be found in the *Revue de métaphysique et de morale* in 1956, and in the reprints of the *Cahiers pour l'analyse* in 1966 and 1967;

"Thérapeutique, expérimentation, responsabilité" (pp. 383-91), published in 1959.

This book has been reprinted many times. It was translated into Japanese in 1991. Extracts from the fifth edition (1989) of this book are included in this reader.

"Biologie et philosophie: Publications européennes," in Raymond Klibansky, ed., *La Philosophie contemporaine, Chroniques*, vol. 2: *Philosophie des sciences* (Florence: La Nuova Italia Editrice, 1968), pp. 387-94.

A review of works published between 1956 and 1966 in biology and on the history of biology.

[Comments in] "Objectivité et historicité de la pensée scientifique," *Raison présente* 8 (1968), pp. 24-54.

Canguilhem's comment can be found on pages 39-41, 46-47 and 51-52. Reprinted in J.-M. Auzias et al., *Structuralisme et marxisme* (Paris: 10/18, 1970), pp. 205-65; Canguilhem's comments there are on pages 235-39 and 260-62.

"Régulation (epistémologie)" *Encyclopaedia universalis* 14 (Paris: Encyclopaedia Universalis France, 1968), pp. 1-3.

Reprinted in following editions.

"La Recherche expérimentale," *Revue de l'enseignement philosophique* 18.2 (Dec. 1967-Jan. 1968), pp. 58-64.

An exchange with Charles Mazières on experimental research, broadcast on French educational television, February 6, 1967.

"Le Vivant," *Revue de l'enseignement philosophique* 18.2 (Dec. 1967-Jan. 1968), pp. 65-72.

An exchange with François Dagognet, broadcast on French educational television, February 20, 1968.

"Un Modèle n'est rien d'autre que sa fonction," in Ministère de l'Education Nationale, *Entretiens philosophiques: A l'usage des professeurs de philosophie de l'enseignement secondaire* (Paris: Institut pédagogique national, 1968), pp. 133-36.

1969

"Jean Hyppolite (1907-1968)," *Revue de métaphysique et de morale* 74 (April-June 1969), pp. 129-30.

Tribute to Jean Hyppolite, the respected scholar and translator of Hegel, at the Ecole Normale Supérieure on January 19, 1969. Canguilhem and Hyppolite had been students at the Ecole Normale and became colleagues at the University of Strasbourg and, later, the Sorbonne.

"Avant-propos," in Dominique Lecourt, *L'Epistémologie historique de Gaston Bachelard* (Paris: Vrin, 1969), p. 7.

This book is Lecourt's master's thesis, prepared under the supervision of Canguilhem.

L'Epistemologia di Gaston Bachelard: Scritti di Canguilhem e Lecourt, trans. Riccardo Lanza and Magni (Milan: Jaca Book, 1969).

Italian translation of Dominique Lecourt's *L'Epistémologie historique de Gaston Bachelard* (with Canguilhem's "Premessa" on p. 11), to which a second part is added comprised of three articles by Canguilhem on Bachelard: "La storia delle scienze nel'opera epistemologica di Gaston Bachelard," pp. 87-98; "Gaston Bachelard e filosofi," pp. 99-105; "La dialettica e la filosofia del 'non' in Gaston Bachelard," pp. 107-16. These three articles first appeared in French in 1963.

1970

With S. Bachelard, J.-C. Cadieux, Y. Conry, O. Ducrot, J. Guillerme, P.G.

Hamamdjian, R. Rashed, C. Salomon-Bayet, J. Sebestik, *Introduction à l'histoire des sciences*, vol. 1: *Eléments et instruments. Textes choisis* (Paris: Hachette, 1970).

"Avant-propos" (pp. iii–v) by Georges Canguilhem. Published in Canguilhem's collection "Textes et documents philosophique," it is aimed mainly at students in the final years of the lycées. At the time of publication, the authors were all participating in Canguilhem's weekly seminars at the Institut d'histoire des sciences et des techniques.

"Qu'est-ce qu'une idéologie scientifique?" *Organon* 7 (1970), pp. 3–13.

Based on an invited lecture given at the Institute for the History of Science and Technology of the Polish Academy of Sciences, in Warsaw and Cracow, in October 1969. Reprinted in *Idéologie et rationalité dans l'histoire des sciences de la vie* (1977). Extracts from this article are included in this reader.

"Bichat, Marie, François-Xavier," in Charles C. Gillispie, ed., *Dictionary of Scientific Biography* (New York: Scribner, 1970), vol. 2, pp. 122–23.

"Présentation," in Gaston Bachelard, *Etudes* (Paris: Vrin, 1970), pp. 7–10.

Canguilhem edited this collection of articles, which Bachelard published between 1931 and 1934.

"Judith Swazey, *Reflexes and Motor Integration: Sherrington's Concept of Integrative Action*, Harvard University Press," *Clio Medica* 5 (1970), pp. 364–65.

Review.

[Introduction] "Georges Cuvier: Journées d'études organisées par l'Institut d'histoire des sciences de l'Université de Paris, les 30 et 31 mai 1969 pour le bicentenaire de la naissance de G. Cuvier," *Revue d'histoire des sciences* 23.1 (1970), pp. 7–8.

1971

Canguilhem retired that year from his professorship at the Sorbonne, and from the direction of the Institut d'histoire des sciences et des techniques.

"C. Konczewski, *La Psychologie dynamique et la pensée vécue* (Paris: Flammarion, 1970)," *Revue philosophique* 161 (1971), pp. 119-20.
Review.
"Logique du vivant et histoire de la biologie," *Sciences* 71 (March-April 1971), pp. 20-25.
An essay review of François Jacob's *La Logique du vivant* (Paris: Gallimard, 1970).
"Cabanis, Pierre-Jean-Georges," in Charles C. Gillispie, ed., *Dictionary of Scientific Biography* (New York: Scribner, 1971), vol. 3, pp. 1-3.
"De la Science et de la contre-science," in S. Bachelard et al., *Hommage à Jean Hyppolite* (Paris: Presses Universitaires de France, 1971), pp. 173-80.
A contribution to a book published in honor of Jean Hyppolite, three years after his death.
With S. Bachelard, Y. Conry, J. Guillerme, P.G. Hamamdjian, R. Rashed, C. Salomon-Bayet, J. Sebestik, *Introduction à l'histoire des sciences*, vol. 2: *Objet, méthode, exemples. Textes choisis* (Paris: Hachette, 1971).
Second and final volume, following the one published the previous year, with a new "Avant-propos" (pp. 3-4).
Lo normal y lo patológico (Mexico: Siglo veintiuno editores, 1971).
A second edition was published in 1978; this translation was made from the French edition of 1966, including its new second part.

1972

"Préface," in *Inédits de Lamarck*, Présentés par Max Vachon, G. Rousseau, Y. Laissus (Paris: Masson, 1972), pp. 1-2.
"Préface," in Gaston Bachelard, *L'Engagement rationaliste* (Paris: Presses Universitaires de France, 1972), pp. 5-6.
"Physiologie animale: Histoire," *Encyclopaedia universalis* 12 (Paris: Encyclopaedia Universalis France, 1972), pp. 1075-77.
Reprinted in the new edition of 1989 under a slightly different title; included in this reader.

"L'Idée de nature dans la théorie et la pratique médicales," *Médecine de l'homme* 43 (March 1972), pp. 6–12.
    An extract is included in this reader.

Le Normal et le pathologique (2nd rev. ed., Paris: Presses Universitaires de France, 1972).
    Reprint of the 1966 edition, with some "rectifications de détails et quelques notes complémentaires" (addendum to the "Avertissement"). This edition has since gone through several printings. Extracts from this edition are included in this reader.

La Mathématisation des doctrines informes: Colloque tenu à l'Institut d'histoire des sciences de l'Université de Paris, sous la direction de Georges Canguilhem (Paris: Hermann, 1972).
    "Avant-propos" (pp. 7–9) and comments on pages 67–68, 69, and 133–34 by Canguilhem. This colloquium was held June 24–26, 1970.

1973

"Vie," *Encyclopaedia universalis* 16 (Paris: Encyclopaedia Universalis France, 1973), pp. 764–69.
    Reprinted in the second edition of 1989; included in this reader.

1974

"Sur l'Histoire des sciences de la vie depuis Darwin," *Actes du XIIIe Congrès international d'histoire des sciences (1971), Conférences plénières* (Moscow: Nauka, 1974), pp. 41–63.
    Translated into German in 1979.

"John Brown (1735–1788). La Théorie de l'incitabilité de l'organisme et son importance historique," *Actes du XIIIe Congrès International d'histoire des sciences (1971)* (Moscow: Nauka, 1974), Section IX, pp. 141–46.
    Reprinted, with modifications, and under a different title, in *Idéologie et rationalité dans l'histoire des sciences de la vie* (1977).

"Histoire de l'homme et nature des choses selon Auguste Comte dans le *Plan des travaux scientifiques pour réorganiser la société*, 1822," *Les Etudes philoso-*

phiques (July-Sept. 1974), pp. 293-97.

Based on a paper given at a colloquium held at the house of Auguste Comte, in Paris, on June 27, 1972; included in this reader.

"La Question de l'écologie: La Technique ou la vie?" *Dialogue* (Bruxelles) 22 (March 1974), pp. 37-44.

Based on a lecture given at the "Journées du protestantisme libéral," in Sète, on November 11, 1973.

"Gaston Bachelard," in *Scienziati e tecnologi contemporanei* (Milan: Mondadori, 1974), vol. 1, pp. 65-67.

Das Normale und das Pathologische, trans. Monika Noll and Rolf Schubert (Frankfurt, Berlin, Vienna: Ullstein, 1974).

Translation of the 1972 second, revised French edition. This translation was reprinted in 1977.

1975

"Auguste Comte," in *Scienziati e tecnologi dalle origini al 1875* (Milan: Mondadori, 1975), vol. 1, pp. 325-28.

[Comments in] *Actes de la journée Maupertuis* (Paris: Vrin, 1975), pp. 180-81.

On Anne Fagot, "Le 'Transformisme' de Maupertuis," pp. 163-78.

These were the proceedings of a colloquium held in Créteil in December 1973.

"Pour la philosophie," *La Nouvelle critique* (May 1975), p. 29.

A short letter by Canguilhem answering questions regarding opposition to reform of the national programs of the lycées, which would affect the teaching of philosophy at that level. The title is not Canguilhem's; all answers given by French philosophers whom the journal contacted were published under this name.

Il normale e il patologico (Rimini: Guaraldi, 1975).

Translation of the 1972 second, revised French edition.

La formación del concepto de reflejo en los siglos XVII y XVIII (Valencia, Barcelona: Juan Lliteras, 1975).

Translation of the 1955 first French edition.

1976

Vie et mort de Jean Cavaillès (Ambialet [Tarn]: Pierre Laleure, 1976).

The texts included had not been previously published:

"Avant-propos" (pp. 7-8);

"Inauguration de l'Amphithéâtre Jean Cavaillès à la nouvelle Faculté des Lettres de Strasbourg (9 mai 1967)" (pp. 9-34);

"Commémoration à l'O.R.T.F., France-Culture (28 octobre 1969)" (pp. 35-39);

"Commémoration à la Sorbonne, Salle Cavaillès (19 janvier 1974)" (pp. 41-53);

"Bibliographie: Publications de Jean Cavaillès" (pp. 57-61).

A new edition was published in 1984.

"Qualité de la vie, dignité de la mort," *Actes du colloque mondial Biologie et devenir de l'homme*, Université de Paris, 1976 (New York: McGraw Hill, 1976), pp. 527-32.

Final report of a commission presented at an international colloquium held at the Sorbonne in Paris, September 19-24, 1974 (it is followed by an English translation of the text, pp. 532-37). Canguilhem was a member of the French organizing and reception committee of the colloquium.

"Nature dénaturée et Nature naturante (à propos de l'oeuvre de François Dagognet)," in *Savoir, espérer, les limites de la raison* (Brussels: Faculté Universitaire Saint-Louis, 1976), pp. 71-88.

"Il ruolo dell'epistemologia nella storiografia scientifica contemporanea," *Scienza & Tecnica '76: Annuario della Enciclopedia della Scienza e della Tecnica* (Milan: Mondadori, 1976), pp. 427-36.

Reprinted in *Idéologie et rationalité dans l'histoire des sciences de la vie* (1977). Translated into German in 1979. Extracts from this article are included in this reader.

"Marc Klein, 1905-1975," *Archives internationales d'histoire des sciences* 26.98 (1976), pp. 163-64.

Klein had spent his career as a professor at the University of Strasbourg's medical school where Canguilhem had completed his degree in medicine. Klein continued to teach at the university when it was moved to Clermont-Ferrand during the German occupation. In 1944, the Gestapo arrested and deported him to the concentration camps of Auschwitz, Grossrosen and Buchenwald, from where he was liberated in 1945. He published widely on histology, endocrinology and on history of biomedical sciences. In this obituary, Canguilhem suggested that Klein's historical papers be collected and published as a book; the book was in fact published in 1980, and Canguilhem wrote the introduction (see below, second entry under 1980).

La conoscenza della vita (Bologna: Il Mulino, 1976).

El conocimiento de la vida (Barcelona: Editorial Anagrama, 1976).

1977

Idéologie et rationalité dans l'histoire des sciences de la vie: Nouvelles études d'histoire et de philosophie des sciences (Paris: Vrin, 1977).

Includes:

"Avant-propos" (pp. 9-10);

"Le Rôle de l'épistémologie biologique dans l'historiographie scientifique contemporaine" (pp. 11-29), published in Italian in 1976;

"Qu'est-ce qu'une idéologie scientifique?" (pp. 33-45), published in 1970;

"Une Idéologie médicale exemplaire, le système de Brown" (pp. 47-54), based on the paper published in 1974 under a different title in the Proceedings of the XIIIth International Congress for the History of Science in Moscow, August 18-24, 1971;

"L'Effet de la bactériologie dans la fin des 'Théories médicales' au XIXe siècle" (pp. 55-77), based on a lecture presented in Barcelona in April 1975, translated into German in 1979;

"La Formation du concept de régulation biologique aux XVIIIe et XIXe siècles" (pp. 81-99), an extended version of the paper published, also in 1977, in the proceedings of a conference held in 1974;

"Sur l'Histoire des sciences de la vie depuis Darwin" (pp. 101-119), published in 1974 in the proceedings of the XIIIth International Congress for the History of Science in Moscow;

"La Question de la normalité dans l'histoire de la pensée biologique" (pp. 121-39), based on a paper presented at a colloquium organized by the International Union of the History and Philosophy of Science, in Jyväskylä, Finland, in June-July 1973.

The book was translated into German in 1979, Portuguese in 1981, English in 1988 and Italian in 1992.

A second edition appeared in 1981. Extracts from the 1988 translation of the first edition are included in this reader.

"La Formation du concept de régulation biologique aux XVIIe et XVIIIe siècles," in André Lichnerowicz, Jacques Lions, François Perroux, Gilbert Gadoffre, eds., *L'Idée de régulation dans les sciences* (Paris: Maloine-Doin, 1977), pp. 25-39.

Paper presented at the Collège de France in December 1974, at a colloquium organized by the editors of the proceedings, on the idea of regulation in science. An extended version was published the same year in *Idéologie et rationalité dans l'histoire des sciences de la vie*.

La Formation du concept de réflexe aux XVIIe et XVIIIe siècles (2nd ed., Paris: Vrin, 1977).

The first edition had been published by the Presses Universitaires de France in 1955. This new edition, "révisée et augmentée," includes a short "Avertissement de la deuxième édition," corrections of misprints and a "Complément bibliographique" (p. 202).

Extracts from this article are included in this reader.

"Jacques Ruffié, *De la Biologie à la culture* (Paris: 1976)," *Encyclopaedia universalis* (Paris: Encyclopaedia Universalis France, 1977), pp. 378-79.

Review.

"J. Schiller et T. Schiller, *Henri Dutrochet*," *Archives internationales d'histoire des sciences* 27 (1977), p. 340.

Review.

"Souvenir de Lucien Herr," *Bulletin de la Société des amis de l'Ecole Normale*

Supérieure 138 (March 1977), pp. 12-13.

On the occasion of the fiftieth anniversary of Herr's death. In 1932, Canguilhem had published a review of a collection of Herr's writings as well as of his biography by Charles Andler.

"Les machines à guérir," *Le Monde* (April 6, 1977).

Review of Michel Foucault, Blandine Barret Kriegel, Anne Thalamy, François Beguin and Bruno Fortier, *Les Machines à guérir (aux origines de l'hôpital moderne)* (Paris: Institut de l'environnement, 1976). Canguilhem is incorrectly identified at the bottom of the review as "*Professeur au Collège de France.*"

1978

"Une Pédagogie de la guérison est-elle possible?" *Nouvelle revue de psychanalyse* 17 (1978), pp. 13-26.

"Le Concept d'idéologie scientifique: Entretien avec Georges Canguilhem," *Raison présente* 46 (1978), pp. 55-68.

Following the previous year's publication of *Idéologie et rationalité dans les sciences de la vie*, which includes the article "Qu'est-ce qu'une idéologie scientifique?" On pages 55-58, Gabriel Gohau comments on that article and raises five questions, which Canguilhem answers (pp. 58-60).

"Célestin Bouglé," *Annuaire de l'Association des anciens élèves de l'Ecole Normale Supérieure* (1978), pp. 29-32.

Canguilhem had written his "Diplôme d'études supérieures" under the supervision of Célestin Bouglé in 1926 (see above, first entry under 1926).

On the Normal and the Pathological, trans. Carolyn B. Fawcett, with the editorial collaboration of Robert S. Cohen. Introduction by Michel Foucault (Dordrecht: Reidel, 1978).

Translation of the 1972 second, revised French edition. Reprinted by Zone Books in 1989; extracts included in this reader.

O normal e o patologico (Rio de Janeiro: Forense-Universitaria, 1978).
Portuguese translation of the 1972 second, revised French edition.

1979

"L'Histoire des sciences de l'organisation de Blainville et l'Abbé Maupied," *Revue d'histoire des sciences* 32 (1979), pp. 73-91.
Included in this reader.

"Préface," in Othmar Keel, *La Généalogie de l'histopathologie* (Paris: Vrin, 1979), pp. i-ii.

"Préface," in François Delaporte, *Le Second règne de la nature* (Paris: Flammarion, 1979), pp. 7-10.
Translated into English in 1982 and German in 1983.

"L'oggetto della storia delle scienze," in Gaspare Polizzi, ed., *Scienza ed epistemologia in Francia (1900-1970)* (Turin: Loescher Editore, 1979), pp. 200-16.
Translation of "L'Objet de l'histoire des sciences," published in *Etudes d'histoire et de philosophie des sciences* (1968).

Wissenschaftsgeschichte und Epistemologie: Gesammelte Aufsätze, Wolf Lepenies, ed., trans. Michael Bischoff and Walter Seitter (Frankfurt am Main: Surkhamp Verlag, 1979).
A reader of Canguilhem's works, including:

"Die Geschichte der Wissenschaften im epistemologischen Werk Gaston Bachelard" (pp. 7-21), first published in French in 1963;

"Der Gegenstand der Wissenschaftsgeschichte" (pp. 22-37), first published in French in 1968;

"Die Rolle der Epistemologie in der heutigen Historiograhie der Wissenschaften" (pp. 38-58), first published in Italian in 1976 and in French in 1977;

"Die Epistemologische Funktion des 'Einzigartigen' in der Wissenschaft vom Leben" (pp. 59-74), first published in French in 1968;

"Theorie und Technik des Experimentierens bei Claude Bernard" (pp. 75-88), first published in French in 1967;

"Die Herausbildung des Konzeptes der biologischen Regulation im 18. und 19. Jahrhundert" (pp. 89–109), first published in French in 1977;
"Der Beitrag der Bakteriologie zum Untergang der 'medizinischen Theorien' im 19. Jahrhundert" (pp. 89–109), first published in French in 1977;
"Zur Geschichte der Wissenschaften vom Leben seit Darwin" (pp. 134–53), first published in French in 1974.

1980

"Le Cerveau et la pensée," *Prospective et Santé* 14 (Summer 1980), pp. 81–98.
Based on a lecture delivered on February 20, 1980, at a conference organized by the "Mouvement universel de la responsabilité scientifique," in Paris. Reprinted with some corrections in 1993.
"Marc Klein, historien de la biologie," in Marc Klein, *Regards d'un biologiste: Evolution de l'approche scientifique. L'Enseignement médical strasbourgeois* (Paris: Hermann, 1980), pp. vii–xii.
See Canguilhem's obituary of Klein above, fifth entry under 1976.
"Préface," in André Pichot, *Eléments pour une théorie de la biologie* (Paris: Maloine, 1980), p. 7–10.
"Conditions de l'objectivité scientifique," *Raison présente* 55 (1980), pp. 81–83.
"What is Psychology?" *Ideology and Consciousness* 7 (1980), pp. 37–50.
Translation by Howard Davies of the text first published in 1958.

1981

Ideologia e racionalidade nas ciências da vida (Lisbon: Ediçôes 70, 1981).
Translation of the first French edition (1977).
Idéologie et rationalité dans les sciences de la vie: Nouvelles études d'histoire et de philosophie des sciences (2nd rev. ed., with corrections, Paris: Vrin, 1981).
An Italian translation of this edition was published in 1992, and an English translation in 1988.

"Préface," in Henri Péquignot, *Vieillir et être vieux* (Paris: Vrin, 1981), pp. i-v.

This text was also included in the second edition, newly entitled *Vieillesses de demain: Vieillir et être vieux* (Vrin, 1986, pp. i-v, with a "Complément pour une nouvelle édition," p. vi).

"Gustave Monod, philosophe, pédagogue," in Louis Cros, ed., *Gustave Monod: Un Pionnier en éducation. Les Classes nouvelles de la Libération* (Paris: Comité universitaire d'information pédagogique, 1981), pp. 15-19.

"What is a Scientific Ideology?" *Radical Philosophy* 29 (1981), pp. 20-25.

Translation and an introduction by Mike Shortland, pp. 19-20.

1982

With G. Lapassade, J. Piquemal, J. Ulmann, *Du Développement à l'évolution au XIXe siècle* (Paris: Presses Universitaires de France, 1982).

Reprint of the study in *Thalès* (1960), published in 1962; with a "Présentation" by Etienne Balibar and Dominique Lecourt, pp. v-vi. A new, identical edition appeared in 1985.

"Foreword," in François Delaporte, *Nature's Second Kingdom* (Cambridge, MA: MIT Press, 1982), pp. ix-xii.

Translation of the book first published in French in 1979.

"Emile Littré, philosophe de la biologie et de la médecine," Centre international de synthèse, *Actes du Colloque Emile Littré 1801-1881. Paris, 7-9 octobre 1981* (Paris: Albin Michel, 1982), pp. 271-83.

These proceedings also constitute a special issue of the *Revue de synthèse* 106-108 (April-Dec. 1982); included in this reader.

1983

Canguilhem was awarded in 1983, in absentia, the Sarton Medal, the highest honor of the History of Science Society (see below, in Part Two, entry under 1984, for the reference to the citation).

Etudes d'histoire et de philosophie des sciences (5th ed., Paris: Vrin, 1983).

Includes all the texts published in the 1968 edition, plus "Puissance et limites de la rationalité en médecine," also published in the proceedings of a conference in 1984 (pp. 392-411).

Extracts included in this reader.

"Vorwort," in François Delaporte, *Das zweite Naturreich, über die Fragen des Vegetabilischen im 18 Jahrhundert* (Frankfurt: Ullstein Materialen, 1983), pp. 7-9.

Translation of the text first published in French in 1979.

1984

"Présentation de l' Anatomie," in G. Canguilhem, C. Debru, G. Escat, F. Guéry, J. Lambert, Y. Michaud, A.-M. Moulin, *Anatomie d'un épistémologue: François Dagognet* (Paris: Vrin, 1984), pp. 7-10.

An introduction to the proceedings of a conference, organized by Canguilhem, and held on May 14, 1983, at the Musée Claude Bernard in Saint-Julien en Beaujolais, to discuss the works of François Dagognet. Dagognet had written his dissertation, *La Raison et les remèdes* (Paris: Presses Universitaires de France, 1964), under Canguilhem's supervision.

"Puissance et limites de la rationalité en médecine," in Charles Marx, ed., *Médecine, science et technique: Recueil d'études rédigées à l'occasion du centenaire de la mort de Claude Bernard (1813-1878)* (Paris: Editions du Centre national de la recherche scientifique, 1984), pp. 109-30.

Reprinted in the fifth edition of *Etudes d'histoire et de philosophie des sciences* (1983); included in this reader.

"Gaston Bachelard, psychanalyste dans la cité scientifique?" *Il Protagora* 24.5 (Jan.-June 1984), pp. 19-26.

Published in an issue of the journal devoted to "Gaston Bachelard. Bilancio critico di una epistemologia."

"Entretien avec Georges Canguilhem" (with Jean-Pierre Chretien-Goni and Christian Lazzeri), in *Indisciplines: Cahiers S.T.S.* 1 (1984), pp. 21-34.

1985

[Comments in] Comité consultatif national d'éthique pour les sciences de la vie et de la santé, *Rapport 1984* (Paris: La Documentation française, 1985), pp. 182-84.

Comments on three papers presented by F. Quéré, M. Glowinski and M. Pelicier at a roundtable on the "Problèmes d'éthiques posés par la recherche sur le système nerveux humain," organized by the French National Committee on Ethics in the Life Sciences and Medicine, December 6, 1984.

Emile Boutroux, *Des Vérités éternelles chez Descartes*, Thèse latine traduite par M. Georges Canguilhem, élève de l'Ecole Normale Supérieure (Paris: Vrin, 1985).

Reprint of the 1927 edition, then published by Félix Alcan, lacking the preface by Léon Brunschvicg; with a short "Avant-Propos" by Jean-Luc Marion.

"Fragments," in *Revue de métaphysique et de morale* 90.1 (1985), pp. 93-98.

"Striking fragments" selected from the works of Canguilhem, by Dina Dreyfus, Claire Salomon-Bayet and Jean-Jacques Salomon.

"Descartes et la technique," *Cahiers S.T.S.* 7 (1985), pp. 87-93.

Reprint of the paper first published in 1937.

1986

"Sur l''Histoire de la folie' en tant qu'événement," *Le Débat* 41 (Sept./Nov. 1986), pp. 37-40.

Note on the circumstances surrounding Canguilhem's report on Foucault's doctoral dissertation. Didier Eribon published the report in 1991 (see below, first entry under 1991).

1987

"La Décadence de l'idée de progrès," *Revue de métaphysique et de morale* 92 (1987), pp. 437-54.

"Lecture et souvenir de Jean Brun," in François Dagognet et al., *Une philosophie du seuil: Hommage à Jean Brun* (Dijon: Editions Universitaires de Dijon, 1987), pp. 1-7.

Published in a Festschrift presented to Jean Brun, a French philosopher who had been a student of Canguilhem at the Lycée Fermat, in Toulouse, in 1937.

"Discours de Monsieur Georges Canguilhem prononcé le 1er décembre 1987 à l'occasion de la remise de la Médaille d'or du CNRS," *Médaille d'or du CNRS 1987* (Paris: Centre National de la Recherche Scientifique, 1987).

A two-page printed text of Canguilhem's acceptance speech of the CNRS's gold medal for scientific achievements.

"Avertissement des éditeurs à la première édition," in Jean Cavaillès, *Sur la Logique et la théorie de la science* (4th ed., Paris: Vrin, 1987), pp. ix–xiii.

The first three editions, beginning in 1947, had been published by the Presses Universitaires de France.

Seijou to Byouri (Tokyo: Hosei University Press, 1987).

Japanese translation, by Takehisa Takizama, of *Le Normal et le Pathologique*.

"Preface," *History and Technology* 4 (1987), pp. 7-10.

This text was Canguilhem's contribution to "Science: la renaissance d'une histoire," a colloquium held in memory of Alexandre Koyré in Paris on June 10-14, 1986. It is printed here as the introduction to a special journal issue of the proceedings of that colloquium.

1988

Ideology and Rationality in the History of the Life Sciences, trans. Arthur Goldhammer (Cambridge, MA: MIT Press, 1988).

Translation of the second, revised French edition (1981); extracts included in this reader.

"Présentation," in Yves Schwartz, *Expérience et connaissance du travail* (Paris: Editions Sociales, 1988), pp. 19-22.

"Le Statut épistémologique de la médecine," *History and Philosophy of the Life*

Sciences 10 (suppl., 1988), pp. 15-29.
    Included in this reader.
Hanshagainen no rekishi (Tokyo: Hosei University Press, 1988).
    Japanese translation, by Osamu Kanamori, of *La formation du concept de réflexe*.
"La santé, concept vulgaire et question philosophique," *Cahiers du séminaire de philosophie* 8: *La santé* (Strasbourg: Editions Centre de Documentation en Histoire de la Philosophie, 1988), pp. 119-33.
    The text of a lecture given at the University of Strasbourg in May 1988. Published as a booklet in 1990, and again, in part, as the introduction to a book in 1992, under the title "La santé, vérité du corps."

1989

"Les Maladies," in André Jacob, ed., *Encyclopédie philosophique universelle: L'Univers philosophique*, vol. 1 (Paris: Presses Universitaires de France, 1989), pp. 1233-36.
    Included in this reader.
"Physiologie, 1: Physiologie animale – Objectifs et méthode," *Encyclopaedia universalis* 18 (2nd ed., Paris: Encyclopaedia Universalis France, 1989), pp. 244-46.
    Reprint from the first edition.
"Régulation (epistémologie)," *Encyclopaedia universalis* 23 (2nd ed., Paris: Encyclopaedia Universalis France, 1989), pp. 711-13.
    Reprint from the first edition.
"Vie," *Encyclopaedia universalis* 23 (2nd ed., Paris: Encyclopaedia Universalis France, 1989), pp. 546-53.
    Reprint from the first edition; excerpts from the first edition are included in this reader.
The Normal and the Pathological (New York: Zone Books, 1989).
    Reprint of the translation published by Reidel in 1978; extracts included in this reader.

"Présentation," in François Delaporte, *Histoire de la fièvre jaune* (Paris: Payot, 1989), pp. 11-13.

Translated into Spanish in 1989, English in 1991 and Japanese (in press).

"Préface," in Anne Fagot-Largeault, *Les Causes de la mort: Histoire naturelle et facteurs de risque* (Paris: Vrin / Lyon: Institut interdisciplinaire d'études épistémologiques, 1989), p. xiii.

"Présentation," in *Michel Foucault philosophe: Rencontre internationale, Paris 9, 10, 11 janvier 1988* (Paris: Seuil, 1989), pp. 11-12.

Based on a speech for the colloquium organized by the Association pour le Centre Michel Foucault.

"Prefacio," in François Delaporte, *Historia de la fiebre amarilla* (Cemca: IIH-UNAM, 1989), pp. 13-14.

Spanish translation of the text published first in French (see above, three entries up).

1990

"Philosophie d'une éviction: l'objet contre la chose," *Revue de métaphysique et de morale* 95.1 (1990), pp. 125-29.

Review of François Dagognet, *Eloge de l'objet* (Paris: Vrin, 1989).

La Santé, concept vulgaire et question philosophique (Pin-Balma: Sables, 1990).

A thirty-six-page booklet reprinting the text first published in 1988.

1991

"Rapport de M. Canguilhem sur le manuscrit déposé par M. Michel Foucault, directeur de l'Institut français de Hambourg, en vue de l'obtention du permis d'imprimer comme thèse principale de doctorat ès lettres."

Canguilhem's report (April 19, 1960) on Foucault's doctoral dissertation published under the title *Folie et déraison: Histoire de la folie à l'âge classique* (Paris: Plon, 1961), in Didier Eribon, *Michel Foucault* (2nd ed., Paris: Flammarion, 1991), pp. 358-61.

"Qu'est-ce qu'un philosophe en France aujourd'hui?" *Commentaire* 14.53 (Spring 1991), pp. 107–12.

Occasioned by the awarding of the Jean Cavaillès Prize to Jean-Pierre Séris for his book *Machine et communication* (Paris: Vrin, 1987), at the Ecole Normale Supérieure, March 10, 1990.

"Hegel en France," *Magazine littéraire* 293 (Nov. 1991), pp. 26–29.

Extracts from the article published in 1949.

L'Homme de Vésale dans le monde de Copernic (Paris: Laboratoires Delagrange, 1991).

A reprint, as a booklet, of the article first published in 1965.

"Témoignage," in Société des Amis de l'Ecole normale supérieure, *Bulletin* 186 (Dec. 1991), pp. 20–23.

On Jean Hyppolite.

"Preface," in François Delaporte, *The History of Yellow Fever* (Cambridge, MA: MIT Press, 1991), pp. ix–xi.

English translation of the text first published in French in 1989.

Kagakushi Kagakutetsugaku Kenkyu (Tokyo: Hosei University Press, 1991).

Japanese translation, by Osamu Kanamori, Shunsuke Matsuura, Shoujirou Koga, Muneyoshi Hyoudou, Yasuko Moriwaki and Kiiko Hiramatsu, of *Etudes d'histoire et de philosophie des sciences*.

1992

"Postface," in Jean Gayon, ed., *Buffon 88: Actes du Colloque international Paris-Montbard-Dijon* (Paris: Librairie Philosophique Vrin / Lyon: Institut interdisciplinaire d'études épistémologiques, 1992), pp. 745–49.

"Ouverture," in Elisabeth Roudinesco, ed., *Penser la folie: Essais sur Michel Foucault* (Paris: Galilée, 1992), pp. 39–42.

Opening address given at the colloquium on the "Histoire de la folie trente ans après," held by the Société d'histoire de la psychiatrie et de la psychanalyse, in Paris, on November 23, 1991.

"La santé, vérité du corps," in Marie-Agnès Bernardis, ed., *L'homme et la santé* (Paris: Seuil, 1992), pp. 9–15.

Partial reprint of the text published twice before, under the title "La santé, concept vulgaire et question philosophique," in 1988 and 1990.

Ideologia e razionalità nella storia delle scienze della vita: Nuovi studi di storia e filosofie delle scienze (Florence: La Nuova Italia Editrice, 1992).
Translation, with an introduction by Jacques Guillerme (see below, Part Two), by Paola Jervis of the 1988 French revised edition.

1993

"Le Cerveau et la pensée," in *Georges Canguilhem: Philosophe, historien des sciences. Actes du colloque (6-7-8 décembre 1990)* (Paris: Albin Michel, 1993), pp. 11-33.
Reprint of the article originally published in 1980; the subtitles that had been added by the journal are omitted, and some of the original paragraphing has been reestablished (see p. 32 n.1).
"Préface," in Jacques Piquemal, *Essais et leçons d'histoire de la médecine et de la biologie* (Paris: Presses Universitaires de France, 1993), pp. 7-8.
"Preface," in François Delaporte, *Ounetsu no rekishi* (Tokyo: Misuzu Shobo, 1993).
Japanese translation of the text first published in French in 1989.

Part Two

A SELECTION OF REVIEWS AND COMMENTS ON CANGUILHEM'S WORKS

1933

Raymond Aron, "Réflexions sur le 'pacifisme intégral'," *Libres propos* (Feb. 1933), pp. 96-99.
On "La Paix sans réserve" (1932).

1946

Daniel Lagache, "Le Normal et le pathologique d'après Georges Canguilhem," *Bulletin de la Faculté des lettres de Strasbourg* 24 (1946), pp. 117-30.

A review of Canguilhem's 1943 study. Lagache, who was one of the early proponents of psychoanalysis in France, had entered the Ecole Normale Supérieure in 1924, the same year as Canguilhem. He also taught at the University of Strasbourg when he wrote this article. This review was also published, in a slightly shorter form, in the *Revue de métaphysique et de morale* 51 (1946), pp. 355-70.

1956

P. Delaunay, Review of *La Formation du concept de réflexe* (Paris: Presses Universitaires de France, 1955), *Archives internationales d'histoire des sciences* 9 (1956), pp. 161-62.

F.B., Review of *La Formation du concept de réflexe aux XVIIe et XVIIIe siècles* (Paris: Presses Universitaires de France, 1955), *L'Année psychologique* 56 (1956), p. 329.

Only the author's initials are given.

1957

[Anonymous], Review of *La Formation du concept de réflexe aux XVIIe et XVIIIe siècles* (Paris: Presses Universitaires de France, 1955), *Revue de métaphysique et de morale* 62 (1957), pp. 99-101.

1958

Alvin P. Dobsevage, Review of *La Formation du concept de réflexe aux XVIIe et XVIIIe siècles* (Paris: Presses Universitaires de France, 1955), *Philosophy and Phenomenological Research* 18 (Sept. 1957-June 1958), pp. 568-69.

1959

Jean Théodoridès, Review of "Les Concepts de 'lutte pour l'existence' et de 'sélection naturelle' (1959)," *Archives internationales d'histoire des sciences* 12 (1959), pp. 32-33.

1964

Pierre Macherey, "La Philosophie de la science de Georges Canguilhem. Epistémologie et histoire des sciences," *La Pensée* 113 (1964), pp. 50-74.
With a foreword by Louis Althusser, pp. 50-54.

1967

Jean Lacroix, "Le Normal et le pathologique," *Le Monde*, Jan. 8-9 (1967), p. 13.
Review of the 1966 edition of the book published under that title.

1968

Frederic L. Holmes, Review of Claude Bernard, *Leçons sur les phénomènes de la vie communs aux animaux et aux végétaux* (Paris: Vrin, 1966), in *Isis* 59.3 (1968), pp. 349-50.

G. Rudolph, Review of Claude Bernard, *Leçons sur les phénomènes de la vie communs aux animaux et aux végétaux* (Paris: Vrin, 1966), in *Archives internationales d'histoire des sciences* 21.82-83 (1968), pp. 196-97.

1970

Mauro Di Giandomenico, Review of *Etudes d'histoire et de philosophie des sciences* (Paris: Vrin, 1968), *Episteme* 4 (1970), pp. 113-14.
In Italian.

Annette Lavers, "For a 'Committed' History of Science," *History of Science* 9 (1970), pp. 101-105.
Review of *Etudes d'histoire et de philosophie des sciences* (Paris: Vrin, 1968).

1971

F. Courtès, Review of *Introduction à l'histoire des sciences*, vol. 1 (Paris: Hachette, 1971), *Etudes philosophiques* 26 (1971), pp. 124-25.

1972

Dominique Lecourt, *Pour une Critique de l'épistémologie* (Paris: Maspéro, 1972). Chapter 3: "L'Histoire épistémologique de Georges Canguilhem," pp. 64-97.

François Russo, "Chronique des sciences de la vie," *Archives de philosophie* 35 (1972), pp. 469-508.

An essay review, including comments on numerous works of Canguilhem.

Jean Starobinski, Review of *Etudes d'histoire et de philosophie des sciences* (Paris: Vrin, 1968), *Bulletin of the History of Medicine* 46 (1972), pp. 88-89.

1973

James L. Larson, Review of *Etudes d'histoire et de philosophie des sciences* (Paris: Vrin, 1968), *Isis* 64 (1973), pp. 115-16.

M. Eck, "Le Normal et le pathologique," *La Nouvelle presse médicale* 2.1 (Jan. 1973), pp. 51-56.

A defense of Canguilhem's viewpoint against attacks made by F. Duyckaerts in his book *La Notion de normal en psychologie clinique* (Paris: Vrin, 1954).

Michel Fichant, "L'épistémologie en France," in François Chatelet, ed., *La philosophie au 20ᵉ siècle* (Paris: Hachette, 1973), pp. 129-72.

Part Four of this essay (pp. 161-70), under the title "Epistémologie et histoire des sciences; le rationalisme appliqué des sciences biologiques," discusses Canguilhem's epistemological views. A second edition was published in 1979.

1974

Yvon Gauthier, Review of *La mathématisation des doctrines informes* (Paris: Vrin, 1968), *Isis* 65 (1974), pp. 527-28.

François Russo, "Epistémologie et histoire des sciences," *Archives de philosophie* 37 (1974), pp. 617-57.
　　A review essay that comments on many of Canguilhem's works.

G. Quarta, "G. Canguilhem, storico della scienza," *Il Protagora* 14 (1974), pp. 95-96.

1977

J.A. Schuster, Review of *La Mathématisation des doctrines informes* (Paris: Hermann, 1972), *Annals of Science* 34 (1977), pp. 78-81.

1978

Michel Foucault, "Introduction," in Georges Canguilhem, *On the Normal and the Pathological* (Dordrecht: Reidel, 1978), pp. ix-xx.
　　For a slightly different translation of the same text, see second entry under 1980.

Everett Mendelsohn, "Editorial Note," in Georges Canguilhem, *On the Normal and the Pathological* (Dordrecht: Reidel, 1978), pp. xxiii-xxiv.

Giuseppe Quarta, "Ideologia e storia delle scienze in G. Canguilhem," *Bolletino di storia della filosofia* 6 (1978), pp. 239-51.

1979

Wolf Lepenies, "Vorbemerkung des Herausgebers," in Georges Canguilhem, *Wis-*

senschaftsgeschichte und Epistemologie: Gesammelte Aufsätze, Wolf Lepenies, ed. (Frankfurt am Main: Surkhamp Verlag, 1979).
Introduction to this reader.

S. Marcucci, Review of *La conoscenza della vita* (Bologna: Il Mulino, 1976), *Rivista critica di storia della filosofia* 34 (1979), pp. 226-33.

1980

Ornella Costa, Review of *Idéologie et rationalité dans les sciences de la vie* (Paris: Vrin, 1977), *Scientia* 115 (1980), pp. 227-35.

Michel Foucault, "Georges Canguilhem: Philosopher of Error," *Ideology and Consciousness* 7 (1980), pp. 51-62.

Translation by Graham Burchell based on the same French original used by Carolyn R. Fawcett in 1978 for the introduction to her translation of *The Normal and the Pathological*. The French text used by the two translators differs in many ways from that published in French in 1985 under the title "La Vie, l'expérience et la science" (see below, under 1985).

Colin Gordon, "The Normal and the Pathological: A Note on Georges Canguilhem," *Ideology and Consciousness* 7 (1980), pp. 33-36.

Russell Maulitz, Review of *On the Normal and the Pathological* (Dordrecht: Reidel, 1978), *Isis* 71 (1980), p. 674.

1981

W.A. Albury, Review of *On the Normal and the Pathological* (Dordrecht: Reidel, 1978), *Clio Medica* 15 (1981), pp. 115-16.

Mike Shortland, "Introduction to Georges Canguilhem," *Radical Philosophy* 29 (1981), pp. 19-20.

A note on Canguilhem, introducing an English translation of "Qu'est-ce qu'une idéologie scientifique?"

Martin Staum, Review of *On the Normal and the Pathological* (Dordrecht: Reidel, 1978), *Journal of the History of Medicine and Allied Sciences* 36 (1981), pp. 88-89.

1982

M. Shortland, "Disease as a Way of Life," *Ideology and Consciousness* 9 (1981-82), pp. 113-22.
A review of *On the Normal and the Pathological* (Dordrecht: Reidel, 1978).

Tamayo, Ruy Pérez, *Triptico* (Mexico: El Colegio Nacional, 1982).
A critical evaluation of Canguilhem's work, in particular *The Normal and the Pathological*, on pp. 15-41.

1983

Christopher Lawrence, Review of Georges Canguilhem, *On the Normal and the Pathological* (Dordrecht: Reidel, 1978), and of F. Kräupl Taylor, *The Concepts of Illness, Disease and Morbus* (Cambridge: Cambridge University Press, 1979), *British Journal for the History of Science* 16 (1983), pp. 95-96.

1984

William Coleman, [Extracts from the citation written and read by William Coleman, on the occasion of the award of the Sarton Medal of the History of Science Society to Georges Canguilhem, on 28 October 1983], "Prize announcements," *Isis* 75.2 (1984), p. 357.

Jean-Pierre Chrétien-Goni, "Georges Canguilhem, 1904-," in Denis Huysmans, ed., *Dictionnaire des philosophes*, 2 vols. (Paris: Presses Universitaires de France, 1984), vol. 1, pp. 460-65.
An analysis of Canguilhem's main works.

1985

The following articles were published in a special issue of *Revue de métaphysique et de morale* 90.1 (1985) devoted to Canguilhem:

François Dagognet, "Une oeuvre en trois temps," pp. 29-38.
    Canguilhem had been Dagognet's dissertation supervisor.
Michel Foucault, "La Vie, l'expérience et la science," pp. 3-14.
    The French version of Foucault's introduction to the English translation of *Le Normal et le pathologique*.
Henri Péquignot, "Georges Canguilhem et la médecine," pp. 39-50.
    Canguilhem had written a preface to Péquignot's book, *Vieillir et être vieux*, in 1981.
Jacques Piquemal, "G. Canguilhem, professeur de Terminale (1937-1938): Un Essai de témoignage," pp. 63-83.
    Jacques Piquemal had been a student of Canguilhem in Toulouse and later in Paris.
Jean-Jacques Salomon, "Georges Canguilhem ou la modernité," pp. 52-62.
    Salomon had been a student of Canguilhem in Paris.
Bertrand Saint-Sernin, "Georges Canguilhem à la Sorbonne," pp. 84-92.
    Saint-Sernin had been a student of Canguilhem at the Sorbonne.
[Anonymous], "Bibliographie des travaux de Georges Canguilhem," pp. 99-105.
    This bibliography partially covers Canguilhem's writings and gives a list of Canguilhem's courses at the Faculté des lettres of the Université de Strasbourg, at the Sorbonne and at the Institut d'histoire des sciences.
G.H. Brieger, Review of *On the Normal and the Pathological* (Dordrecht: Reidel, 1978), *Bulletin of the History of Medicine* 59 (1985), pp. 132-33.

1986

C.M.P.M. Hertogh, *Bachelard en Canguilhem: epistemologische Discontinuiteit en het medisch normbergrip* (Amsterdam: VU Uitgeverij, 1986).
D. Chevroton, Review of *Du Développement à l'évolution* (Paris: Presses Universitaires de France, 1985), *L'Année Psychologique* 86 (1986), pp. 275-76.

1987

The journal *Prospective et Santé* published a special issue 40 (Winter 1986-87)

on "Le Normal et le pathologique en question" in honor of Canguilhem:
François Dagognet, "Le Normal et le pathologique," pp. 7–10;
Jean-Claude Beaulne, "Canguilhem, Foucault et les autres," pp. 11–20;
Christiane Sinding, "Relire Canguilhem. De la Normativité à la normalité," pp. 21–25;
Henri Péquignot, "La Clinique face au défi technique," pp. 27–31;
Anne Fagot-Largeault, "Vers un nouveau naturalisme," pp. 33–38;
Denis Versant, "Epistémologie de l'incertain," pp. 39–46;
Hervé Le Bras, "La 'norme' démographique: Politique et idéologie dans les sciences sociales," pp. 47–50;

Gilles Errieau, "Un Praticien face aux concepts: 'Normal' et 'pathologique' pour le généraliste," pp. 51–52;

Stuart F. Spicker, "L'Un et le multiple: L'Epistémologie médicale française vue des USA," pp. 53–59;

François Raveau, "Pour un dialogue nature/culture: Les Vues de l'anthropologie médicale," pp. 61–62;

Charles Brisset, "La 'Double' histoire de la folie: Avant et après la psychiatrie...," pp. 63–66;

Marcel Colin and Thierry Guichard, "Déviance, psychiatrie et société: Les Impuissances du corps social," pp. 67–69;

Mireille Delmas-Marty, "Normes et droit: Repères pour une 'mise en compatibilité,'" pp. 71–76;

Françoise Gaill, "Exemplaire océanographie...Le 'Normal' en révolution permanente," pp. 77–79.

Stuart F. Spicker, "An Introduction to the Medical Epistemology of Georges Canguilhem: Moving beyond Michel Foucault," *The Journal of Medicine and Philosophy* 12 (1987), pp. 397–411.

An analysis of *On the Normal and the Pathological*.

F. Vásquez García, "La crítica de la historia dogmática de las ciencias en la epistemología de Georges Canguilhem," in Angel M. Lorenzo, José L. Tasset and Francisco Vásquez, *Estudios de historia de las ideas*, vol. 1: *Locke, Hume, Canguilhem* (Los Palacios, Villafranca: A.M. Lorenzo, 1987), pp. 95–126.

Pichot, André, Review of *Du Développement à l'évolution* (Paris: Presses Universitaires de France, 1985), *Etudes philosophiques* 42 (1987), pp. 329-30.

1988

François Azouvi, "Canguilhem, Georges," *Le Débat* 50 (May-Aug. 1988), p. 236.
A short biographical notice.

1989

Gary Gutting, *Michel Foucault's Archaeology of Scientific Reason* (Cambridge: Cambridge University Press, 1989).
Includes a discussion of Canguilhem's work, pp. 32-54.

Kenneth A. Long, Review of *Ideology and Rationality in the History of the Life Sciences* (Cambridge, MA: MIT Press, 1988), *Clio* 18.4 (1989), p. 407.

1990

David Brain, "From the History of Science to the Sociology of the Normal," *Contemporary Sociology* 19 (1990), pp. 902-906.
A review of *The Normal and the Pathological* (New York: Zone Books, 1989) as well as of those works of Michel Foucault translated into English.

M. Ereshefsky, Review of *Ideology and Rationality in the History of the Life Sciences* (Cambridge, MA: MIT Press, 1988), *Quarterly Review of Biology* 65 (1990), pp. 58-59.

S. Gilman, Review of *The Normal and the Pathological* (New York: Zone Books, 1989), *Isis* 81 (1990), pp. 746-48.

Howard L. Kaye, Review of *The Normal and the Pathological* (New York: Zone Books, 1989), *Journal of Interdisciplinary History* 21 (1990), pp. 141-43.

C. Lawrence, Review of *Ideology and Rationality in the History of the Life Sciences* (Cambridge, MA: MIT Press, 1988), and of Bruno Latour, *The Pasteuriza-*

tion of France (Cambridge, MA: Harvard University Press, 1988), *Medical History* 34 (1990), pp. 113-14.

Roger Smith, Review of *Ideology and Rationality in the History of the Life Sciences* (Cambridge, MA: MIT Press, 1988) and of *The Normal and the Pathological* (New York: Zone Books, 1989), *Annals of Science* 47.2 (1990), pp. 199-201.

1991

Joy Harvey, Review of *Ideology and Rationality in the History of the Life Sciences* (Cambridge, MA: MIT Press, 1988), *Isis* 82 (1991), p. 610.

Paul Jorion, Review of *Idéologie et rationalité* (Paris: Vrin, 1977), *Revue de l'Institut de sociologie* 1-2 (1991), pp. 167-70.

G. Kearns, Review of *Ideology and Rationality in the History of the Life Sciences* (Cambridge, MA: MIT Press, 1988), *Environment and Planning D-Society and Space* 9 (1991), pp. 373-74.

M. Nicolson, "The Social and the Cognitive: Resources for the Sociology of Scientific Knowledge," *History and Philosophy of Science* 22 (1991), pp. 347-69.

An essay review of *The Normal and the Pathological* (New York: Zone Books, 1989).

R. Olby, Review of *Ideology and Rationality in the History of the Life Sciences* (Cambridge, MA: MIT Press, 1988), *British Journal for the History of Science* 24 (1991), pp. 494-96.

D. Porter, Review of *The Normal and the Pathological* (New York: Zone Books, 1989), *Journal of the History of Biology* 21 (1991), pp. 542-45.

1992

Peter Ostwald, Review of *The Normal and the Pathological* (New York: Zone Books, 1989), *Journal of the History of the Behavioral Sciences* 28 (Oct. 1992), pp. 422-24.

Jacques Guillerme, "Presentazione dell'edizione italiana: Georges Canguilhem, un eròe moderno?," in Canguilhem, *Ideologia e razionalità nella storia delle scienze della vita: Nuovi studi di storia e filosofia delle scienze* (Florence: La Nuova Italia Editrice, 1992), pp. vii–xvi.

Jacques Guillerme had been a student of Canguilhem in Paris.

1993

Georges Canguilhem: Philosophe, historien des sciences. Actes du colloque (6–7–8 décembre 1990) (Paris: Albin Michel, 1993).

Edited by the organizers of the colloquium: Etienne Balibar, Mireille Cardot, Françoise Duroux, Michel Fichant, Dominique Lecourt and Jacques Roubaud.

Includes:

Michel Fichant, "Georges Canguilhem et l'Idée de la philosophie," pp. 37–48;

Françoise Duroux, "L'Imaginaire biologique du politique," pp. 49–57;

Etienne Balibar, "Science et vérité dans la philosophie de Georges Canguilhem," pp. 58–76;

Hélène Vérin, "Georges Canguilhem et le génie," pp. 77–89;

Jean-Pierre Séris, "L'Histoire et la vie," pp. 90–103;

François Gros, "Hommage à Canguilhem," pp. 104–109;

Claude Debru, "Georges Canguilhem et la normativité du pathologique: Dimensions épistémologiques et éthiques," pp. 110–20;

Anne Marie Moulin, "La Médecine moderne selon Georges Canguilhem. 'Concepts en attente'," pp. 121–34;

Elisabeth Roudinesco, "Situation d'un texte: Qu'est-ce que la psychologie?" pp. 135–44;

Yvette Conry, "La Formation du concept de métamorphose: Un Essai d'application de la problématique canguilhémienne du normal et du pathologique," pp. 145–57;

Gérard Molina, " 'Darwin et Wallace...', trente ans après," pp. 158–74;

Pascal Tassy, "Développement et temps généalogique," pp. 175–93;

Jean Mathiot, "Génétique et connaissance de la vie," pp. 194–207;

Gérard Lebrun, "De la supériorité du vivant humain dans *L'Evolution créatrice*," pp. 208-22;

François Delaporte, "La Problématique historique de la vie," pp. 223-32;

Alfonso M. Iacono, "Georges Canguilhem et l'histoire du concept de fétichisme," pp. 233-42;

Jan Sebestik, "Le Rôle de la technique dans l'oeuvre de Georges Canguilhem," pp. 243-50;

Marc Jeannerod, "Sur le Concept de mouvement volontaire," pp. 251-61;

Dominique Lecourt, "La Question de l'individu d'après Georges Canguilhem," pp. 262-70;

Alain Prochiantz, "Le Matérialisme de Georges Canguilhem," pp. 271-78;

Francisco J. Varela, " 'Le Cerveau et la pensée'," pp. 279-85;

Pierre Macherey, "De Canguilhem à Canguilhem en passant par Foucault," pp. 286-94;

Alain Badiou, "Y a-t-il une théorie du sujet chez Canguilhem?" pp. 295-304;

Yves Schwartz, "Une Remontée en trois temps: Georges Canguilhem, la vie, le travail," pp. 305-21; and

Michel Deguy, "Allocution de clôture," pp. 324-30.

Includes a letter received from Canguilhem, on p. 324.

François Azouvi, "Un Maître influent et discret," *Le Monde*, May 27, 1993.

Review of *Georges Canguilhem: Philosophe, historien des sciences. Actes du colloque* (Paris: Albin Michel, 1993).

Didier Eribon, "Canguilhem le patron," *Le Nouvel Observateur* (March 18-24, 1993), p. 56.

Review of *Georges Canguilhem: Philosophe, historien des sciences. Actes du colloque* (Paris: Albin Michel, 1993).

Marc Regon, "King Cang," *Libération* (Feb. 4, 1993), pp. 19-21.

Information on Canguilhem's life and work, on the occasion of the publication of *Georges Canguilhem: Philosophe, historien des sciences. Actes du colloque* (Paris: Albin Michel, 1993).

Acknowledgements

The compilation of this bibliography benefited substantially from the generous assistance of many people to whom I am greatly indebted and very grateful. These include: François Delaporte, Claude Ménard, Monique David-Ménard, Pietro Corsi, Hélène Vérin, Jacques Guillerme and Yves Schwartz, who provided photocopies and references of titles difficult to locate, and Georges Canguilhem who gave me copies of rare items. I also wish to thank two of my research assistants, Stéphane Castonguay who conducted thorough searches in bibliographical data banks, and Vincent Paquette who transcribed different versions of the bibliography, as well as the documentalists of our research center, Pierre Di Campo and Marie-Pierre Ippersiel, who dedicated much time and were remarkably ingenuous in procuring texts. I also want to thank the authorities of the Bibliothèque nationale in Paris, who provided me with a microfilm of the complete *Libres propos d'Alain*. Finally, I want to stress the exceptional competence, thoroughness and dedication of the editors at Zone Books, particularly Meighan Gale, without whom I would have made numerous inconsistencies and mistakes. Of course, any remaining deficiency is fully mine.

Notes

INTRODUCTION

1. Jean-François Sirinelli, *Génération intellectuelle: Khâgneux et normaliens dans l'entre-deux-guerres* (Paris: Fayard, 1988), p. 465.
2. Ibid., p. 599.
3. Georges Canguilhem, *Le Normal et le pathologique* (Paris: Presses Universitaires de France, 1966); *The Normal and the Pathological*, trans. Carolyn R. Fawcett (New York: Zone Books, 1989).
4. Jean-Jacques Salomon, "Georges Canguilhem ou la modernité," *Revue de métaphysique et de morale* 1 (1985).
5. Louis Althusser, "Présentation," in Pierre Machery, "La Philosophie de la science de Georges Canguilhem," *La Pensée* 113 (1964), p. 51.
6. Canguilhem, "Introduction: The Role of Epistemology in Contemporary History of Science," in *Ideology and Rationality in the History of the Life Sciences* (Cambridge, MA: MIT Press, 1988), p. 9.
7. Ibid., p. 3.
8. Bruno Latour and Geof Bowker, "A Booming Discipline Short of Discipline: (Social) Studies of Science in France," *Social Studies of Science* 17 (1987).
9. Canguilhem, "L'Objet de l'histoire des sciences" (1968), in *Etudes d'histoire et de philosophie des sciences* (5th ed., Paris: Vrin, 1983), p. 11.
10. Ibid., p. 16 and see pp. 25–26 of this reader.
11. Ibid., p. 18.

12. Canguilhem's Doctorat d'Etat, *La Formation du concept de réflexe aux XVII et XVIII siècles* (Paris: Presses Universitaires de France, 1955; Vrin, 1977).

13. François Dagognet, "Une Oeuvre en trois temps," *Revue de métaphysique et de morale* 1 (1985), p. 30.

14. Ibid., p. 31.

15. Ibid., p. 37.

16. Canguilhem, "The Question of Normality in the History of Biological Thought" (1973), in *Ideology and Rationality*, p. 128 and see p. 205 of this reader.

17. Canguilhem, *The Normal and the Pathological*, p. 131 and see p. 343 of this reader.

18. Ibid., pp. 196-97.

19. Canguilhem, "Le Concept et la vie," in *Etudes d'histoire et de philosophie des sciences*, p. 335.

20. Dagognet, "Oeuvre," p. 32.

21. "Le Concept," p. 360.

22. Ibid., p. 362.

23. Michel Foucault, "La Vie, l'expérience et la science," *Revue de métaphysique et de morale* 1 (1985), translated as the Introduction to *The Normal and the Pathological*.

24. Jean Cavaillès, *Méthode axiomatique et formalisme: Essai sur le problème du fondement des mathématiques* (Paris: Hermann, 1938), *Remarques sur la formation de la théorie abstraite des ensembles* (Paris: Hermann, 1939); Canguilhem, *Vie et Mort de Jean Cavaillès*, in *Les Carnets de Baudesar* (Ambialet: Pierre Laleure, 1976).

25. Cavaillès, "Protestantisme et Hitlerisme: La crise du Protestantisme allemand," *Esprit* (Nov. 1933).

PART ONE: METHODOLOGY

1. Hélène Metzger, *La Genèse de la science des cristaux* (Paris: Alcan, 1918).

2. No doubt, a "natural object" is not naturally natural but rather the object of common experience and perception within a culture. For example, the object "mineral" and the object "crystal" have no significant existence apart from the

activity of the quarryman or miner, from work in a quarry or mine. To dwell on this commonplace here would take us too far afield.

3. Quoted in Metzger, *La Genèse*, p. 195.

4. This is, in part, the subject of a study by Jacques Piquemal.

5. "Theoretical practice falls within the general definition of practice. It works on a raw material (representations, concepts, facts) which is given by other practices, whether 'empirical,' 'technical,' or 'ideological.'... The theoretical practice of a science is always completely distinct from the ideological theoretical practice of its prehistory" (Louis Althusser, *For Marx*, trans. Ben Brewster [New York: Vintage, 1970], p. 167).

6. See my "Gaston Bachelard," *Scienziati e tecnologi contemporanei* 1, pp. 65–67. [Bachelard's work in the history of science and epistemology is much better known in Europe than in the United States, where his reputation is primarily as a literary critic. Interested readers without French may wish to consult my translation of *The New Scientific Spirit* (Boston: Beacon, 1985), which contains biographical and other information. – TRANS.]

7. Gaston Bachelard, *Le Matérialisme rationnel* (Paris: Presses Universitaires de France, 1953).

8. Ibid., p. 86.

9. See Anne Fagot's paper, "Le 'Transformisme' de Maupertuis," and my remarks in the ensuing discussion in *Actes de la Journée Maupertuis*, Créteil, December 1, 1973 (Paris: Vrin, 1975). Emile Guyénot in *L'Evolution de la pensée scientifique, les sciences de la vie aux XVIIe et XVIIIe siècles* (Paris: A. Michel, 1941) goes so far as to call Maupertuis "a geneticist" (p. 389).

10. Jean Cavaillès, *Sur la logique et la théorie de la science* (3rd ed., Paris: Vrin, 1976), p. 70.

11. Ibid., p. 78.

12. [In French: *fracture*. The word, which is to be compared with the notions of an epistemological break (*rupture*) or "tear" (*déchirure*) used by Bachelard, is borrowed from Jean Cavaillès: "...ces fractures d'indépendance successives qui chaque fois détachent sur l'antérieur le profil impérieux de ce qui vient après nécessairement et pour le dépasser" (*Sur la logique et la théorie*, p. 28).]

13. The response to Darwin in France has been studied from the standpoint of critical epistemology by Yvette Conry, *L'Introduction du darwinisme en France au XIXe siècle* (Paris: Vrin, 1974).

14. See Alexandre Koyré, "Galilée et Platon," *Etudes d'histoire de la pensée scientifique* (Paris: Gallimard, 1971), pp. 166-95, and *Etudes galiléennes* (Paris: Hermann, 1940). At the beginning of the latter work, Koyré states that he borrowed the term *mutation* from Bachelard. It is true that in *Le Nouvel esprit scientifique* (1934) and *La Philosophie du non* (1940), epistemological discontinuity is described using metaphors borrowed from biology. This early Bachelardian vocabulary was eliminated in favor of "epistemological break" in *Le Rationalisme appliqué* (1949).

15. Maurice Clavelin, *La Philosophie naturelle de Galilée* (Paris: Armand Colin, 1968), confirms the validity of the Archimedean model and challenges the usefulness of the Platonist affiliation.

16. Ludovico Geymonat, *Galileo Galilei* (Turin: Einaudi, 1957).

17. Koyré, *Etudes galiléennes*, pp. 171-72.

18. Jacques Piquemal, "Aspects de la pensée de Mendel," lecture delivered at the Palais de la Découverte, Paris, 1965.

19. In this case, the name of the science was transferred post hoc to the ideology; in the case of atomism, it was the other way around.

20. Gerd Buchdahl, "On the Presuppositions of Historians of Science," in Alistair Cameron Crombie and Michael Hoskins, eds., *History of Science* 1 (1967), pp. 67-77.

21. See the inaugural lecture in a course on the general history of science, Collège de France (March 26, 1892), printed in *Revue occidentale*, May 1, 1892, p. 24.

22. Bachelard, *L'Activité rationaliste de la physique contemporaine* (Paris: Presses Universitaires de France, 1951), p. 25. See also "L'Actualité de l'histoire des sciences," lecture delivered at the Palais de la Découverte, Paris.

23. Bachelard, *L'Activité rationaliste*, p. 3. See also *Le Rationalisme appliqué*, p. 112: "Rationalist thinking does not 'begin.' It corrects. It *regularizes*. It *normalizes*."

24. Thomas Kuhn, *The Structure of Scientific Revolutions* (2nd ed., Chicago: University of Chicago Press, 1970); *The Copernican Revolution* (New York: Vintage, 1959).

25. François Russo, "Epistémologie et histoire des sciences," *Archives de philosophie* 37.4 (1974). Father Russo frequently refers to the important work *Criticism and the Growth of Knowledge*, edited by Imre Lakatos and Alan Musgrave (Cambridge, UK: Cambridge University Press, 1970), in which Kuhn's ideas are discussed at length and at times severely criticized by Lakatos, Karl Popper and Paul Feyerabend.

26. See Buchdahl, "On the Presuppositions of Historians of Science."

27. For a critique of externalism, see Koyré, "Perspectives sur l'histoire des sciences," in *Etudes d'histoire de la pensée scientifique*. This text is a comment on a paper by Henri Guerlac, "Some Historical Assumptions of the History of Science," in A.C. Crombie, ed., *Scientific Change* (London: Heinemann, 1963).

28. J.T. Clark, "The Philosophy of Science and History of Science," in Marshall Claget, ed., *Critical Problems in the History of Science* (2nd ed., Madison: University of Wisconsin Press, 1962), p. 103.

29. Koyré, *From the Closed World to the Infinite Universe* (Baltimore: Johns Hopkins University Press, 1957).

30. See Koyré, *The Astronomical Revolution: Copernicus, Kepler, Borelli*, trans. R.E.W. Maddison (Ithaca: Cornell University Press, 1973), p. 40.

31. For a critique, see Michel Foucault, *The Order of Things: An Archaeology of the Human Sciences* [1966] (New York: Vintage, 1973), pp. 145-65.

32. See Piquemal, "Aspects de la pensée de Mendel."

33. Koyré, *The Astronomical Revolution*, p. 77.

34. A.L. Jeitteles, "Wer ist der Begründer der Lehre von den Reflexbewegungen?" *Vierteljahrschrift für die praktische Heilkunde* (Prague, 1858), vol. 4, pp. 50-72.

35. Du Bois-Reymond is known for the concluding word, "Ignorabimus!," of his *Über die Grenze des Naturerkennens* (1874).

36. Du Bois-Reymond, "Gedächtnisrede auf Johannes Müller," in *Reden* (Leipzig, 1887), vol. 2, p. 204. The text of this address was first published

in the *Abhandlungen der Akademie der Wissenschaften* (Berlin, 1859).

37. See Du Bois-Reymond's lecture on La Mettrie in *Reden* (Leipzig, 1886), vol. 1, p. 178. Du Bois-Reymond surely was not unaware that La Mettrie had sought and found asylum at the court of Frederick II in 1748.

38. Du Bois-Reymond neglected a point that was clear to Konrad Eckhard, namely, the relation between the problem of sympathies and that of reflexes.

39. I have not been able to consult this work, but I have already noted [earlier in the work from which this excerpt is taken – TRANS.] that Prochaska used the concept of reflection in his *Physiologie* of 1797.

40. *Reden*, 2nd series, vol. 2, p. 205.

41. Ibid, p. 317.

42. According to Fritz Lejeune, *Leitfaden zur Geschichte der Medizin* (1943), p. 123, Prochaska performed more than three thousand operations to remove cataracts.

43. Du Bois-Reymond's lecture is not mentioned in Fearing's text or bibliography.

44. Blainville, *Cuvier et Geoffroy Saint-Hilaire* (Paris: Librarie J.B. Baillière, 1890), p. 436; compare Duvernoy, *Notice historique sur les ouvrages et la vie de M. Le B'on Cuvier* (Paris: F.G. Levrault, 1833). The "science of finance," in which Cuvier took courses at the Caroline Academy, included economic theory and practice, "policy science" and technology.

45. Blainville, *Cuvier et Geoffroy Saint-Hilaire*, pp. 48-49.

46. Georges Cuvier, *Histoire des sciences naturelles* (Paris: Fortin, Masson, 1841-45), vol. 3, pp. 14-15.

47. Blainville and Maupied, *Histoire des sciences de l'organisation et de leurs progrès, comme base de la philosophie* (Paris: J. Lecoffre, 1847), vol. 2, p. 65.

48. Ibid., vol. 2, pp. 253, 273, 280.

49. Ibid.

50. Ibid., p. 295.

51. Cuvier, *Histoire*, vol. 3, pp. 55-56.

52. Ibid., p. 61.

53. Blainville and Maupied, *Histoire*, vol. 1, pp. xiii-xiv.

54. Ibid., p. xvi.

55. Ibid., vol. 3, p. 14.

56. Ibid., vol. 1, p. 246.

57. Ibid., p. 284.

58. The concept of measure in comparative anatomy appears in Claude Perrault, architect and anatomist. In the preface to *Mémoires pour servir à l'histoire naturelle des animaux* (1671-76), he wrote, "It has been necessary to agree on a Measure or a Module, as one does in architecture...so that, when one says, for example, that a dog has an elongated head, a small ventricle, and an uncomplicated leg, it is only by comparison of all these parts with all the parts of the human body." Quoted in François Dagognet, *Pour une théorie générale des formes* (Paris: Vrin, 1975), p. 178. But for Blainville and Maupied, the man-measure is the more-than-animal-man: that is the criterion of perfection in the series.

59. See especially Blainville and Maupied's *Histoire*, vol. 3, pp. 15 and 337; and *Cuvier et Geoffroy Saint-Hilaire*, p. 431: "M. Cuvier, in my view, is one of the most eminent examples of political philosophy in action."

60. Ibid., vol. 1, p. vii.

61. Ibid., vol. 3, p. 16.

62. Ibid., p. 529.

63. Ibid.

64. Ibid., vol. 2, p. 58.

PART TWO: EPISTEMOLOGY

1. Aristotle, *De anima*, trans. Kenelm Foster and Silvester Humphries (New Haven, CT: Yale University Press, 1951), II.1, art. 217-19, p. 163.

2. Ibid., II.2, art. 254, p. 184.

3. Ibid., II.3, pp. 196-203.

4. Jean-Baptiste Lamarck, *Recherches sur l'organisation des corps vivant* (Paris: Fayard, 1986).

5. Lamarck, *Philosophie zoologique* (Paris: Chez Dentu et L'Auteur, 1809), vol. 2, p. 6.

6. Georges Cuvier, *Rapport Historique sur les progrès des sciences naturelles*

depuis 1789 jusqu'à ce jour (Paris: De L'imprimerie impériale, 1810).

7. Ibid.

8. Michel Foucault, *The Order of Things: An Archaeology of the Human Sciences* [1966] (New York: Vintage, 1973), ch. 8.

9. Otto Rank, *The Trauma of Birth* (New York: Harcourt, Brace, 1929).

10. Rank, *The Myth of the Birth of the Hero*, trans. F. Robbins and Smith Ely Jelliffe (New York: R. Brenner, 1952).

11. Claude Bernard, *Introduction à l'étude de la médecine expérimentale* (Paris and New York: Librairie J.B. Baillière, 1865), vol. 2, p. 1.

12. René Descartes, "Treatise on Man" (AT XI.201-202), in *Descartes: Selected Philosophical Writings*, trans. John Cottingham, Robert Stoothoff and Dugald Murdoch (Cambridge, UK: Cambridge University Press, 1984-91), vol. 1, p. 108.

13. Descartes, "Passions of the Soul" (AT XI.364-65), in ibid., vol. 1, art. 47, p. 346.

14. Marcello Malpighi, *De formatione pulli in ovo* (Londini: Apud Joannem Martyn, 1673 [1669]).

15. Caspar Friedrich Wolff, *Theoria generationis* (Halae ad Salam: Litteris Hendelianis, 1759), and *De formatione intestinorum* (1768-69).

16. Gottfried Wilhelm Leibniz, *The Monadology of Leibniz*, trans. Herbert Wildon Carr (Los Angeles: University of Southern California Press, 1930), p. 112.

17. Leibniz, "Letter to Arnauld, Nov. 28, 1686," in *G.W. Leibniz; Philosophical Essays*, ed. and trans. Roger Ariew and Daniel Garber (Indianapolis, IN: Hackett, 1989), p. 80.

18. Daniel Duncan, *Histoire de l'animal, ou la connaissance du corps animé par la mécanique et la pout chimie* (1686).

19. Charles Bonnet, "Tableau des considerations sur des corps organisés," in *La Palingénésie philosophique* (Geneva: C. Philibert and B. Chirol, 1769).

20. Immanuel Kant, *Critique of Judgment*, trans. James Creed Meredith (Oxford, UK: Clarendon Press, 1952), art. 65, pp. 20-22.

21. Auguste Comte, *Cours de philosophie positive* (Paris: Schleicher frères, 1907-24), vol. 3, lessons 40-44.

22. Paul-Joseph Barthez, *Nouveaux éléments de la science de l'homme* (Paris: Goujon et Brunot, 1806), vol. 9.

23. Comte, *Cours de philosophie positive*, vol. 4, lesson 48.

24. Bernard, *Pensées, notes détachées* (Paris: Librairie J.B. Baillière, 1937).

25. Henri Atlan, "Mort ou vif?" in *L'Organisation biologique et la théorie de l'information* (Paris: Hermann, 1972).

26. Jorge Luis Borges, "The Aleph" (1962), in *A Personal Anthology* (New York: Grove, 1967), pp. 138-54.

27. Antoine Augustin Cournot, *Considérations sur la marche des idées et des événements dans les temps modernes* (Buenos Aires: El Ateneo, 1964), vol. 2, p. 136.

28. Bernard, *Rapport sur les progrès et la marche de la physiologie générale en France* (Paris: L'Imprimerie Imperiale, 1867), p. 221 n.209.

29. Blondlot, born in 1810, was a professor of chemistry at the faculty of Nancy. His fistulation technique is discussed by Bernard in lesson 26 of *Leçons de physiologie opératoire* (Paris: Librairie J.B. Baillière, 1879).

30. Albrecht von Haller, *Elementa physiologiae* (1762), vol. 4, p. 26.

31. See Erwin Heinz Ackerknecht, *Therapie* (Stuttgart: Enke, 1970); "Die Therapie in Fegefeuer während des 19.Jarhunderts," *Osterreichische Arztezeitung* 24 (March 1969); "Aspects of the History of Therapeutics," *Bulletin of the History of Medicine* 36.5 (1962).

32. Bernard, *Principes de médecine expérimentale* (Geneva: Alliance Culturelle du livre, 1963), p. 211.

33. Louis Peisse, *La Médecine et les médecins* (Paris: Librairie J.B. Baillière, 1857), vol. 2, p. 401.

34. Jean Baptiste Bouillaud, *Essai sur la philosophie médicale et sur les généralités de la clinique médicale* (Paris: J. Rouvier et E. Le Bouvier, 1836), p. 75.

35. Foucault, *The Birth of the Clinic*, trans. A.M. Sheridan Smith (New York: Vintage, 1975), p. 192.

36. Littré is quoted in Peisse, *La Médecine et les médecins*, vol. 2, p. 362.

37. Bernard, *Principes de médecine expérimentale*, p. 442.

38. Cited by Bouillaud in *Essai sur la philosophie médicale et sur les généralités de la clinique médicale* (1836), p. 69.

39. Bernard, *Introduction à l'étude de la médecine expérimentale*, vol. 2, ch. 2, sec. 3: "Vivisection."

40. Bernard, *Principes de médecine expérimentale*, p. 440.

41. Ibid., pp. 179-80.

42. In his way, Bernard remained faithful to Cuvier's view that the nervous system is the animal and essentially the only organic regulator.

43. See Mirko Drazen Grmek, *Raisonnement expérimental et recherches toxicologiques chez Cl. Bernard* (Geneva, Paris: Droz, 1973), esp. pp. 408-16.

44. René-Théophile Hyacinthe Laënnec, *De l'Auscultation médiate* (Paris, 1819), p. 57.

45. François Dagognet, "L'Immunité, historique et méthode," lectures at the Palais de la Découverte, Paris, January 4, 1964.

46. On Ehrlich and his work, see Hans Loewe, *Paul Ehrlich, Schöpfer der Chemotherapie* (Stuttgart: Wissenschaft Verlagsgesellschaft, 1950); Felix Marti Ibanez, *The Mind and the World of Paul Ehrlich* (New York: 1958), pp. 257-69; Léon Vogel, "Paul Ehrlich," *Revue d'histoire de la médecine hébraique* 84-85 (1969); and Pauline M.H. Mazumdar, "The Antigen-Antibody Reaction and the Physics and Chemistry of Life," *Bulletin of the History of Medicine* 48 (1974), pp. 1-21.

47. On these matters, see Dagognet, *La Raison et les remèdes* (Paris: Presses Universitaires de France, 1964), and *Surréalisme thérapeutique et formation des concepts médicaux*, in homage to Gaston Bachelard (Paris: Presses Universitaires de France, 1957).

48. Bachelard, *Le Matérialisme rationnel* (Paris: Presses Universitaires de France, 1953), p. 202.

49. Dagognet, *Méthodes et doctrine dans l'oeuvre de Pasteur* (Paris: Presses Universitaires de France, 1967).

50. Ibid., p. 67.

PART THREE: HISTORY

1. My understanding of cell theory owes a great deal to Marc Klein, *Histoire des origines de la théorie cellulaire* (Paris: Hermann, 1936).

2. Robert Hooke, *Micrographia, or Some Physiological Descriptions of Minute Bodies Made by Magnifying Glass, with Observations and Inquiries Thereupon* (London, 1667).

3. See, for example, P. Bouin, Auguste Prenant and L. Maillard, *Traité d'histologie* (Paris: Schleicher, 1904–11), vol. 1, p. 95, fig. 84; or Max Aron and Pierre Grasse, *Précis de biologie animale* (Paris: Masson, 1939), p. 525, fig. 245.

4. Ernst Heinrich Haeckel, *Gemeinverständliche Werke* (Leipzig: Kröner, 1924), vol. 4, p. 174.

5. Comte Buffon, *Histoire naturelle des animaux* (1748), ch. 10.

6. Buffon, *Des Eléments*, in ibid., pt. 1: on light, heat and fire.

7. Ibid.

8. Ibid.

9. Ibid.

10. On Oken as a nature philosopher, see Jean Strohl, *Lorenz Oken und Georg Büchner* (Zurich: Verlag der Corona, 1936).

11. On Schwann and cell theory, see the fundamental work of Marcel Florin, *Naissance et déviation de la théorie cellulaire dans l'oeuvre de Théodore Schwann* (Paris: Hermann, 1960).

12. Marc Klein, *Histoire des origines de la théorie cellulaire*, p. 19.

13. On the origins of cell theory, see J. Walter Wilson, "Cellular Tissue and the Dawn of the Cell Theory," *Isis* 100 (August 1944), p. 168, and "Dutrochet and the Cell Theory," *Isis* 107–108 (May 1947), p. 14.

14. Haeckel, *Die Welträtzel*, in *Gemeinverständliche Werke*, vol. 3, p. 33.

15. Klein offers additional information on this point in his "Sur les débuts de la théorie cellulaire en France," *Thalès* 6 (1951), pp. 25–36.

16. Jean Rostand, "Les Virus Protéines," in *Biologie et médecine* (Paris: Gallimard, 1939). For a summary of later work, see Rostand's "La Conception particulaire de la cellule," in *Les Grands courants de la biologie* (Paris: Gallimard, 1951).

17. The lines that follow were added to an article first written in 1945. The addition seemed natural. I do not say this in order to claim any prophetic gift but, rather, to call attention to the fact that certain innovations are really

somewhat older than advocates more eager to use than to understand them are willing to admit.

18. Paul Busse Grawitz's *Experimentelle Grundlagen zu einer modernen Pathologie: Von Cellular zur Molecular-pathologie* (Basel: Schwabe, 1946) is the German version of a work first published in Spanish.

19. Charles Naudin, "Les Espèces affines et la théorie de l'évolution," *Revue scientifique de la France et de l'étranger*, ser. 2, vol. 8 (1875).

20. Article 10 of "The Passions of the Soul" is entitled "How the Animal Spirits Are Produced in the Brain," but in fact Descartes shows how the spirits come from the heart in the form of "very fine parts of the blood." They undergo "no change in the brain" other than to be separated from "other, less fine parts of the blood" (AT XI.335) in *Descartes: Selected Philosophical Writings*, trans. John Cottingham, Robert Stoothoff and Dugald Murdoch (Cambridge, UK: Cambridge University Press, 1988), vol. 1, p. 331. Thus it is not incorrect to say that the heart is the "source" of the spirits in "Treatise on Man" (AT XI.166), in ibid., vol. 1, p. 104.

21. Descartes, "Treatise on Man" (AT XI.132), in ibid., p. 101 n.1.

22. Descartes, Discourse Four in "Optics" (AT VI.109-14), in ibid., pp. 164-66.

23. Ibid.

24. Descartes, "Treatise on Man" (AT XI.129-31), in ibid., p. 100.

25. Descartes, Discourse Four in "Optics" (AT VI.109-14), in ibid., pp. 164-66. Rabelais's friend Guillaume Rondelet (1507-1566) of Montpellier appears to have been the first to hypothesize that nerves consist of independent bundles of centripetal and centrifugal conductors.

26. In the 1664 preface to Descartes's "Treatise on Man," Clerselier points out that the nerve's insertion into the muscle, and therefore the muscle's expansion by the animal spirits, were poorly represented by Louis de La Forge, who believed that the nerves conducted the flow of spirits into the muscles, whereas Descartes taught that "the nerve fibers and branches ramify in the muscles themselves, and as those fibers swell or collapse, their arrangement causes the muscles to swell or collapse accordingly, producing various effects" (AT XI.119-202),

in ibid., pp. 99-108. Clerselier is undoubtedly right on this point.

27. "It is important to know the true cause of the heart's movement that without such knowledge it is impossible to know anything which relates to the theory of medicine. For all the other functions of the animal are dependent on this, as will be clearly seen in what follows" ("Description of the Human Body" [AT XI.245], in ibid., p. 319).

28. Poisson took this argument from Descartes himself (cf. "To Plempius, 23 March 1638" [AT II.67], in *The Philosophical Writings of Descartes*, vol. 3, p. 93), to defend the Cartesian view against Father Fabri's objections: see *Remarques sur la méthode de Monsieur Descartes*, part 5, second observation, p. 293 of the second volume of the 1724 edition. The example of the frog whose heart is excised or head severed clearly embarrassed Descartes when the question was put to him by a correspondent. He escaped the difficulty by arguing that life is defined not by muscular movement but by cardiac heat. See "Letters to Boswell (?), 1646 (?)" (AT IV.686 and 695), in Charles Adam and Paul Tannery, eds., *Oeuvres de Descartes* (Paris: Vrin, 1974), vol. 4, pp. 686, 695.

29. "(Cor) enim non viscus nobile et princeps est ut usque adeo uti perhibetur sed merus musculus, carne tantum et tendinibus more coeterorum constans, et sanguini circumpellendo inserviens" (Willis, *Pharmaceutices rationalis* [1673], pt. I, sec. 6, ch. 1). See Appendix, p. 174. See also *De sanguinis incalescentia* (1670), in *Opera omnia*, vol. 1, p. 663 ("ex quo liquet cor esse merum musculum") and *De nervorum descriptio et usus* (1664), in ibid., vol. 1, p. 368 ("Dicendum erit quod ipsius cordis compages, carne valde fibrosa constans, potius musculus quam parenchyma appellari debet").

30. "In corde, sicut in toto praeterea musculoso genere, spirituum insitorum particulis spirituosalinis copula sulphurea a sangine suggesta adjungitur; quae materies, dum spiritus agitantur, denuo elisa, ac velut explosa (non secus a pluveris pyrii particulae accensae ac rarefactae) musculum, sive cor ipsum, pro nixu motivo efficiendo inflant ac intumefaciunt" (Willis, *De nervorum descriptio et usus*). On the comparison of the heart to a hydraulic machine: "Circa motum sanguinis naturalem, non hic inquirimus de circulatione ejus, sed quali cordis

et vasorum structura velut in machina hydraulica constanti ritu circumgyretur" (*De febribus* [1659], in *Opera omnia*, vol. 1, p. 71).

31. "Calorem tamen cor omnino a sanguine et non sanguis a corde mutuatur" (Willis, *De sanguinis incalescentia*, in ibid., vol. 1, p. 663).

32. Cf. ibid., vol. 1, p. 661.

33. Willis, *Cerebri anatome* (1664), chs. 9, 10 and 14, in *Opera omnia*, vol. 1, pp. 289ff., 320. In *De fermentatione* (ibid.), p. 4, Willis describes a still and explains how it works. The hierarchical arrangement of the terms "distillation," "purification," "sublimation" and "spiritualization" provides remarkably precise corroboration of an idea of Gaston Bachelard's: "Imagination necessarily ascribes value.... Consider the alchemists. For them, to transmute is to perfect...for an alchemist, a distillation is a purification that ennobles a substance by removing its impurities." See *L'Air et les songes* (Paris: J. Corti, 1943), pp. 296, 298.

34. Willis, *De motu musculari* (Londini: Apud Jacobum Martyn, 1670), in *Opera omnia*, vol. 1, pp. 680-84.

35. *De fermentatione* contains all the physical and chemical preliminaries to Willis's physiological theories; see esp. ch. 10, "De natura ignis et obiter de colore et luce: Ex preamissis non difficile erit pulveris pyrii in tormentis bellicis usitati naturam explicare."

36. On the anatomy and physiology of the nerve, see *Cerebri anatome*, ch. 19.

37. "Quippe spiritus animales a Cerebro et Cerebello, cum medullari utriusque appendice, velut a gemino luminari affluentes, Systema nervosum irradiant" (*Cerebri anatome*, vol. 1, p. 336).

38. "Quapropier longe melius juxta hypothesim nostram, hos spiritus e sanguinis flamma emissos, lucis radiis, sallem iis aurae aerique intertextis, similes dicamus" (*De anima brutorum*, in *Opera omnia*, vol. 2, p. 31). "Spiritus animales, velut lucis radios, per totum systema nervosum diffundi supponimus" (*Cerebri anatome*, p. 338).

39. "Pari fere mode ac si quisquam pulveris pyrii acervos per funem ignarium, ad distance accenderet" (*Pharmaceutices rationalis*, pt. 2, p. 149). See also *De motu*

musculari, in vol. 2, p. 681.

40. "We call 'reflex movements' movements due to a stimulo-motor nervous force produced by the unconscious functional activity of the sensory nerves. It would be more correct to call them movements produced by a nervous reflex action, for it is not the movement whose direction changes but the nervous force, which we regard as having been somehow reflected inside the organism so that a centripetal motion becomes a centrifugal one. But the first expression is convenient, and its use is sanctioned by custom" (Henri Milne-Edwards, *Leçons sur la physiologie comparée de l'homme et des animaux* [Paris, 1878–79], vol. 13, p. 112).

41. See Willis's *De anima brutorum* (1672).

42. J.A. Unzer, *Erste Gründe einer Physiologie der eigentlichen thierischen Natur thierischen Körper* (1771), sec. 495.

43. It is cheating a little to include the name of Legallois, whose first paper on his experiments with cutting the spinal cord dates from 1809. As for Whytt, I mention him only insofar as his conceptions coincide at various points with those of authors who made explicit use of the notion of reflection.

44. Johannes Müller, *Handbuch der Physiologie des Menschen* (Coblenz: J. Holscher, 1833–37), bk. 3, ch. 3, sec. 3.

45. François Jacob, *La Logique du vivant* (Paris: Gallimard, 1970), p. 302.

46. Aristotle, *De anima*, trans. J.A. Smith (Oxford, UK: Clarendon Press, 1908–52), II.1.

47. Aristotle, *De partibus animalium*, trans. William Ogle (London: K. Paul, French, 1882), I.5.

48. Aristotle, *De generatione animalium*, trans. David Balme (Oxford, UK: Clarendon Press, 1972), IV.10.

49. Descartes, "Principles of Philosophy" (AT VIIIA.326), in *The Philosophical Writings of Descartes*, vol. 1, p. 288.

50. Descartes, "Meditations on First Philosophy" (AT VII.85), in ibid., vol. 2, art. 6, p. 58.

51. Ibid.

52. Ibid.

53. See E. Aziza Shuster, *Le Médecin de soi-même* (Paris: Presses Universitaires de France, 1972), ch. 1.

54. Georg Ernst Stahl, *De autocratia naturae* (1696).

55. See, for example, Buffon's article on "The Ass" in *Histoire naturelle des animaux*.

56. See *On the Origin of the Species*, ch. 14.

57. Salvador Edward Luria, *Life: The Unfinished Experiment* (New York: Scribner, 1973).

58. Xavier Bichat, *Anatomie générale appliquée à la physiologie et la médecine* (1801), vol. 1, pp. 20-21.

59. See my preface to the modern edition of Claude Bernard's *Leçons sur les phénomènes de la vie communs aux animaux et aux végétaux* (Paris: Vrin, 1966).

60. See my *The Normal and the Pathological* (New York: Zone Books, 1989), pp. 275-89.

61. Marjorie Greene, *Approaches to a Philosophical Biology* (New York: Basic Books, 1965).

PART FOUR: INTERPRETATIONS

1. René Descartes, "To [The Marquess of Newcastle], October 1645" (AT IV.329), in *The Philosophical Writings of Descartes*, trans. John Cottingham, Robert Stoothoff and Dugald Murdoch (Cambridge, UK: Cambridge University Press, 1984-91), vol. 3, p. 275.

2. "Rules for the Direction of the Mind" (AT X.380), in ibid., vol. 1, pp. 20-21.

3. Descartes wrote, "For there is within us but one soul, and this soul has within it no diversity of parts: it is at once sensitive and rational, and all its appetites are volitions" ("The Passions of the Soul" [AT XI.364], in ibid., vol. 1, art. 47, p. 346).

4. Part 5 of "Discourse on the Method" (AT VI.40-60), in ibid., vol. 1, pp. 131-41; "To the Marquis of Newcastle, 23 November 1646" (AT IV.570-76), in ibid., vol. 3, pp. 302-304.

5. "To More, 5 February 1649" (AT V.267-70), in ibid., pp. 360-67. On the

relation between sensibility and the disposition of the organs, see Descartes's theory of the "degrees of the senses" in "Author's Replies to the Sixth Set of Objections" (AT VII.436-39), in ibid., vol. 2, sec. 9, pp. 294-96.

6. "Letter of 19 March 1678, to Conring," in *Gottfried Wilhelm Leibniz: Sämtliche Schriften und Briefe* (Darmstadt: Reidel, 1926), 2nd ser., vol. 1, pp. 397-401. Compare Leibniz's criteria for distinguishing animals from automata with Descartes's arguments as well as with Edgar Allan Poe's profound reflections on the same question in "Maelzel's Chessplayer." On Leibniz's distinction between machine and organism, see "A New System of Nature and the Communication of Substances," in *Leibniz: Philosophical Letters and Papers*, trans. and ed. Leroy Loemker (Chicago: University of Chicago Press, 1956), sec. 10, and "Monadology," in *Monadology and Other Philosophical Essays*, trans. Paul Schrecker and Anne Martin Schrecker (New York: Macmillan, 1985), secs. 63-66.

7. Leibniz too was interested in the fabrication of machines and automata. See, for example, his correspondence with the Duke of Hanover (1676-79) in *Sämtliche Schriften und Briefe* (Darmstadt: Reidel, 1927), 1st ser., vol. 2. In *Bedenken von Aufrichtung einer Akademie oder Societät in Deutschland zu Aufnehmen der Künste und Wissenschaften*, Leibniz praised the superiority of German art, which had always been interested in the fabrication of moving machines (monsters, clocks, hydraulic machinery and so on), over Italian art, which concentrated almost exclusively on making static, lifeless objects to be contemplated from without. See ibid., p. 544. This passage was cited by Jacques Maritain in *Art et scolastique* (Paris: Librairie de l'art catholique, 1920), p. 123.

8. "Treatise on Man" (AT XI.119-20), in *The Philosophical Writings of Descartes*, vol. 1, p. 99.

9. What is more, Descartes cannot explain God's construction of animal-machines without invoking a purpose: "considering the machine of the human body as having been formed by God in order to have in itself all the movements usually manifested there" ("Sixth Meditation," in *The Philosophical Works of Descartes*, trans. E.S. Haldane and G.R.T. Ross [Cambridge, UK: Cambridge University Press, 1911], vol. 1, p. 195).

10. "Description of the Human Body and All of Its Functions" 1 (AT II.225), in *The Philosophical Writings of Descartes*, vol. 1, p. 315.

11. See Raymond Ruyer, *Eléments de psycho-biologie* (Paris: Presses Universitaires de France, 1947), pp. 46-47.

12. "It is so important to know the true cause of the heart's movement that without such knowledge it is impossible to know anything which relates to the theory of medicine. For all the other functions of the animal are dependent on this" ("Description of the Human Body and All of Its Functions" 2 [AT XI.245], in *The Philosophical Writings of Descartes*, vol. 1, p. 319).

13. "Treatise on Man" (AT XI.165), in *The Philosophical Writings of Descartes*, vol. 1, p. 104.

14. "Traité de l'homme" (AT XI.173-90), in Charles Adam and Paul Tannery, eds., *Oeuvres de Descartes* (Paris: Vrin, 1974), vol. 11, pp. 173-90; also in André Bridoux, ed., *Oeuvres et lettres* (Paris: Gallimard, 1953). [This passage is omitted from the English translation of "Treatise on Man" in *The Philosophical Writings of Descartes*.]

15. "Traité de l'homme" (AT XI.193), in ibid., vol. 11, p. 193; Bridoux, p. 867.

16. Ibid.

17. "Traité de l'homme" (AT XI.192), in ibid., vol. 11, p. 192; Bridoux, p. 866.

18. "Primae cogitationes circa generationem animalium" (AT XI.519), in ibid., vol. 11, p. 519.

19. Martial Gueroult, *Descartes selon l'ordre des raisons*, vol. 2: *L'Ame et le corps* (Paris: Anbier, 1953), p. 248.

20. See section 85 of this volume, above.

21. Descartes, "To Mersenne, 28 October 1640" (AT III.213), in *The Philosophical Writings of Descartes*, vol. 1, p. 155; Bridoux, p. 1088.

22. "Bruta nullam habent notitiam commodi vel incommodi, sed quaedam ipsis in utero existentibus obvia fuerunt, quorum ope creverunt et a quibus ad certos motus impulsa sunt: unde, quoties illis postea simile quid occurit, semper eosdem motus edunt" ("Primae cogitationes circa generationem animalium"

[(AT XI.520], in *Oeuvres de Descartes*, vol. 11, p. 520).

23. "Accretio duplex est: alia mortuorum et quae non nutriuntur, fitque per simplicem partium oppositionem, sine ulla earum immutatione, vel sallem sine magna...Allia accretio est viventium, sive eorum quae nutriuntur, et fit semper cum aliqua partium immutatione...Perfecta nutritio sive accretio simul generationem sive seminis productionem continet" ("Excerpta anatomica: de accretione et nutritione" [AT XI.596], in ibid., vol. 11, p. 596). Descartes here contrasts the growth of an aggregate whose parts remain unchanged with that of an individual through transformation of its parts.

24. "To More, 5 February 1649" (AT V.277-78), in *The Philosophical Writings of Descartes*, vol. 3, p. 366.

25. "Principles of Philosophy" (AT VIIIA.326), in ibid., vol. 1, p. 288.

26. "Meditations on First Philosophy" (AT VII.84), in ibid., vol. 2, art. 6, p. 58.

27. Gueroult, *Descartes selon l'ordre des raisons*, p. 181.

28. Ibid, p. 193.

29. Ibid.

30. Ibid., p. 194.

31. Auguste Comte, *Cours de philosophie positive* (Paris: Schleicher, 1908), vol. 6, pp. 150-51.

32. Albrecht von Haller, *Bibliothèque anatomique*, vol. 2, p. 583.

33. Auguste Comte, *Système de politique positive* (Paris: Presses Universitaires de France, 1975), vol. 1, p. 584.

34. Comte, *Cours*, vol. 6, Preface, p. xvii.

35. Comte, *Cours*, fortieth lesson, vol. 3, p. 151.

36. Comte, *Système de politique positive*, vol. 1, pp. 574, 592, 650.

37. Comte, *Cours*, forty-first lesson, vol. 3, p. 280.

38. Comte, *Système de politique positive*, vol. 1, p. 440.

39. Ibid., vol. 1, pp. 578-80.

40. Ibid., vol. 1, p. 602.

41. Ibid.

42. Comte, *Cours*, fortieth lesson, vol. 3, p. 243n.

43. Comte, *Système de politique positive*, vol. 1, p. 661.
44. Comte, *Cours*, fortieth lesson, vol. 3, p. 171.
45. Ibid., forty-first lesson, vol. 3, p. 281.
46. Comte, *Système de politique positive*, vol. 1, p. 641.
47. Comte, *Cours*, fortieth lesson, vol. 3, p. 163.
48. Ibid.
49. See the *Comptes rendus de la Société de Biologie* 40 (1899). The report is reprinted in Emile Gley, *Essais de philosophie et d'histoire de la biologie* (Paris: Masson, 1900).

50. Georges Pouchet published an interesting biographical note and bibliography of Robin's work in the *Journal de l'anatomie et de la physiologie* in 1886.

51. Emile Littré, *Médecine et médecins* (2nd ed., Paris: Didier, 1872), p. 433; and *La Science au point de vue philosophique* (Paris: Didier, 1873), p. iii.

52. Littré, *Médecine et médecins*, p. 486.
53. Ibid., p. 487.
54. *La Philosophie positive*, 2nd ser., vol. 27, Jul.-Dec. 1881.
55. Littré, *La Science au point de vue philosophique*, p. vii.
56. John Stuart Mill, *Auguste Comte and Positivism*, in J.M. Robson, ed., *Collected Works of John Stuart Mill* (Toronto: University of Toronto Press, 1969), vol. 10, pp. 284-92.

57. Littré, *La Science*, p. 230.
58. Ibid., p. 234.
59. Ibid., p. 260.
60. Ibid., p. 261.
61. Littré, *Médecine et médecins*, p. 170.
62. Charles Robin, "De la biologie, son objet et son but, ses relations avec les autres sciences," *La Philosophie positive* 4 (May-June 1869), p. 331.

63. Littré, *La Science*, p. 340; this text originally appeared as an article in the January 1870 issue of *La Philosophie positive*.

64. Littré, *Médecine et médecins*, p. 148.
65. Ibid., p. 246.

66. "Transrationalisme," *La Philosophie positive* 4 (1880), p. 35ff.
67. Littré, *Médecine et médecins*, pp. 269-70.
68. Ibid., pp. 276-77.
69. In German in 1866, in French in 1874. [The *Oxford English Dictionary* gives 1896 as the date of the first use of "ecology" in English. – TRANS.]
70. Ibid., p. 284.
71. Gaston Bachelard, *The New Scientific Spirit*, trans. Arthur Goldhammer (Boston: Beacon, 1984), p. 136.
72. The first two formulations are to be found in Bernard's *La Science expérimentale* (Paris: Librairie J.B. Baillière, 1878), p. 45, and the third in his *Pensées: notes détachées*, ed. L. Delhoume (Paris: Librairie J.B. Baillière, 1937), p. 36.
73. Bernard, *Leçons sur les phénomènes de la vie communs aux animaux et aux végétaux* (Paris: Librairie J.B. Baillière, 1879), vol. 1, p. 40.
74. Bernard, *Rapport sur les progrès et la marche de la physiologie générale en France* (Paris: Imprimerie Imperiale, 1867), n. 211.
75. Bernard, *Introduction à l'étude de la médecine expérimentale* (Paris: Librairie J.B. Baillière, 1865), p. 142.
76. Ibid., p. 143.
77. Bernard, *Leçons sur les phénomènes de la vie*, vol. 1, p. 342.
78. Ibid., vol. 2, p. 524.
79. Bernard, *Principes*, p. 71.
80. Ibid., p. 52.
81. Bernard, *Introduction*, p. 70.
82. Bernard, *Principes*, p. 26.
83. Ibid., p. 152 n.2.
84. Ibid., p. 156.
85. Bernard, *Introduction*, p. 365.
86. Bernard, *Principes*, p. 171.
87. Bernard, *Introduction*, p. 401.
88. Bernard, *Principes*, p. 139.
89. Ibid., p. 165.
90. The full title of the work is "La Nature opprimée par la médecine

moderne, ou la nécessité de recourir à la méthode ancienne et hippocratique dans le traitement des maladies" (Paris: Debure, 1768).

91. Bernard, *Principes de médecine expérimentale* (Geneva: Alliance Culturelle du livre, 1963), p. 181n.

92. Bernard, *Introduction*, p. 252.

93. Bernard, *Principes*, pp. 51ff.

94. Ibid., p. 53.

95. Ibid., p. 392.

96. See the paper by Marc Klein and Mme Sifferlen in the *Comptes rendus* of the Congrès National des Sociétés Savantes, Strasbourg and Colmar, 1967, Section des Sciences, vol. 1, pp. 111-21.

97. Bernard, *Principes*, pp. 95 and 125.

98. Mirko Drazen Grmek, "Réflexions inédites de Claude Bernard sur la médecine pratique," *Médecine de France* 150 (1964), p. 7.

99. Bernard, *Cahier de notes*, ed. Mirko Drazen Grmek (Paris: Gallimard, 1965), p. 126.

100. Published in Paris by V. Masson and Son. The work first appeared as an article in the *Gazette hébdomadaire de Médecine et de Chirurgie*.

101. Ibid., p. 117.

102. Bernard, *Principes*, p. 117.

103. Bernard, *Pensées: notes détachées*, p. 76.

PART FIVE: PROBLEMS

1. Emanuel Radl, *Geschichte der biologischen Theorien in der Neuzeit* (2nd ed., Leipzig: W. Engelmann, 1913), vol. 1, chap. 4, sec. 1.

2. Walther Riese, *L'Idée de l'homme dans la neurobiologie contemporaine* (Paris: Alcan, 1938), p. 8; see also p. 9.

3. Aristotle, *Politics*, in *The Basic Works of Aristotle*, ed. Richard McKeon (New York: Random House, 1947), I.ii.11.

4. Théophile de Bordeu, *Recherches anatomiques sur les positions des glandes* (Paris: G.F. Quillau, 1751), sec. 64, quoted in Charles Victor Daremberg, *Histoire des sciences médicales* (Paris: Librairie J.B. Baillière, 1870), vol. 2, p. 1157 n.2.

5. Julien Pacotte, *La Pensée technique* (Paris: Alcan, 1931).

6. Franz Reuleaux, *Theoretische Kinematik: Grundzüge einer Theorie des Maschinwesen* (Braunschweig: Vieweg, 1875).

7. According to Marx, tools are moved by human strength, whereas machines are moved by natural forces; see his *Capital*, trans. Samuel Moore and Edward Aveling (New York: International Publishers, 1967), vol. 1, pp. 374-79.

8. For example, trochlea (from the Greek for a block of pulleys), thyroid (from the Greek for shield), scaphoid (boatshaped), hammer (in the ear), sac, duct, *trompe* (the French for fallopian tube, so called because of its resemblance to a horn), thorax (from the Greek for chest), tibia (originally, a kind of flute), cell – and so on.

9. See my "Modèles et analogies dans la découverte en biologie," in *Etudes d'histoire et de philosophie des sciences* (Paris: Vrin, 1968), p. 306.

10. Aristotle explained the flexing and extension of the limbs by analogy with a catapult: see *De motu animalium*, trans. Martha Craven Nussbaum (Princeton, NJ: Princeton University Press, 1978), 701 b9.

11. Descartes, "To Mersenne, 20 February 1639" in (AT II.525), *The Philosophical Writings of Descartes*, trans. John Cottingham, Robert Stoothoff and Dugald Murdoch (Cambridge, UK: Cambridge University Press, 1984-91), vol. 3, p. 134.

12. Claude Bernard, *Introduction* (1865), pt. 2, ch. 2, sec. 1.

13. Kant, *Critique of Judgment*, trans. J.H. Bernard (New York: Hafner, 1951), sec. 65.

14. Bernard, *Introduction*, pp. 356-57.

15. Ibid., pp. 359-60.

16. Auguste Comte, *Cours de philosophie positive* (Paris: Schleicher, 1907-24), vol. 3, forty-first lesson.

17. See my *La Connaissance de la vie* (Paris: Vrin, 1965), on cell theory. Appendix II of that work treats the relations between cell theory and the philosophy of Leibniz.

18. Etienne Wolff, "Les Cultures d'organes embryonnaires 'in vitro,' " *Revue scientifique* (May-June 1952), p. 189.

19. Bernard, *Cahier de notes*, ed. Mirko Drazen Grmek (Paris: Gallimard, 1965), p. 171.

20. Aristotle, *Metaphysics*, in *The Basic Works of Aristotle*, art. 966a, p. 718.

21. See Kant's Appendix to the Transcendental Dialectic in the *Critique of Pure Reason* (New York: Doubleday, 1966), p. 425ff.

22. Henry E. Sigerist, *Man and Medicine: An Introduction to Medical Knowledge*, trans. Margaret Galt Boise (New York: Norton, 1932), p. 102.

23. Ibid., pp. 117-42.

24. Comte, "Considérations philosophiques sur l'ensemble de la science biologique" (1838), fortieth lecture of the *Cours de philosophie positive* (Paris: Schleicher, 1908), vol. 3, p. 169.

25. Ibid., p. 175.

26. Ibid., p. 179.

27. Sigerist, *Man and Medicine*, p. 109.

28. Comte, *Cours*, pp. 175, 176.

29. Ibid., p. 169.

30. Claude Bernard, *Leçons sur le diabète et la glycogenèse animale* (Paris: Librairie J.B. Baillière, 1877), p. 56.

31. Ibid.

32. Ibid., pp. 65-66.

33. Ibid., p. 181.

34. Ibid., p. 132.

35. Ibid., p. 360.

36. Bernard, *Leçons sur la chaleur animale* (Paris: Librairie J.B. Baillière, 1876), p. 391.

37. J.M. Guardia, *Histoire de la médecine d'Hippocrate à Broussais et ses successeurs* (Paris: Doin, 1884), p. 311.

38. Victor Prus, *De l'Irritation et de la phlegmasie, ou nouvelle doctrine médicale* (Paris: Panckoucke, 1825), L.

39. Georges Teissier, "Intervention," *Une Controverse sur l'évolution*, *Revue trimestrielle de l'Encyclopédie française* 3.2 (1938).

40. Emile Guyénot, *La Variation et l'évolution*, 2 vols. (Paris: Doin, 1930).

41. Jean Rostand, *Hommes de vérité: Pasteur, Bernard, Fontenelle, La Rochefoucauld* (Paris: Stock, 1942), p. 96.

42. Bernard, *Introduction à l'étude de la médecine expérimentale* (Paris: Librairie J.B. Baillière, 1865), trans. by Henry Copley Greene as *Introduction to the Study of Experimental Medicine* (New York: Macmillan, 1927; Collier, 1961), p. 96.

43. Alfred North Whitehead, *Nature and Life* (Cambridge, UK: University Press, 1934), p. 5. Quoted by Alexandre Koyré in a report in *Recherches philosophiques* 4 (1934-35), p. 398.

44. Xavier Bichat, *Anatomie générale appliquée à la physiologie et à la médecine* (Paris: Brosson and Chaudé, 1801); new ed. by Béclard, 1821. Trans. by George Hayward as *General Anatomy, Applied to Physiology and Medicine*, 2 vols. (Boston: Richardson and Lord, 1822), vol. 1, pp. 20-21.

45. Besides, Hegel understood this perfectly well: see *Wissenschaft der Logik*, chs. 1 and 3.

46. Teissier, "Intervention."

47. Théodore de Saussure, *Le Miracle grec* (Paris: De noël, 1939).

48. Plato, *The Sophist*, 239b, in *The Sophist and the Statesman*, trans. and intro. A.E. Taylor, ed. R. Klibansky and E. Anscombe (London: Nelson, 1961). [Orthology: the art of using words correctly (*Webster's New International Dictionary*, 2nd ed., 1958) – TRANS.]

49. See Pierre Guiraud, *La Grammaire* (Paris: Presses Universitaires de France, 1958), p. 109.

50. Claude Favre de Vaugelas, *Remarques sur la langue française* (1647), preface.

51. Establishment of conscription and the medical examination of conscripts; establishment of national studfarms and remount depots.

52. Guiraud, *La Grammaire*, p. 109.

53. See Jacques Maily, *La Normalisation* (Paris: Dunod, 1946), pp. 157ff. My brief account of normalization owes much to this work, which is useful for its clarity of analysis and historical information as well as for its references to a study of Dr. Hellmich, *Vom Wese der Normung* (1427).

54. Jean de la Fontaine, *Fables*, 6.4, "Jupiter et le Métayer" (Jupiter and

the Share Cropper).

55. Hans Kelsen, *Reinen Rechtslehre* (Leipzig: F. Deuticke, 1934; 2nd ed., 1960), trans. as *Pure Theory of Law* (Berkeley: University of California Press, 1967).

56. Julien Freund, *L'Essence du politique* (Paris: Sirey, 1965), p. 332.

57. Ibid., p. 293.

58. See Henri Bergson, *The Two Sources of Morality and Religion*: "Whether human or animal, a society is an organization; it implies a coordination and generally also a subordination of elements; it therefore exhibits, whether merely embodied in life or, in addition, specifically formulated, a collection of rules and laws" (trans. R. Ashley Adura and Cloudesley Breton [Garden City, NY: Doubleday, 1954], p. 27).

59. Claude Lévi-Strauss, *Tristes Tropiques*, trans. John and Doreen Weightman (New York: Atheneum, 1981), p. 387.

60. Friedrich Nietzsche, Letter of February 1870 to Paul Deussen, in *Nietzsche Briefwechsel* (Berlin: Walter de Gruyter, 1977), p. 100.

61. Nietzsche, *The Birth of Tragedy*, trans. Walter Kaufman (New York: Vintage, 1967).

Sources

The following publishers have granted permission to use excerpts from copyrighted works:

Etudes d'histoire et de philosophie des sciences (5th ed., Paris: Vrin, 1989), pp. 12-23, 55, 63-73, 75-79, 131, 135-41, 144-46, 147-51, 158-60, 226-27, 231-38, 260-71, 296-304, 323-27, 329-33, 336-45.

Ideology and Rationality in the History of the Life Sciences (Cambridge, MA: MIT Press, 1988), pp. 10-37, 52-55, 58-63, 65-70, 125-44.

La Formation du concept de réflexe aux XVIIe et XVIIIe siècles (2nd ed., Paris: Vrin, 1977), pp. 3-6, 30-32, 34-35, 41, 52-56, 60-63, 65-66, 68-69, 130-31, 138-42, 155-56.

"L'Histoire des sciences de l'organisation de Blainville et l'Abbé Maupied," *Revue d'histoire des sciences* 32 (1979), pp. 75-82 and 90-91.

"Vie," *Encyclopaedia universalis* 16 (1973), pp. 762a-66b, 767a-69c.

"Physiologie animale: Histoire," *Encyclopaedia universalis* 12 (1972), pp. 1075-77a.

"La Physiologie animale au XVIIIe siècle," in René Taton, ed., *Histoire générale des sciences*, vol. 2 (Paris: Presses Universitaires de France, 1958), pp. 593-98, 601-603, 618-19.

"La Physiologie en Allemagne," in Taton, ed., *Histoire générale des sciences*, tome III: *La Science contemporaine*, vol. 1, *Le XIXe siècle* (Paris: Presses Universitaires de France, 1961), pp. 482-84.

"L'Idée de nature dans la théorie et la pratique médicales," *Médecine de l'homme* 43 (March 1972), pp. 6-7.

"Les Maladies," in André Jacob, ed., *Encyclopédie philosophique universelle: L'Univers philosophique*, vol. 1 (Paris: Presses Universitaires de France, 1989), p. 1235a.

"Le Statut épistémologique de la médecine," *History and Philosophy of the Life Sciences* 10 (suppl., 1988), pp. 15-29.

La Connaissance de la vie (Paris: Vrin, 1989), pp. 47-50, 52-56, 58-63, 69-71, 73-76, 79, 86-87, 88-89, 91-92, 102-104, 110-15.

"Descartes et la technique," *Travaux du IXe Congrès international de philosophie Congrès Descartes*, tome II (Paris: Hermann, 1937), pp. 79-85.

"Histoire de l'homme et nature des choses selon Auguste Comte dans le *Plan des travaux scientifiques nécessaires pour réorganiser la société, 1882*," *Les Etudes philosophiques* (1974), pp. 294-97.

"Emile Littré, philosophe de la biologie et de la médecine," Centre international de synthèse, *Actes du Colloque Emile Littré 1801-1881. Paris, 7-9 octobre 1981* (Paris: Albin Michel, 1982), pp. 271-77 and 279-80.

"Un Physiologiste philosophe: Claude Bernard," *Dialogue* 5.4 (1967), pp. 556-57, 560-62, 566-68.

"Préface," in Claude Bernard, *Leçons sur les phénomènes de la vie communs aux animaux et aux végétaux* (Paris: Vrin, 1966), p. 9.

This edition designed by Bruce Mau
with Greg Van Alstyne
Type composed by Archetype
Printed and bound Smythe-sewn by Maple-Vail
using Sebago acid-free paper